Physics

Second Edition

by Johnnie T. Dennis and Gary Moring

ALPHA

A member of Penguin Group (USA) Inc.

This book is dedicated to everyone who wishes to gain an understanding of the inner workings of nature. May it be a stepping-stone in developing a deep appreciation for the beautiful and fragile world in which we live.

ALPHA BOOKS

Published by the Penguin Group

Penguin Group (USA) Inc., 375 Hudson Street, New York, New York 10014, USA

Penguin Group (Canada), 90 Eglinton Avenue East, Suite 700, Toronto, Ontario M4P 2Y3, Canada (a division of Pearson Penguin Canada Inc.)

Penguin Books Ltd., 80 Strand, London WC2R 0RL, England

Penguin Ireland, 25 St. Stephen's Green, Dublin 2, Ireland (a division of Penguin Books Ltd.)

Penguin Group (Australia), 250 Camberwell Road, Camberwell, Victoria 3124, Australia (a division of Pearson Australia Group Pty. Ltd.)

Penguin Books India Pvt. Ltd., 11 Community Centre, Panchsheel Park, New Delhi—110 017, India

Penguin Group (NZ), 67 Apollo Drive, Rosedale, North Shore, Auckland 1311, New Zealand (a division of Pearson New Zealand Ltd.)

Penguin Books (South Africa) (Pty.) Ltd., 24 Sturdee Avenue, Rosebank, Johannesburg 2196, South Africa

Penguin Books Ltd., Registered Offices: 80 Strand, London WC2R 0RL, England

International Standard Book Number: 978-1-59257-531-2
Library of Congress Catalog Card Number: 2006922332

13 10

Interpretation of the printing code: The rightmost number of the first series of numbers is the year of the book's printing; the rightmost number of the second series of numbers is the number of the book's printing. For example, a printing code of 06-1 shows that the first printing occurred in 2006.

Printed in the United States of America

Publisher: *Marie Butler-Knight*
Editorial Director/Acquiring Editor: *Mike Sanders*
Managing Editor: *Billy Fields*
Development Editor: *Michael Thomas*
Production Editor: *Megan Douglass*
Copy Editor: *Keith Cline*
Cartoonist: *Shannon Wheeler*
Cover Designer: *Bill Thomas*
Book Designers: *Trina Wurst/Kurt Owens*
Indexer: *Julie Bess*
Layout: *Chad Dressler*
Proofreader: *John Etchison*

Contents at a Glance

Appendixes

Contents

Appendixes

Foreword

This book is about physics. But what is physics? Why would anyone want to learn some physics? Isn't it terribly hard and mathematical? Don't you have to be an Einstein to understand any of it? Let's talk about this.

Physics is the science that has carried out the basic research that has produced electric motors and generators, radio, radar, television, nuclear reactors, lasers, x-rays, magnetic resonance imaging, transistors, radio carbon dating, electron microscopes, quantum mechanics, superconductivity …. And physics is the science that has developed high-powered mathematical descriptions of the natural universe and hopes one day to develop one single theory or principle to describe all observable natural phenomena.

I think there are a multitude of reasons for learning physics. It's fun, beautiful, and some would say even elegant. From a more practical viewpoint, it is a wonderful foundation to have before plunging into other sciences and technologies such as chemistry, geology, astronomy, electronics, engineering of all varieties, medical technology, and so on.

Okay, so it's interesting and useful, but isn't it really hard to understand? You might find this difficult to believe, but compared to high school–level molecular biology or chemistry or precalculus, it's pretty simple stuff. Eventually, the planners of high school science curricula will wake up to the idea that students should first learn physics, then chemistry, and then molecular biology instead of in the reverse order, the usual practice today. Don't be frightened by the math in this book; you worked with more complicated math in the eighth, ninth, and tenth grades.

How should you go about reading this physics book? The answer depends on how much physics you already know. I'll assume you don't know much physics. You shouldn't read physics the way you do a novel or a newspaper. It won't work. Of course, it won't damage you in any way to read rapidly or to skim through a few pages of this book, but I can guarantee you won't understand or retain much. Here is how I suggest you read.

First, start at the beginning of the book, and don't skip around. I confess, I often read biographies and magazines starting at the back or in the middle, or I just jump around, and this seems to work fine. But this won't work in physics, at least not until you learn a lot more physics. Physics is sequential. The later topics require that you know the material that came earlier in the book. So start at the beginning.

Next, read slowly and carefully. Take notes summarizing all you have read (including text, math, diagrams, and graphs). Give special attention to definitions of technical words and to general principles. You will find that preparing these notes will take you a long way down the path leading to a deep understanding of physics. Many technical words used in physics have been borrowed from everyday English. They have been given a specialized or restricted meaning when used in a physics context. Consider the following four sentences:

Max is a forceful speaker.

Amy is an energetic speaker.

Ivy is a powerful speaker.

Allison is a dynamic speaker.

In everyday language, the four sentences have essentially the same meaning. But in physics each of the four nouns—force, energy, power, and dynamics—from which the adjectives in the previous sentences were derived has a meaning distinctly different from the others. So the exact meaning of key words used in physics is of great importance, and we have to be careful not to carry over from ordinary English a looser, more general meaning.

Next, make sure you work out the problems. Try your hand at solving the problems in the book without looking at the author's solutions. If you get stuck, take a peek at his solution. Make up your own problems similar to the book's, but based on situations you see in the real world. Or look in other physics books for problems. You will be amazed at how interesting and different the world looks from the perspective of the new physics you are learning. Practice, practice, practice. It's fun.

And finally, review before starting a new session; look over all the material from previous sessions. Work a few typical problems. Overlearn the material. Use either this book or your own summary notes for the review. This will give you a mastery that will help you to retain your knowledge of physics the rest of your life.

Johnnie Dennis, the author of this book and recipient of the National Teacher of the Year award, is a gifted physics teacher. The many years he spent honing his teaching skills will become readily apparent as you start your tour through his carefully crafted presentation of the world of physics. Bon voyage!

—Gilbert Ford, Ph.D., Nuclear Physics, Harvard University

Vice President for Academic Affairs, Emeritus, Northwest Nazarene University

Introduction

At one point in the movie *The Matrix*, Keanu Reeves's character Neo is given the option of taking either a red pill or a blue pill. He is told that if he takes the blue pill everything will remain the same and he will go on with his life as if nothing had happened. But if he takes the red pill, he will be given the opportunity to enter Wonderland and to see how far down the rabbit hole really goes. Reading this book will be like taking the red pill. For in studying and learning about physics, you'll get a chance to see behind the everyday world in which you live. The mysteries of nature will be explained to you in a way that will not only reveal its fundamental mathematical beauty, but will also show how it is connected to everything you do, from throwing a baseball, to listening to music, to flying in an airplane.

Why study physics? For many students of physics, it is simply a means to an end. A large number of premed students are required to show proficiency in physics to become medical doctors. Others need physics under their belt to become successful engineers. Yet others study physics on their way to being awarded degrees in biology, chemistry, architecture, environmental science, or astronomy; and there is the occasional student who, upon studying physics, decides that there is no other subject that will hold their interest like understanding fundamentally how the universe works.

So what is physics? In ancient times, the term *physics* referred to the study of the natural world and the phenomena that took place in it. The study was sometimes called "natural philosophy." A more modern definition of physics is that it is the study of matter and energy and the interactions between them, and the discipline of physics embodies a set of techniques to observe, model, and understand the natural world.

Many aspects of physics tie it to any number of nonscientific disciplines. The physics of sound is critical to the understanding of musical instruments and the manipulation of sound by musicians. Artists, for example, often use the physics of light to work with colored glass to create sculptures. The fundamental nature of the questions posed by physicists about the way the world works places them in contact, from time to time, with theologians asking many of the same questions: How did we get here? Are there universal laws, and if so, how did they arise? Does the presence of universal laws imply the presence of a creator, or are the laws just a fundamental part of the universe? In addition, many mathematical techniques that are useful to physicists interested in modeling complex physical systems are just as useful to economists modeling complex systems in the marketplace.

Our understanding of the world derives from the sciences and humanities equally. Therefore, many chapters also include some historical background information to provide you with a time frame or context in which some of the ideas and concepts that you will be studying were born. As important as it is to understand the concepts and applications of physics, it is also valuable to know why certain discoveries were made, who made them, and why it is significant for you to know this.

You'll be able to check your progress by working a few problems at the end of each of the chapters. You can find these problems under the section called "Problems for the Budding Rocket Scientist." Within the chapters themselves, you will find several examples that are outlined for you in words, solved algebraically, and calculated numerically. You can use the examples as models for solving the problems and check your results by looking at the answers in Appendix B. You are encouraged to follow the outline for solving problems in the examples and find your own solution before checking in the book. That way you can make the several checks on your work outlined in this book, such as correct units of measurement and the correct number of significant figures.

This book is designed as an introduction to physics and not an advanced book. So consequently, the highest form of mathematics used to present the material and to solve problems is algebra, although in the last two chapters we use just a touch of trigonometry. Calculus and all the rest of trigonometry, although extremely useful for solving all types of physics problems, are not incorporated in this book because they go beyond and assume a higher level of aptitude than the level at which this book is written.

How This Book Is Organized

The book is divided into five parts:

Part 1, "Motion Near the Surface of the Earth," takes you from motion along a straight line that may be described with a linear mathematical model to motion described with a quadratic model. You'll understand why a new invention called a vector was needed to explain what happens when an object moves off a straight line into a plane.

Part 2, "How and Why Things Move," shows you that there must be a cause for motion. Not only is there a cause, you'll learn to identify it as a force and measure it. The realization that there is a force of attraction between every two bodies in the universe leads to an understanding of the motions of an artificial satellite of the Earth as well as natural satellites of the Earth and the sun.

Part 3, "Work and Energy," guides you to the identification and definition of work. You'll see how work is related to kinetic energy and kinetic energy to potential energy. If you do not already know what an *erg* is, you will find out and use that word in the same way that you use the words *joule* or *watt*. By developing an understanding of the concepts of power, momentum, and impulse, you'll be able to predict what happens when two objects collide, either by actually touching or interacting at a distance.

Part 4, "Heat, Sound, and States of Matter," explores the motions of particles in solids, liquids, and gases. By having a clear comprehension of what pressure is using the idea of force, you'll learn to recognize that there is a distinct difference between force and pressure. The discovery of a different type of energy in heat leads to an understanding of temperature that you measure in different systems of measurement. Heat influences the energy of sound and how fast it travels. Sound is found to have wavelength and frequency that enables you to explain the change in pitch of the motor of a speeding car when it passes you.

Part 5, "The Anatomy of Atoms, Electricity, and Light," takes you inside the atom by using a model to name the parts of the atom. Some parts of the atom lead to the identification of charge. Charged particles are generated and detected. Charge is associated with the electron, the proton, and the ion. Charge in motion leads to an exploration of current in an electrical circuit. From motion of charge to motion of light particles leads to the explanation of images formed by mirrors and lenses. The dual nature of light is explored, and we examine the properties of light both as a particle and as a wave.

Extras

The sidebars in this book offer guideposts to focus your attention by providing suggestions, definitions, and valuable information about possible difficulties to be avoided. A word of encouragement is always in order.

Physics Phun

Problems are posed for mastery of complicated notions where practice with applications may promote a deeper understanding.

def•i•ni•tion

Definitions of words used in the book to help clarify the meaning of subject matter being discussed.

Newton's Figs

Suggestions and tips on how best to master an idea or to remember relationships among concepts. Ideas that are apt to turn up later are emphasized.

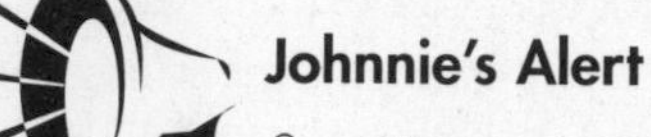

Johnnie's Alert

Sometimes serves as a warning about possible difficulties that can be avoided and sometimes offers an encouraging word.

Acknowledgments

The first group of individuals who inspired this book is all my former students. Many took the time to comment or make suggestions about what worked and what did not make sense in my presentations. It is my desire that you are able to recognize the characters and examples used throughout the book. Thanks to each of you and especially to those who secretly changed the spelling of words on my bulletin boards.

To my wife, Shirley Dennis, who caught typos in the manuscript and kept me on my treatment recovery diet throughout the project. To Charles and Karen, Deanna and Kevin, Maureen and Steve, and Kevin and Mary, who encouraged me at every turn.

To Sarah, Caitlin and John, Nicholas and Christopher, and Summer and Seth, who have urged me to do a good job because they have high expectations for a book they plan to use.

I will always be grateful to Mike Sanders and the staff at Alpha for making this project possible. Nancy Lewis is a devoted editor who helped me to gather all of the parts together in a meaningful way.

Trademarks

Part 1

Motion Near the Surface of the Earth

Everything around us is in constant motion. Airplanes fly overhead; cars whisk down the street; children throw a ball back and forth in a park; even molecules of oxygen, nitrogen, and carbon dioxide move about in the atmosphere in a constant jittery dance. This first part of your introduction to physics will help you better understand what it means to move. You will learn that sometimes it makes a difference as to which direction you want to go. Even though the objects discussed are everyday kinds of things, you may find that some brand-new ideas are needed to describe how they move.

Chapter 1

Mathematics, the Language of Physics

In This Chapter

- It's all about measurement
- Standard units in physics
- Significant figures
- Scientific notation
- How to round off numbers

Since ancient times humans have taken measures of their world. How many sheep do I own? How large is my tract of land? How many days until the winter festival? How much does this steer weigh? Of course, measurement is fundamental to any science, but particularly to physics, the science that probes the physical properties of the natural world, from the scale of the atom to the scale of clusters of galaxies.

One of the great discoveries in the history of science is the connection between observable phenomena and mathematics. The discovery of the connection between mathematics and the workings of the natural world was made by the Pythagoreans—followers of the Greek philosopher

Pythagoras, in the fifth century B.C.E., a time when the *Iliad* and the *Odyssey* were cast into their final form, when Confucius walked the earth, and when the Greeks began to question nature rather than oracles for answers.

Mathematics, the language of physics, is useful for several reasons: mathematics expresses the underlying assumption that nature is logical—in other words, that every effect has a cause—and quantitative (meaning that things can be described in mathematical terms); mathematics provides a way to generalize the results of a small number of particular experiments and thus to make predictions about new experiments; mathematics is usually briefer and more convenient than words; and mathematics provides a way to proceed step by step from one result to the next.

When taking measurements of the physical world, there are important choices to be made. The size of the unit chosen should be appropriate to the object being measured. For example, astronomers wouldn't use the centimeter to discuss the distances to other stars; instead, they use the light-year, which is the distance a photon moving at the speed of light can travel in a year. (A photon is a "packet" of radiation. Light or other transmitted electromagnetic radiation, such as radio waves, microwaves, or x-rays, are all transmitted as photons.) Likewise, nuclear physicists do not use acres or hectares to measure the apparent size of atomic nuclei; they use barns. And not the kind you keep your cows and horses in, but a very, very, very small unit.

This chapter contains information that you will refer to again and again as you read this book. You do not need to master every idea presented here, but you'll know that you have it at your fingertips any time you need it.

Why All of This Is Important

The rise of modern science grows out of quantification. But quantification means much more than expressing observations in mathematical form. It is also a turning away from natural philosophy—grand schemes based on aesthetic preference—to detailed and precise observations and measurement. In other words, it is a turning to the accumulation of knowledge by small, detailed increments. Before Copernicus, many speculated on a sun-centered universe, but Copernicus took the trouble to do the detailed calculations and produce astronomical tables that made his system a serious competitor for the very successful Ptolemaic system that came before it. Others before Galileo speculated on the properties of matter in motion, but Galileo based his arguments on detailed observations. And just in case these names don't mean anything to you, don't worry; by the time you're done with this book, they will.

Quantification of physics tends to condense its ideas into mathematical formulas. As Galileo put it, the great book of nature lies ever open before our eyes, but it is written in mathematical characters. History teaches us that mathematics helps to advance physics, but it also shows that, like the tides, ideas flow in both directions. New discoveries in one field often lead to improvements in another.

After Copernicus and his revolution and the events that led Europe through the years of Kepler, Galileo, and Newton, our view of the universe was, and still is today, that we live on a speck of dust in a lost corner, somewhere in the universe. The early Greek natural philosophers tried to teach us humility by placing us in a lowly sphere, isolated from the serene perfection of the heavens, but never in their wildest dreams could they have imagined the impact of the psychological change that occurred when we first realized that we are not at the center of the universe.

Making Measurements

Throughout history, establishing common units of time, distance, and weight for the sake of orderly agriculture and commerce has been one of the principle responsibilities of government. But units of measure were seldom the same in different countries and were generally based on some convenient or traditional magnitude. For example, still preserved by use in the United States today, the mile (from the Latin *milia*, "thousand") was once 1,000 standard paces of a Roman legion; the yard was the distance from one's nose to his outstretched fingers; the foot was the length of the reigning king's foot; and the inch was one thumb, from joint to tip.

But finally, thanks to one of the legacies of Napoleon's conquests in Europe, a new system of units, not based on tradition or whimsy, but on precise French logic and the decimal system, took hold. Yet it is also firmly based on human magnitudes and on the properties of water, an essential ingredient of life. For example, the fundamental unit of mass, the gram, is the mass of 1 cubic centimeter of water; the unit of volume, the liter, is 1,000 cubic centimeters, so that a liter of water has a mass of 1 kilogram.

The definitions of these quantities no longer vary from one country to another; they are fixed by international treaty, and are used everywhere except in the United States. This system is formally known as the International System of Units, or SI for short. Americans who want to learn science or engineering, or just to shop in Europe or Canada, must learn to convert from units based on the size of a dead king's foot to units based on the length of a bar of platinum alloy kept in a refrigerated vault in France.

These *fundamental units* for length and mass, along with the second, the fundamental unit for measuring time, are all part of the metric system. And within the metric system you will find reference to MKS and CGS. MKS is shorthand for meter, kilogram, and second; CGS is shorthand for centimeter, gram, and second. Another system that is used, the British system, is called the FPS system; FPS is just a short way of labeling foot, pound, and second.

The utility and simplicity of the metric system can again be seen in the conversions from one multiple or mass to another. For example, there are 1,000 grams (g) in a kilogram (kg). There are 100 centimeters (cm) in a meter. A nanosecond is one billionth of a second. The following table indicates some of the most common prefixes used in the metric system. Multiples of any fundamental units are simply named by placing a prefix before the word *meter*, *gram*, or *second*.

Common Metric Prefixes

Multiple	Prefix
1,000,000	mega (M)
1,000	kilo (k)
0.01	centi (c)
0.001	milli (m)
0.000,001	micro (μ)
0.000,000,001	nano (n)

To give you a clear idea of just how the letters CGS, MKS, and FPS are used to express the three fundamental quantities, we've listed them in a table you can refer to throughout your reading of this book. You will find that the table will be extended to include *derived quantities*, which are also very important in our effort to understand the language of physics.

Units of Measurement of Physical Quantities

Fundamental Quantities	MKS	CGS	FPS
Time	s	s	s
Length	m	cm	ft
Mass	kg	g	slug

As you can see, the fundamental quantity of time is measured in the same units in all systems. You will see later that there is a nice mathematical relationship between units in the CGS system and the MKS system. In countries that still use the British system, people are often confused between kilograms and pounds. These units refer to different physical quantities. Yet labels list the weight of an item in pounds along with its mass in kilograms and do not specify that one is a weight and the other is mass. The problems in this book are devoted mostly to the MKS and CGS systems.

def•i•ni•tion

Fundamental quantities or **units** are the building blocks for the foundation of physics. Time, length, and mass are the quantities required to study the area of physics known as mechanics. **Derived quantities** or **derived units** are simply combinations of fundamental units such as meters per second, written m/s. Or they can be expressed as one fundamental unit used more than once, as in a second squared.

Recording Measurements

Let's use this new language to become familiar with some other important concepts in physics. First let's consider *significant figures.*

Significant Figures

Significant figures are digits that have physical meaning. If we are measuring quantities, the number of significant figures is determined by the accuracy of our measurements. For example, a ruler with centimeter divisions would not give us as accurate a measure of length as a ruler with millimeter divisions. When you make a measurement, you should record and report the best observation you can make. To do that, you should record every digit that you are sure of, given the measuring instrument that you are using, plus one more digit that you estimate. That way you and anyone who reads your report will know that any error that may occur will be recorded in that last digit you write down.

Even when you take all of this care, not all digits are significant when you record measurements using the conventions of writing numerals. Following are some rules you can follow to make sure your records communicate the same information to anyone who reads your work. It's a good idea to keep this list handy until you know the rules:

1. All nonzero digits are significant. Example: In the number 134 m, there are three significant digits.

2. All zeros occurring between nonzero digits are significant. Example: In the number 104 m, there are three significant digits.
3. All zeros occurring beyond nonzero digits in a decimal fraction are significant. Example: In the number 0.1340 m, there are four significant digits. They are the last four digits.
4. Zeros used to locate the decimal point are not significant. Example: In the number 0.0104 m, there are only three significant digits. They are the last three digits. Also in the number 25,000 mi, there are two significant digits because the three zeros serve only to locate the decimal point.

These rules apply whenever you make a measurement in physics. And it's helpful to know that a unit of measurement will always be a part of such a record. So whenever you see a numeral with a unit of measurement, you will know that significant figures must be observed. This is true all of the time—with one exception. If the numeral with units results from a *defining equation*, then all digits are significant. One final thought before you're overloaded: it's good to be aware of the rule that all numerals without units of measurement contain digits all of which are significant.

If these rules are a little confusing, like anything else, they will become easier as you use them in practice by making measurements or working problems.

Defining Equations

Before we move on to a couple of other new mathematical concepts, let's talk briefly about defining equations. You're probably already familiar with defining equations, but you may not have called them by that name. We'll list some of the most common ones used in physics, and afterward see whether there are any you remember that aren't listed: 1 hr = 60 min, 1 min = 60 s, 1 mi = 5,280 ft, 1 kg = 1,000 g, 1 m = 100 cm.

Physics Phun

How many significant figures are recorded in each of the following? (1) 2,500 km, (2) 3.14, (3) 0.005 m, (4) 3.100 m, (5) 300,000,000 m/s

Are all of these familiar to you? Maybe some more so than others. A little bit later we will talk about *unit analysis* and at that time show you how to use these equations to define something called unity. Then you will be able to transition comfortably from one measurement to another. There are many other defining equations; we will introduce some of them to you and demonstrate how they are used.

So for now that's it. We've covered a ton of material and that's a lot to remember, so again, it's a good idea to keep your list handy. You'll soon have a chance to apply everything we've covered so far.

Counting Significant Figures

Recorded Measurement	Number of Significant Figures
102.33 m	5 (by rules 1 and 2)
10.0500 s	6 (by rules 1, 2, and 3)
92,000,000 mi	2 (by rules 1 and 4)

Scientific Notation

Because of the enormous range in scales that physicists deal with, they (and you) need to become familiar not only with measurement but with the use of a shorthand known as *scientific notation*, a way to represent very large and very small numbers in a compact way. This shorthand is also sometimes called *powers of 10 notation*, because it entails thinking of large and small numbers as multiples of 10. Referring to multiples of 10 for various measurements fits perfectly with units of measurement in the metric system. For example, there are 1,000 (or 3 powers of 10, 10^3) millimeters in a meter.

def•i•ni•tion

Significant figures are those digits an experimenter records that he or she is sure of plus one last digit that is doubtful. A **defining equation** is a statement of a relationship between two units of measure. **Unit analysis** is the process of changing from one unit of measure to a larger or smaller unit of measure without changing the value of the measured quantity.

def•i•ni•tion

Scientific notation or **powers of 10 notation** is a method of writing very large or very small numbers in a special pattern as follows: _._ _ $\times 10^n$, where the first blank is a digit from 1 through 9 inclusive and the remaining blanks are for all remaining significant figures. In other words, a number written in scientific notation has two parts: (1) a coefficient, which is a number between 1 and 9 inclusive, and (2) the power of 10, which is an exponent either positive or negative. The pattern described above could also look like this $A \times 10^n$. If the exponent (the power of 10) is positive, it tells you how many places to the right of the coefficient to move the decimal. If the exponent is negative, it tells you how many places to the left of the coefficient to move the decimal.

Coefficients and Exponents

A number written in scientific notation has two parts: (1) a coefficient, which is a number from 1 to 9 inclusive; and (2) the power of 10, which is an exponent, either positive or negative. A positive exponent means that the number is greater than 1, and a negative exponent means that the number is less than 1. Putting these ideas together, the format of a number in scientific notation is as follows:

> $A \times 10^B$, where A is the coefficient and B is the exponent expressing the power of 10.

Oh, and just in case you run into it, an exponent of zero is equal to 1. It would look like this: 100 = 1. Below is a table that gives some examples of numbers written both as decimals and in scientific notation.

Examples of Writing Numbers in Scientific Notation

Original Number	Written in Scientific Notation
250,000 mi	2.5×10^5 mi
0.003 m	3×10^{-3} m
5.200×10^{-3} g	5.200×10^{-3} g
2.0500 km	2.0500×10^0 km

The fundamental unit of length is the meter. One million meters is 1,000,000 m, or six powers of 10, written as 10^6 m; that is, 10^6 m is a 1 followed by 6 zeroes. The exponent indicates how many places to the right of the coefficient value to move the decimal. For example, 5.38×10^2 m is the same as 538 m, because we moved the decimal two places (two powers of 10) to the right.

The astronomical unit (AU) is the mean distance between the earth and the sun, or about 1.5×10^{11} m. (You are probably more familiar with the distance as 93,000,000 miles, but we're dealing with metric units in physics, so this gives you an opportunity to see it in other terms.) This number is much easier to write and manipulate in scientific notation than the number 150,000,000,000 m. A light-year (LY), or the distance that light travels in a year, is 9.5×10^{15} m. How many times would you want to write out 9,500,000,000,000,000 m?

This same shorthand can be used for numbers that are very small. A millimeter is 0.001 m, or 10^{-3} m. When the number is smaller than 1, the exponent indicates how many places to the left of the coefficient to move the decimal. A micron is a millionth of a meter. You could write this as 0.000001 m or, more simply, as 10^{-6} m.

Multiplying and Dividing in Scientific Notation

Multiplying and dividing numbers in scientific notation is also relatively simple. If you are multiplying two numbers in scientific notation, you multiply the coefficients and add the exponents. For example, take the two numbers below and see how they are multiplied.

$$(3.0 \times 10^4) \times (2.5 \times 10^2) = 7.5 \times 10^6$$

Dividing two numbers in scientific notation involves dividing the coefficients and subtracting the exponents. For example, see how we deal with the numbers here.

$$(6.4 \times 10^5)/(2.0 \times 10^3) = 3.2 \times 10^2$$

Physics Phun

Write the following numbers in scientific notation: (1) 250,000 mi, (2) 4,830,000 m, (3) 23,000,000 s, (4) 0.000876 kg, (5) 1.0003 cm

Newton's Figs

The number 6.02×10^{23} is known as Avogadro's number. It is used in physics and chemistry and is defined as the number of carbon 12 atoms in 12 grams of carbon 12.

It's About Time, Too!

Earlier in the chapter, you learned that the fundamental units for length and mass are the meter and gram, respectively. But before we work on some calculations, you should be aware of the third fundamental unit used in physics: units of time.

Mastering the flow of time and dividing it into units seems to have been a part of the growth of every civilization on Earth. Astronomer-priests of agricultural societies were responsible for deciding when to begin the annual cycle of tilling, planting, and harvesting. Smaller divisions of time corresponded, at least roughly, to the death and rebirth of the moon (months) and, of course, the daily cycle of light and dark.

Intermediate clusters of days, 5 or 10, and by Roman times 7 days per week, also came up. Dividing time into units smaller than a day proved more difficult, because it involved inventing time-keeping devices rather than mere counting. Hours, minutes, and seconds are relatively recent inventions, as is the idea that these units should have the same duration all year, regardless of the proportion of daylight and darkness each day had.

Fortunately, unlike the units of length and mass, the units of time are used everywhere, even in the United States. Units of one second and longer have the traditional names (minute, hour, day, week, month, century, and millennium), whereas shorter times get metric-style prefixes such as millisecond (one thousandth of a second), microsecond (one millionth of second), and nanosecond (one billionth of a second). And if you look back to the table of fundamental quantities you'll see that (s) for second is the fundamental unit of time for all three systems. That's one less new term for you to think about.

A Few Additional Rules to Keep in Mind

Now that you know the fundamental units of length, mass, and time, and have significant figures and scientific notation fresh in your mind, let's review some rules for precisely calculating physical quantities using given measurements.

- When adding or subtracting, round off the terms of the sum or difference to the precision of the least precise term in the sum or difference, and then carry out the operation.
- When multiplying or dividing, carry out the operation and then round off your answer to have only as many significant figures as the factor in the product or quotient having the least number of significant figures.

Here are a few simple rules to follow when rounding off numbers:

- If the number being rounded off is greater than five, drop the number and increase the preceding number by one.
- If the number being rounded off is less than five, drop the number and leave the preceding number unchanged.
- If the number being rounded off is exactly five and the number preceding is odd, drop the five and increase the preceding number by one. If the number preceding the five is even, drop the five and leave the preceding number unchanged.

When you multiply and divide, use the same rounding procedures, but remember to put them in scientific notation when necessary.

Got it? If not, this will all make a little more sense as you work through some problems at the end of the chapter. And remember this motto: practice, practice, practice!

Putting It All Together

We know you've been patiently waiting to get to some calculations, so let's wind up this chapter by doing exactly that. Suppose you want to add and subtract some measurements of distance. If you arrange the sum or difference so that the decimal points are in a vertical line, as in the following examples, it makes the work much easier. The first column for each operation below is the original set of terms to be combined. The second column shows how the set of terms is rounded to the next closest number.

Add these measurements:

283.6 cm	283.6 cm
34.621 cm	34.6 cm
91.25 cm	91.2 cm
8.36 cm	8.4 cm
	417.8 cm

Subtract these measurements:

478.348 m	478.3 m
332.1 m	332.1 m
	146.2 m

Multiply these measurements:

151 ft

46 ft

906

604

6,900 ft^2 = 6.9×10^3 ft^2

Physics Phun

1. Add the following measurements: 36.98 cm, 905.6 cm, 8.64 cm. 2. Subtract the smaller of these two measurements from the larger: 348.256 ft, 24.3 ft. 3. Find the product of 671.89 m and 53.4 m. 4. Divide the larger measurement value by the smaller: 3258.25 cm^2, 42.5 cm.

Divide these measurements:

$$\frac{480.6m^2}{47.8m} = 10.054 \text{ m rounded to } 10 \text{ m}$$

Did you put the first answer in scientific notation and round your second answer? Another great thing is that the units of measurement behave just like numerals in calculations—that is, ft × ft = ft^2 and m^2/m = m. This is the last important idea for this chapter. And as you work through the problems, you'll get a chance to work with this concept, too.

Problems for the Budding Rocket Scientist

1. Express each of the following in scientific notation: (a) 627.4, (b) 0.000365, (c) 20,001, (d) 1.0067, (e) 0.0067.
2. (a) How many millimeters are there in a kilometer? (b) How many centimeters are there in a millimeter?
3. The diameter of the earth is about 1.27×10^7 m. Find its diameter in (a) millimeters, (b) kilometers, (c) miles.

The Least You Need to Know

- The three fundamental quantities of physics are length, mass, and time.
- Significant figures are those digits in a final calculation that have physical meaning.
- Scientific notation is a form of scientific shorthand used to write either very large numbers or very small numbers.
- The fundamental units of measurement in the MKS system are meters, kilograms, and seconds; in the CGS system they are centimeters, grams, and seconds; and in the FPS system they are feet, slugs, and seconds.

Chapter 2

Movement in One Dimension

In This Chapter

- Speed and velocity
- Changes in motion
- Uniform motion
- Calculating velocity and speed

Understanding the movement of objects is a great place to start our excursion into the world of physics. Now that you have some idea about the type of mathematics used, this chapter introduces you to the terms and ways in which movement in one dimension is measured and defined. We explore the relationship among the concepts of position, speed, velocity, and acceleration. You get a chance to see how each of these is calculated and solve some basic problems that will clearly show you what it means to describe an object that is moving. This will prepare you for upcoming chapters in which we examine movement in two dimensions, how vectors operate, and why things fall down instead of up.

It All Starts with Position

Before we examine how an object moves, it's important to look at where it is in space. In other words, what is its position?

Position is a "relative" measurement. We might say, "I am 4 miles east of the library," or "I live in a house that is 2 blocks east of the zoo." Measurements of position require a zero point, and the zero point we choose is arbitrary. In a laboratory experiment, you might be measuring the motion of a cart from a resting state. The position of a cart at the top of an inclined plane at the start of the experiment is its zero point; its motion (down the incline) is measured from that arbitrary zero point.

In studies of motion, we are concerned with the measurement of the change in position of an object. For example, we might want to know that in a specified period of time, an object covered a measured distance. The position of an object and how it changes as a function of time are basic measurements that we can make, and we can use these measurements to test our ideas of how objects move in space. Distances are typically measured in meters or multiples of this unit (e.g., centimeters, kilometers).

One of Galileo's (there's that name again) great contributions to physics was the exploration of objects in motion, which was made possible by the accurate measurement of time. After time could be measured accurately, then not only the position but also the velocity and acceleration of an object could be measured. (We get to the concept of acceleration a little later in this chapter.)

Let's begin by taking a look at how an object moves from one place to another. Imagine that you're watching a friend sitting on a couch reading this book. At some point your friend gets a little uncomfortable and decides to shift his position by moving farther along the couch. And let's also imagine that over a period of time he would continue to change his position, eventually getting all the way to the other end of the couch. He had to *move* in order for you to notice that he is no longer in the position he was in just a little while ago. Let's call the first place your friend was sitting his initial position and the last place he was sitting his final position. By thinking in terms of physics, it is reasonable for us to say that he has experienced a change in position. This might seem obvious to you, but what we're doing is taking a simple example and putting it into terms of an observation and breaking it down into its component parts so we can discuss movement in the most fundamental way and build from there.

As we can see, not only is there a change in position, but also during the process that his body was in motion, there was something else that occurred. He exhibited *speed.* (Perhaps not the quick movement of someone sitting on a sharp object, but there was still movement from one position to another.) The amount of speed depends on how long it took to make the change in position—that is, the amount of time required to go from his initial position to his final position on the couch, based on the path he took as he went from one position to the other. And because he moved in one direction only, his direction is the same. (In Chapters 3 and 4, we examine movement in more than one direction.)

def•i•ni•tion

An object is said to **move** when it experiences a change in position. How fast an object moves is its **speed. Velocity** is the rate of change, along with the change in direction (if any) of its position.

In properly measuring the motion of an object, we must specify two quantities: its speed and direction. Speed is the rate of change in an object's position. Common units of speed are miles per hour (mi/h), feet per second (ft/s), meters per second (m/s), and knots. A knot is 1 nautical mile per hour, or 1.15 statute mile (the kind we're used to talking about) per hour.

Speed coupled with direction, then, is *velocity.* So whether you're traveling at 60 mi/h due east or 60 mi/h due south is an important piece of information. These same two speeds, but with different velocities, will land you in very different locations. Because the direction of motion is typically an important piece of information, problems in physics generally involve velocities, which are represented by a vector. A vector is simply any measured quantity that has both a direction and a magnitude, or size. (You'll learn all about vectors in Chapter 3.)

Changes in Movement

We now have the basic concepts for a discussion of motion. And those concepts are …? That's right—position, speed, and velocity. Not much is known about your friend's speed so far. Let's begin by assuming that several speeds were involved in accomplishing the change in position.

It took a lot of words to describe the movement of your friend on the couch, but the language of physics will show how simple such a discussion can be. Algebra plays a very important role in helping you to understand these concepts in physics. So let's break down his movement and put it in mathematical terms so we can use these terms to define his position, speed, and velocity.

By using the following symbols instead of words, we can develop formulas that relate his movement to the concepts that we've been discussing:

- x_0 = initial position, or position when we begin measuring time
- x_f = final position, or position at the end of measured time
- v_0 = initial speed, or velocity when we begin measuring time
- v_f = final velocity, or speed at the end of measured time
- $\bar{v}$ = average velocity, or speed over the time interval
- t_0 = initial time, or the reading on a stopwatch to begin measuring time
- t_f = final time, or the reading on a stopwatch at the end of the measured time interval
- Δ = symbol used to represent change

To begin with, we will use x to represent the position at any time. The change in position will be $\Delta x = x_f - x_0$. The change in time can be expressed as $\Delta t = t_f - t_0$. Similarly, if v represents the velocity at any time, then $v_f - v_0 = \Delta v$, or the change in velocity. If we want to find the *average velocity*, we simply divide the change in position by the change in time: in symbols, $\bar{v} = \frac{\Delta x}{\Delta t}$, which can also be written $\bar{v} = \frac{x_f - x_0}{t_f - t_0}$. Did you see how the different parts in the equation can be substituted for one another? Easy, huh?

Now, imagine that a straight line can represent the couch. The distances that your friend moved, x, can be marked off along the line from one end of the couch to the other. Suppose we measure all distances relative to the first place your friend was sitting so that $x_0 = 0$ and $x_f = x$. And to take it one step further, let's suppose that we started the stopwatch when he moved from the initial position and stopped it when he reached his final position so that $t_0 = 0$ and $t_f = t$. That means that $\bar{v} = \frac{x_f - x_0}{t_f - t_0} = \frac{x - 0}{t - 0} = \frac{x}{t}$. In this instance, we just replaced two of the terms, x_0 and t_0, with their numeric values to arrive at our simplified formula of $\bar{v} = \frac{x}{t}$. In other words, the average velocity is equal to the length of the couch, x, divided by the time, t, required for your friend to travel that length. Therefore, from $\bar{v} = \frac{x}{t}$

we get the expression $x = \bar{v}t$, which represents the position at any time t, assuming that your friend is always traveling with the same motion. (If you missed how we got from one expression to the other, all I did was put the expression in terms of x; in other words, we solved $\bar{v} = \frac{x}{t}$ for x.)

def•i•ni•tion

The **average velocity** is the change in position of an object divided by the change in time.

So the position or distance traveled at the end of a time interval is calculated by multiplying the average speed during that time interval by the time of the interval.

Using Algebra to Calculate Velocity

As you can see, algebra enables us to communicate several ideas clearly and quickly with a few simple mathematical statements. Let's consider the following example using the equation we developed earlier, $\bar{v} = \frac{\Delta x}{\Delta t}$. Can you translate that into English?

That equation is read as "velocity equals the change in x divided by the change in t."

So let's say that you leave for vacation. If you travel a distance of 300 mi in 6 h, then your average velocity was 300 mi per 6 h, or 50 mi/h. (We just plugged the numbers into the equation and solved it.) The typical unit of velocity in physics problems is meters per second (m/s), but any unit of distance divided by a unit of time in a specified direction represents a velocity.

To determine how far an object has traveled (x), from an initial position (x_0), after a set amount of time (t), traveling at a constant velocity (v), you could use the following formula, which is derived from the preceding equation.

$$x = x_0 + vt$$

In the example we just gave, if you were traveling with an average velocity of 50 mi/h, and you drove for 10h on your second travel day, and your starting point (x_0) was 300 mi beyond where you were the day before (because you already drove 300 miles yesterday), then your total distance traveled (at the end of the second day) would be

$$x = (300 \text{ mi}) + (50 \text{ mi/h}) \times (10 \text{ h})$$

$$x = 800 \text{ mi}$$

Did you notice that the units are written as a fraction (mi/h) and are read "miles per hour" in this case? Soon we'll guide you through the details of determining the correct units of measure for a given situation by using unit analysis. You probably knew immediately that the units for the answer to this problem would be miles. It's a familiar problem and one that you probably solve daily without even thinking. You'll soon realize that you know a lot of physics stuff already.

Describing Uniform Motion

Did the preceding problem seem familiar to you? Somewhere in your background you learned that distance is equal to rate times time. That idea was used to introduce the relationship between average speed and distance. We also reminded you that your friend's speed was not the same throughout his movement from one end of the couch to the other. It might have been zero at times. From your observations of his behavior, it probably is true that he moved from one end of the couch to the other in a series of rest stops.

After calculating the average speed, we described his motion as if the speed was the same throughout the journey. Averages are very much like the touch-up paint you use for scratches on your car. The paint makes the surface look nice and smooth, like the original paint. Similarly, the concept of "average speed" covers up irregular changes that occurred during the time of motion.

> **def•i•ni•tion**
>
> **Uniform motion** is motion that is characterized by a constant velocity or speed. Whenever a situation in physics states or implies uniform motion, then you will know that the speed is constant. If the speed or velocity is constant, then you know that the object has uniform motion.

Up until now we've treated average speed as if it was the same at every point, and the results we achieved are fine. But average speed is not the same as constant speed. What happens if the speed is constant? Well, as it turns out, one of the simplest types of motion to describe is *uniform motion*. An object traveling with uniform motion is traveling with a constant speed, neither speeding up nor slowing down. Unlike the motion we reviewed in the preceding section (since you were driving, your car was at times speeding up and slowing down), with uniform motion, speed is always the same. Let's translate this idea from words into the symbols we will use to describe uniform motion:

- v = the constant velocity
- x = the distance traveled
- t = the time the object is in motion

Uniform motion is described simply as $x = vt$. Because the velocity is constant (that is, there is no change in speed or direction), the distance traveled is found by calculating the product of the average velocity and the time. The average velocity is always the same because the velocity is constant.

Johnnie's Alert

Always observe significant figures when calculating results that include measured quantities. Always use scientific notation if measurements of physical quantities are very large or very small.

Suppose you drive along the interstate, set your cruise on 80 km/hr, and continue on your way for 30 min. How far will you travel? You probably know that 40 *km* is the correct answer. You are given a constant speed of 80 km/hr and a time of 30 min, both having one significant figure. That means that your car is traveling with uniform motion, and, because $x = vt$, then $80\frac{km}{hr} \times \frac{1}{2}hr = 40km$.

Unit Analysis

Now is probably a good time to talk a bit about unit analysis—that is, units of measure and the units of our answers.

In the answer to the preceding problem, $80\frac{km}{hr} \times \frac{1}{2}hr = 40km$, we expressed the result in kilometers (km). The question is, "How did we get km?" We were given that the constant speed is 80 km/hr and that the time is 30 min. We know that we want distance in km in this problem. By using the defining equation 60 min = 1 hour, we can translate that in terms of hr/min.

In other words, if we divide both sides of that defining equation by 60 min, we get $1 = \frac{1hr}{60\,\text{min}}$. We can multiply any quantity by 1 without changing the value of the quantity. We just change its appearance. Therefore, $80\frac{km}{hr} \times \frac{1hr}{60\,\text{min}} = \frac{80km}{60\,\text{min}}$ because $\frac{hr}{hr} = 1$.

Remember, units of measurement behave like rational numbers.

When we multiply $\frac{80km}{60\,\text{min}} \times 30\,\text{min}$, we get 40 km because $\frac{30\,\text{min}}{60\,\text{min}} = \frac{1}{2}$ and $80km \times \frac{1}{2} = 40km$.

Let's look at a couple of other examples. Express 6.0×10^1 mi/hr in ft/s. Use the following defining equations: 5,280 ft = 1 mi, 60 s = 1 min, 60 min = 1 hr. Let's put that into an equation in which the defining equations show their correct units:

$$6.0 \times 10^1 \frac{mi}{hr} \times 5280 \frac{ft}{mi} \times \frac{1hr}{60\,\text{min}} \times \frac{1\,\text{min}}{60s} = 88 \frac{ft}{s}$$

If you simply cancel the units out in the numerator and denominator (hr/hr, min/min, mi/mi), you are left with ft/s. Then do the math, and you have 88. Putting a defining equation such as 60 min = 1 hour into the form of hr/min is sometimes called "designing unity to have a disguise." Canceling out the common unit terms is another way to express different disguises of unity.

Here's another example. Express 1 year in seconds. Use the following defining equations: 1 yr = 365 days, 1 day = 24 hr, 1 hr = 60 min, and 1 min = 60 s. Then disguise unity from each of those equations and write the following: $1yr \times 365 \frac{da}{yr} \times 24 \frac{hr}{da} \times 60 \frac{\text{min}}{hr} \times 60 \frac{\text{s}}{\text{min}}$, which is another way of writing $1\,yr \times 1 \times 1 \times 1 \times 1$ (which is still 1 yr), but the disguises of unity result in an answer of 3.1536×10^7 sec, the number of seconds in a year.

This process of unit analysis is an easy way to keep track of units of measurement. All you need are the defining equations that can be found in many sources.

> **Physics Phun**
>
> Use unit analysis and appropriate defining equations to express (1) 9.2×10^7 mi in meters, (2) 1 day in seconds. Hint: remember that all digits are significant in defining equations such as 1 day = 24 hours. Only measured quantities limit the number of significant figures in your calculations.

Understanding Acceleration

So far, we have considered only motions involving constant velocity. In the real world, of course, velocities are always varying and are rarely constant. *Acceleration* is defined as the rate of change of the velocity—that is, the change in velocity divided by the time it takes for the change to occur, or

$$\text{Acceleration} = \frac{\Delta v}{\Delta t}$$

Remember that the Greek letter delta (Δ) stands for "change in" the quantity that it precedes. Because velocity is measured in meters per second (m/s) and time is measured in seconds (s), the typical unit of acceleration is meters per second per second (m/s²). Don't try to visualize a square second. There is no such physical quantity. It simply means that the speed in meters per second changes every second. Take a look at the following table and notice the difference between units of speed and acceleration. What do you see as the one real difference between the two?

def•i•ni•tion

Acceleration is the rate of change of velocity or the change in velocity divided by the time for that change to take place.

Units of Measurement of Physical Quantities

Derived Quantities	MKS	CGS	FPS
Speed	m/s	cm/s	ft/s
Acceleration	m/s^2	cm/s^2	ft/s^2

Acceleration is expressed in terms of seconds squared. That's the only difference.

Let's take a look at our formula of acceleration in just a bit more detail and clarify why acceleration is expressed the way that it is:

- a = acceleration
- v_f = final velocity or the velocity at the end of the time interval
- v_0 = initial velocity or velocity at the beginning of the time interval
- t_0 = initial time or time at the beginning of the interval
- t_f = final time or time at the end of the interval

Using this information we can find Δv and Δt. We can then define acceleration as follows: $a = \frac{\Delta v}{\Delta t} = \frac{v_f - v_0}{t_f - t_0}$; that is, acceleration is the change in velocity divided by the change in time or the rate of change of velocity.

Okay, now that we have a good idea of what acceleration is, at least in mathematical terms, let's look at an example of putting it into use. Imagine that you're sitting at a red light. When it changes to green, you press the gas pedal (the "accelerator," which must have been named by a physicist), and the car moves from a velocity of 0 mi/hr to a velocity of 35 mi/hr. As you should know, this change in velocity is called acceleration, and the rate at which your velocity changes tells you how great your acceleration is.

Accelerations determine the final velocity of an object. To determine an object's final velocity (v_f), given that it started as some initial velocity (v_0) and experienced acceleration (a) over a period of time (t), use the equation

$$v = v_0 + at$$

For example, if you're in a car that starts at rest (with an initial velocity of $v_0 = 0$) and you accelerate for 5 s at an acceleration of 10 m/s², what will be your final velocity? Let's plug those values into our equation. We then get

$$v = 0 \text{ m/s} + (10\text{m/s}^2) \times (5 \text{ s})$$

or

$$v = 50 \text{ m/s}$$

Another interesting notion to consider is that a change in direction, not just a change in speed, also constitutes acceleration. Why is this? Well, our definition of velocity is a speed in a given direction. So a change in direction is a change in velocity, and any change in velocity is acceleration (the way we've been defining it, because slowing down is deceleration—but we'll come to that in a second).

When you leave the freeway and enter a curved off ramp, even if you're moving at a constant speed of 50 mi/hr, you are accelerating. This should come as no surprise because you can usually "feel" acceleration. Whether you are feeling your car accelerate from 0 to 60, or feeling it as you round a curve, you are feeling the acceleration.

Now, if you're driving along and for some reason you slam on your brakes, this would cause another kind of acceleration (a negative acceleration, or slowing down, called deceleration). For example, suppose a car traveling at 10 m/s is brought to rest by its brakes at the uniform rate of 2.5 m/s². How long will it take for the car to stop? If we know that the braking acceleration is –2.5 m/s², then the car will slow by 2.5 m/s every second. So a car moving at 10 m/s will be moving at 7.5 m/s after the 1 s, and 5.0 m/s after 2 s, and will be at a standstill after 4 s.

Describing Uniformly Accelerated Motion

There's one more aspect of acceleration that we need to consider before closing this chapter—the idea of uniformly accelerated motion. An object that travels with uniform accelerated motion is one that simply has constant acceleration. Remember how we were able to describe the motion of your friend on the couch by using his average velocity? Now let's describe your friend's motion in terms of constant acceleration—a very different discussion. We will use the same symbols we used before to describe motion. Along with these symbols, we will develop the three basic ideas you need for the solution to any problem concerning an object that has constant acceleration. Some of the algebraic solutions that you'll want to be able to construct are outlined here, too.

Any time that you consider a problem that states or implies that an object moves with constant acceleration, these three ideas apply:

- $x = \overline{v}t$: Distance is equal to the average velocity or speed times the time.
- $\overline{v} = \frac{v_f + v_0}{2}$: The definition of average velocity where v_f and v_0 are given or implied.

  v_f is the velocity at the end of the time interval and v_0 is the velocity at the beginning of the time interval.
- $a = \frac{v_f - v_0}{t}$: The definition of acceleration, where t is the length of the time interval.

An example will illustrate how you can use these three formulas. You're in a car that was initially at rest but is now traveling with acceleration a and continues to accelerate until it reaches a final speed of v_f. How far did you travel? Use the procedure that we went through earlier in the sections "Changes in Movement" and "Using Algebra to Calculate Velocity."

Express the problem first in general algebraic terms, then substitute the given information from your problem into the equations.

In general $x = \overline{v}t$; that is, distance equals average velocity times the time. Use the definition of $\overline{v}$ (the second equation above) to find that $x = \left(\frac{v_f + v_0}{2}\right)t$. Using the definition of acceleration, we find that $x = \left(\frac{v_f + v_0}{2}\right)\left(\frac{v_f - v_0}{a}\right) = \left(\frac{v_f^2 - v_0^2}{2a}\right)$ for the general solution. The second term in the equation $\frac{v_f - v_0}{a}$ is simply the equation $a = \frac{v_f - v_0}{t}$ solved for (t).

The particular solution is obtained by substituting the information implied in the statement of the problem, $v_0 = 0$, and finding that $x = \frac{v_f^2}{2a}$. We still have a solution in

> **Physics Phun**
>
> An object has an initial velocity v_0 and an acceleration a. How far does it travel in a time t? If $a = 6.0$ ft/s^2, $v_0 = 6.0$ ft/s, and $t = 1.0 \times 10^1$ s, then $x = 360$ ft.

algebraic symbols, but the beauty of this solution is that you have solved all problems of this type that involve uniformly accelerated motion.

So with all of that under your belt, just remember that this is a discussion of motion in one direction along a straight line. Velocity and speed are closely related and may be used interchangeably under these conditions. We can describe uniform motion by the simple statement $x = vt$, and describe uniformly accelerated motion using three ideas: $x = \bar{v}t$, $a = \frac{v_f - v_0}{t}$, and $\bar{v} = \frac{v_f + v_0}{2}$.

Problems for the Budding Rocket Scientist

1. You use an Internet map site to calculate the driving distance from your house to your parents' house for the holidays. The results say that the distance is 250 miles, and the travel time is 5 hours. What speed is the program assuming you will average for the trip?
2. A skydiver reaches a terminal velocity of 10 mi/hr at a height of 1 mile above the earth. How long will it take her to reach the ground in minutes?
3. A spaceship accelerates from rest at $1g$ (9.8 m/s^2) for 10 minutes. What is the final velocity of the spaceship after this 10-minute acceleration in (a) km/s and (b) km/hr?

The Least You Need to Know

- Average speed or average velocity is equal to distance divided by time.
- Uniform motion is defined as $x = vt$.
- Uniformly accelerated motion can be described using these three equations: $x = \bar{v}t$, $a = \frac{v_f - v_0}{t}$, and $\bar{v} = \frac{v_f + v_0}{2}$.
- Changes in units of measurement are accounted for by using defining equations and unit analysis.
- Always observe significant figures when calculating results that include measured quantities.

Chapter 3

Introduction to Vectors

In This Chapter

- Why do we need vectors?
- The difference between vectors and scalars
- Vector addition and subtraction
- Component parts of vectors
- Velocity vectors

In the legendary days of Long John Silver and swashbuckling pirates, when tall ships sailed the high seas, treasure booties were frequently buried on uncharted, out-of-the-way islands. To aid their rum-soaked memories, pirates drew treasure maps, pointing the way to their hidden treasures. These maps usually had a coconut tree or a particularly odd-looking rock as a starting point, and consisted of a series of arrows that would steer the lucky map holder to the treasure. Each arrow was labeled with a number of paces to be taken and a compass direction. Unknown to them, pirates were using vectors and vector addition long before they were invented by mathematicians.

Vectors—Geometry in Motion

Throughout their history, mathematics and physics have been intimately related; a discovery in one field led to an improvement in the other. Early natural philosophers grappling with quantities such as distance, speed, and time used geometry inherited from the Greeks to explore physical problems. But in the tumultuous years of the seventeenth century, physics underwent a transformation—a shift in emphasis from numerical quantities, such as distance and speed, to *vector* quantities, such as displacement and velocity. (If you don't know what those are yet, don't worry, you will very soon.)

This transition was neither abrupt nor confined to that century. It was necessary to invent new mathematical objects—vectors—and new mathematical machinery for manipulating them—vector algebra—to embody the properties of the physical quantities they were to represent.

def•i•ni•tion

A **vector** is a quantity that has both magnitude and direction. Distance and speed are examples of magnitudes of vectors. **Magnitude** is the size of the vector in the proper units of measurement. It is the length of the arrow representing the vector. The length can be drawn to scale to represent the magnitude of any vector.

Geometrically, a vector is represented by an arrow whose length corresponds to the *magnitude* and whose direction, from the tail to the head of the arrow, corresponds to the direction of the vector.

Although a vector has a magnitude and direction, it has no fixed position in space. You can take a vector, like the one shown in Figure 3.1, and slide it parallel to itself anywhere, and you still have the same vector. Two vectors having the same magnitude and direction are considered equal. They can be translated so that one fits exactly on top of the other.

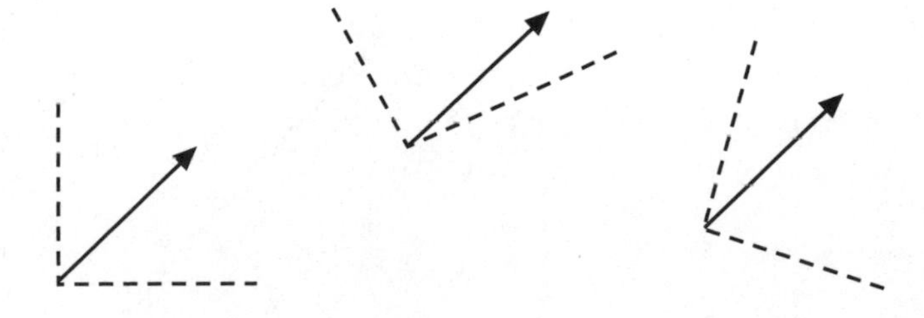

Figure 3.1

The same vector shown at different positions in space.

Vectors exist independent of any coordinate system, because they don't have any definite position. For this reason they are useful in mechanics or the branch of physics that deals with the study of motions and forces produced by moving bodies through direct contact. Why is this, you may ask? Before Copernicus, there was only one conceivable reference frame, the center of the earth. (Before Copernicus, everyone thought that the earth was the center of the universe.) The essence of the Copernican revolution was to move the origin of the coordinate system describing all physics from the center of the earth to the center of the sun. Then, in devising a new mechanics, Galileo discovered through the law of inertia (something else you'll be learning about in Chapters 6 and 8) that there is no preferred frame of reference. There is nothing unique about coordinate systems; no one is better than another. Consequently, coordinate systems or reference frames can be anywhere in space, oriented in any way. Similarly, vectors can be anywhere in space. Vectors are natural devices to describe physical quantities with complete generality and independence from particular reference frames.

Newton's Figs

Equality of vectors is very special in that two vectors are equal if and only if they have equal magnitude and the same direction.

Because a vector has a magnitude and a direction, it is more than a single number, and therefore needs to be represented by a special symbol, different from what we use to represent a single number. The most common symbol used in physics to represent a vector is a letter with an arrow over it, like this: $\vec{A}$. In Figure 3.2, we have given you a representation of two vectors. Included also is a label for each. The symbols $\vec{A}$ and $\vec{B}$ are used to refer to the arrows representing the corresponding vectors so that we may discuss the vectors without drawing the arrows. In addition, these two vectors are equal because they have the same length and the same direction. In terms of the new symbols, we say $\vec{A} = \vec{B}$.

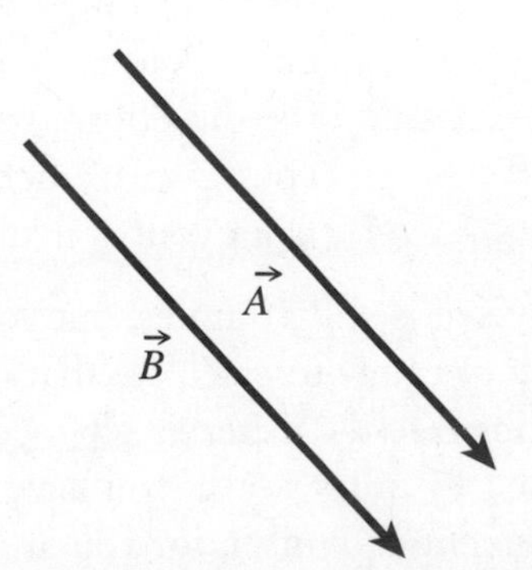

Figure 3.2

Vectors $\vec{A}$ and $\vec{B}$ demonstrate equal vectors. $\vec{A}$ and $\vec{B}$ are algebraic symbols used to name the arrows that represent vectors.

Vectors and Scalars

A quantity that has both size and direction is a vector quantity, or simply a vector. One that has size only is a *scalar* quantity. A scalar is simply an ordinary number such as 3 or 5. The magnitude of a vector is also a scalar. Distance is usually treated as a scalar quantity. A car can be said to have traveled a distance of 15 miles regardless of the direction it was traveling. On the other hand, under certain conditions, direction does make a difference when it is combined with the size of the distance. If town B is 15 miles north of town A, then it is not enough to direct a motorist to travel 15 miles to reach town B. The direction must also be given. If he travels 15 miles north, he'll get there; but if he travels 15 miles east (or any direction other than north), he'll never make it. If we call a combination of size and direction of distance traveled displacement, then we can say that displacement is a vector.

In physics, different names are used to distinguish vector quantities from their scalar magnitudes. One example is velocity versus speed. The velocity v of a moving object is a vector quantity represented by an arrow pointing in the direction of the motion. The length of the velocity vector, v, is called the speed; it tells how fast the object is moving and is a scalar quantity.

Another example is displacement versus distance. If a particle moves a distance s from one point to another, the vector s joining them is called the displacement; its magnitude s (the distance it moved) is a scalar.

Since one letter is oftentimes used for both quantities (to add to your confusion), they are differentiated by one letter being in bold type and the other in italic type.

def•i•ni•tion

A **scalar** is a name for ordinary numbers such as 2 or 7. The magnitude of a vector is also a scalar.

The importance of differentiating between vectors and scalars is that the two are manipulated differently. For example, to add scalars, you use ordinary addition. If you travel 15 miles in one direction, then travel 15 miles in another direction, the total distance you travel is 15 plus 15, or 30 miles. Whatever the directions, the total mileage is 30. But what is your displacement? How far, in other words, are you from your starting point?

The total distance traveled is still 30 miles, but your final displacement is 21.2 miles northeast. (We'll show you how to calculate that displacement in a little while. For now we just wish to illustrate the difference between scalars and vectors.) If you travel 15 miles north and then 15 miles south, you have still traveled 30 miles altogether, but your total displacement is 0 miles, for you are back at your starting point.

As a final way to help clarify the difference between scalars and vectors, we've put together a table that lists both types of quantities. Take a look at it and notice the difference between the bold type and the italic type. Some of the terms we haven't covered yet, but it will all become clear as we progress through the material.

Table 3.1 Common Scalar and Vector Quantities in Physics

Scalars		Vectors	
distance	*s*	displacement	**s**
speed	*v*	velocity	**v**
acceleration	*a*	acceleration	**a**
force	*F*	force	**F**
time	*t*		
mass	*m*		

Vector Addition

All vectors are subject to three simple mathematical operations: addition, subtraction, and multiplication. Let's take a look at the first of these, addition. As you've already seen, adding scalars is straightforward; 15 and 15 is always 30. But in vector addition, 15 and 15 can be anything from 0 to 30, depending on the directions involved. Let's consider the two vectors in Figure 3.3.

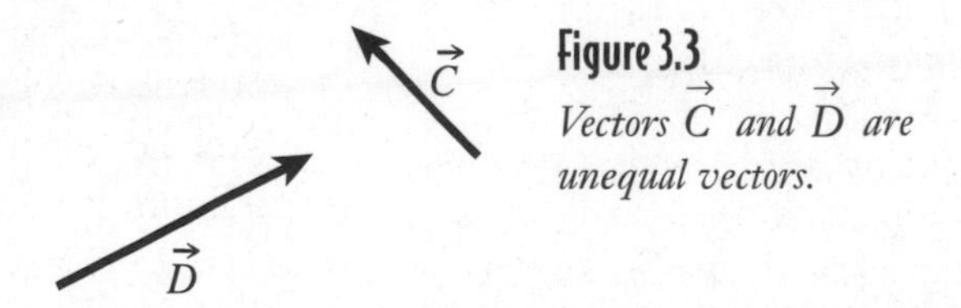

Figure 3.3
Vectors $\vec{C}$ and $\vec{D}$ are unequal vectors.

As you can see in the figure, vectors $\vec{C}$ and $\vec{D}$ are not equal. They have neither the same direction nor the same magnitude. We'll use these vectors to demonstrate a method for adding vectors. And you'll soon see that what is true about adding these two vectors is true for adding any two vectors.

The tools you need to construct vector drawings are a ruler and protractor. The ruler will help you measure magnitude and the protractor will provide you with the angular measurement, which defines the direction. And to make it even easier, you may want to provide yourself with compass headings, such as the top of the paper to represent the direction north and the bottom of the paper as south. (These compass headings aren't really necessary yet, but they will come in handy when we examine movement in two directions in the next chapter.)

Here's one way that you can add the vectors in Figure 3.3. These aren't rules, they're just steps to follow when constructing vectors that you'll be adding together.

1. Construct vector $\vec{C}$ at some convenient place in the plane. (A plane here is represented by a flat sheet of paper.)
2. Construct vector $\vec{D}$ with its foot exactly at the tip of the head of vector $\vec{C}$.
3. Draw a new vector from the foot of $\vec{C}$ to the head of $\vec{D}$. Let's call the new vector $\vec{R}$, which is the sum of $\vec{C}$ and $\vec{D}$. To write what we just did in vector algebra, we can say that $\vec{C} + \vec{D} = \vec{R}$.

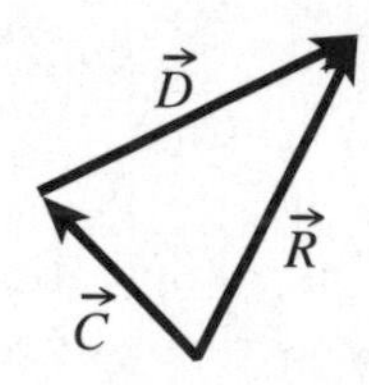

Figure 3.4

The vector sum demonstrates a method of adding vectors.

Newton's Figs

Remember the steps for adding vectors. These will apply any time you add two or more vectors together.

Now, let's add these vectors another way. The only thing you need to always do is be sure that the length and angle of the vectors you're drawing are the same. In other words, if you're checking to see whether two or more vectors always add up to the same *resultant vector*, the magnitude and direction of the vectors that you're comparing must always be exactly alike.

In Figure 3.5, instead of starting with $\vec{C}$ we begin with $\vec{D}$ and add the same two vectors. That is, construct $\vec{D}$ first and then place the foot of $\vec{C}$ exactly at the head of $\vec{D}$. Draw the new vector from the foot of $\vec{D}$ to the head of $\vec{C}$ and compare it with the vector $\vec{R}$ that we found in the previous example. Have you noticed anything special? Well, if you added your vectors in reverse but kept the same magnitude and direction, the resultant vector $\vec{R}$ should be exactly the same magnitude and direction as the previous resultant vector. This leads us to an important axiom in vector addition: no matter what order you add the vectors in, the resultant vector will always be the same. If we want to state this mathematically (or in vector algebra), we have $\vec{D} + \vec{C} = \vec{C} + \vec{D}$.

def•i•ni•tion

The **resultant vector**, or just the resultant, is the name assigned to the vector representing the sum of two or more vectors.

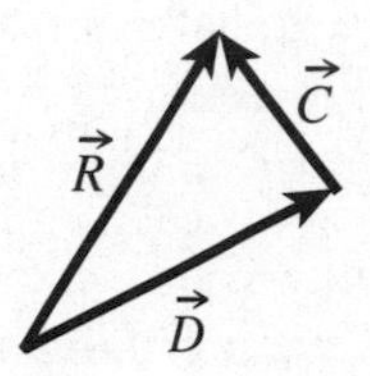

Figure 3.5

The vector sum demonstrates that vectors can be added together in any order and the resultant vector will be the same.

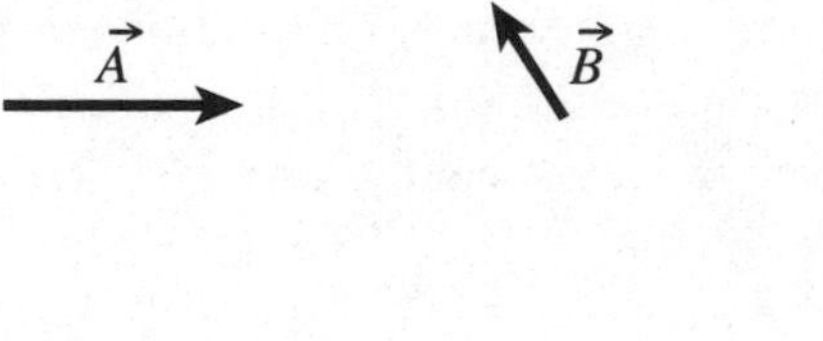

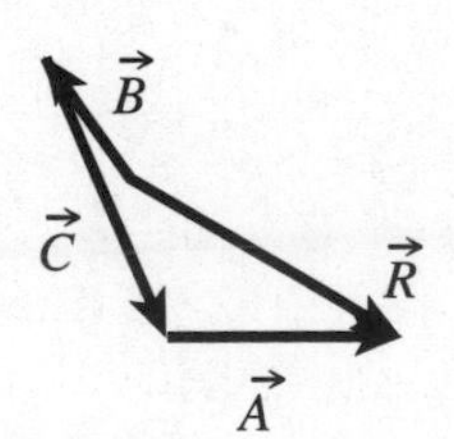

Figure 3.6

The vector sum $\vec{B} + \vec{C} + \vec{A}$ *shows a method of adding more than two vectors together.*

In Figure 3.6, we used three different vectors to illustrate how you can add more than two vectors together. We chose the order $\vec{B} + \vec{C} + \vec{A}$, but as you now know, any order of adding them would produce the same result. And one of the most important points to remember when adding any number of vectors in any order is to always draw the resultant vector from the foot of the first vector you construct to the head of the last vector in your sum.

When we first began drawing vectors in this section, we mentioned that the tools you would need are a ruler and protractor. We also mentioned that to orient yourself, a compass point would be helpful. Thus far we've only used angles to define the direction of a vector. Let's look at how compass points can be used to define direction.

We've already used a ruler to measure magnitude. For example, if your vector is 5 cm long, it has a magnitude of 5. But what is its direction? Up until now, that direction has been defined by the angle read on a protractor in relation to the other vector. It is easier to define a direction, especially if you need something more concrete than an arbitrary starting point, by relating it to one of the directions on a compass. The easiest convention to use is to have north as the vertical upward direction and south as the vertical downward direction. That would then make east to the right of your page and west to the left. Pretty straightforward, don't you think?

Let's suppose that you have an angle that's 60° upward from the horizontal. Refer to Figure 3.7 to see what we're talking about. As you can see, the direction can be either East 60° North or West 60° North. In other words, the direction is upward and to the right of the page (east) or to the left of the page (west). Got it? The only time that you'll be using just one direction in a heading is if the movement is straightforward in that direction, like due east or due south. Otherwise, you'll need two directions to indicate the direction of the vector. Let's see if you've got a handle on this by doing the following Physics Phun problem.

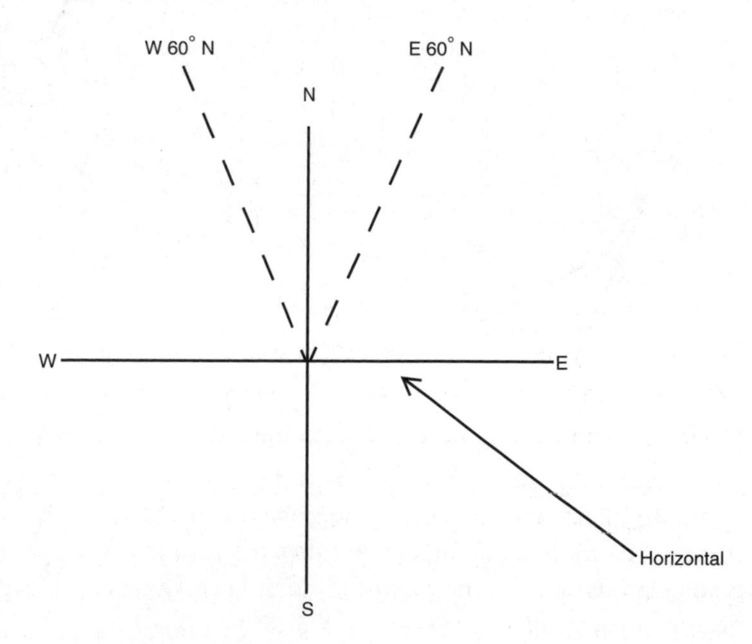

Figure 3.7

Vectors in relation to compass points.

Physics Phun

We measured the vectors in Figure 3.6 using a ruler and a protractor, and got $\vec{A}$ = 1.9 cm E, $\vec{B}$ = 1.1 cm, W55°N, and $\vec{C}$ = 2.4 cm E67°S. Did you notice that both the magnitude and direction must be given to specify a vector? Use those same instruments and make the same measurements to see if you agree with the measurements we recorded. Then add these vectors together using the steps outlined in this section along with the aid of the example for adding three vectors. Now, take a measurement of $\vec{R}$. That's a little different from adding 1.9 + 1.1 + 2.4. As you can see, our vector sum doesn't equal 5.4.

Subtracting Vectors

When we know how to add vectors, subtraction is pretty straightforward because we can always treat subtraction as the addition of a negative vector. But what's a negative vector? Well, it is simply a vector with the same magnitude but opposite direction. Look at Figure 3.8 and you'll see an example of a negative vector, represented by $-\vec{H}$. The figure also shows just what a graphical representation of vector subtraction looks like.

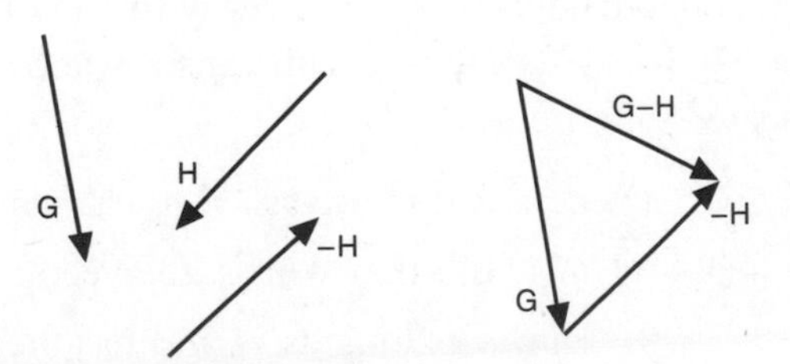

Figure 3.8

A negative vector and vector subtraction.

Let's say that you have two vectors, $\vec{G}$ and $\vec{H}$, and you want $\vec{G}-\vec{H}$. You first form $-\vec{H}$, then add it tail to head to $\vec{G}$. The vector from the tail of $\vec{G}$ to the head of $-\vec{H}$ is $\vec{G}-\vec{H}$. We can express subtraction mathematically as $\vec{G}-\vec{H} = \vec{G} + (-\vec{H})$.

A swimmer in a stream and an airplane in a wind are good physical illustrations of vector addition and subtraction. The swimmer, for example, makes a certain progress with respect to the water he's in, but he's also swept along in some other direction by the current. What is his actual velocity as seen by someone standing on the bank of the stream? Can you see that it is the vector sum of the swimmer's velocity in still

Physics Phun

In Figure 3.8, construct the negative of $\vec{G}$ and add it to $\vec{H}$. How does the difference vector $\vec{H}-\vec{G}$ compare to the difference vector $\vec{G}-\vec{H}$?

water plus the velocity of the stream itself? (We have a problem similar to this example at the end of the chapter, with the answer in Appendix C.)

Before moving on to vector multiplication, there's one last important point to make. If we asked you what the vector sum is called in vector addition, you would say the resultant, right? So what do you think the vector is called in subtraction? No big surprise here—it's simply called the difference vector.

Multiplication in Vector Algebra

So far we've covered some important parts of vector algebra. We've covered (1) the definition of a vector, (2) what equal vectors are, (3) what vector addition is, and (4) how to subtract vectors. The explanation of one more operation, vector multiplication, will provide us with all the vector algebra we need to pursue the development of many ideas in physics.

There are several types of vector multiplication, but we consider only *scalar multiplication*, which is the process of multiplying a vector by a scalar. When you first learned to multiply, it was probably introduced as a quick way to add. That is, 4×3 is just a quick way to write $3 + 3 + 3 + 3$. You get the same answer both ways, but 4×3 is faster and less complicated to write.

Let's suppose you are given a vector $\vec{A}$ and are asked to calculate $4\vec{A}$. One way to look at it would be $\vec{A} + \vec{A} + \vec{A} + \vec{A}$. In other words, the vector $4\vec{A}$ is a vector four times as long as $\vec{A}$ and has the same direction as $\vec{A}$. Did you notice that scalar multiplication produces a vector and not a scalar? A general statement that we can make is that for any scalar, say n, $n\vec{A}$ is a vector having the same direction as $\vec{A}$ but n times as long as $\vec{A}$. That means that $\frac{1}{4}\vec{A}$ is a vector with the same direction as $\vec{A}$ but only $\frac{1}{4}$ the length of $\vec{A}$. So as you can see, multiplication of a scalar times a vector is pretty straightforward.

def•i•ni•tion

Scalar multiplication is the product of a scalar and a vector that results in a vector with a magnitude determined by the scalar. The direction of the product is the same as the original vector if the scalar is positive and opposite the direction of the original vector if the scalar is negative.

Components of Vectors

A handy skill to learn in analyzing vectors is to look at the parts of a vector. The process of identifying the parts of a given vector is called *resolution.* The parts are called the *components* of the vector. This might be the same as asking for change for a dollar. There are many possible combinations that might add up to a dollar. But suppose that we want to use the change in a vending machine. Then pennies would not be very helpful. In the same way, some vector components might prove to be more helpful than others. As we proceed in our review of physics, you will find that resolving a vector into certain components can be very helpful in solving problems.

Two components you may want to be prepared to find are the two that are perpendicular to each other. The easiest way to do this is to construct the projection of a line segment on a plane. A component of a vector is its "shadow" or a perpendicular drop on an axis in a given direction. So drop a perpendicular from the endpoints of a line segment to a plane and connect the two points in the plane with a line segment. We have constructed that bit of geometry for you in Figure 3.9. Take a look at the line segment labeled AB and its projection (shadow) on line l. That projection is a line in a plane perpendicular to the page and is labeled A′B′.

def•i•ni•tion

Resolution is the name of the process of identifying the parts of a vector. The **components** of a vector are those parts whose sum is the given vector.

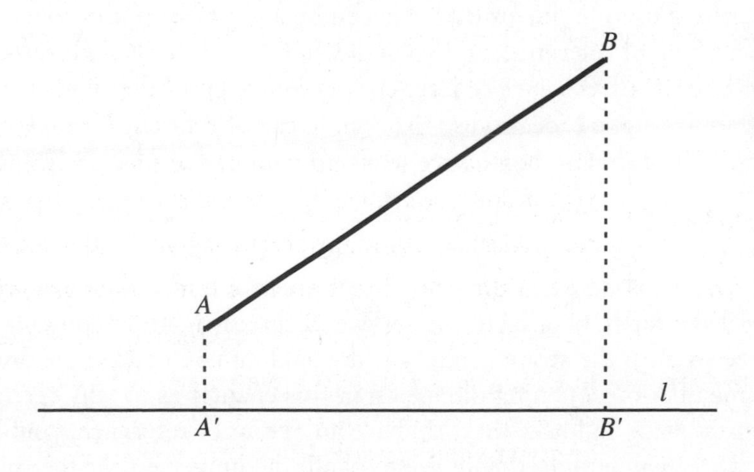

Figure 3.9

The projection of AB on l illustrates the projection of a line segment on a plane.

The information summarized in Figure 3.9 can be used to find two mutually perpendicular components of any vector. Now look at Figure 3.10. Given a vector $\vec{A}$, let's construct two mutually perpendicular lines through the foot of $\vec{A}$. First construct the projection of the vector just like you did in the previous example and get line A′B′. Then rotate your plane 90° and do the same thing for line m and get segment A′C′. Place an arrowhead on each of the components at the points on lines l and m, corresponding to the projections of the tip of the arrow representing $\vec{A}$.

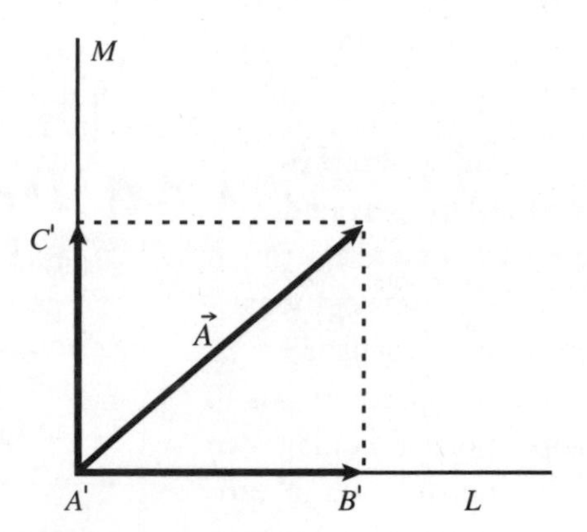

Figure 3.10

The resolution of $\vec{A}$ into two mutually perpendicular components illustrates the identification of two crucial vector components.

Position and Displacement

Our study of motion has made us aware that an object moves in relation to some point or place. Describing motion means that you assume a certain point or place, often referred to as a frame of reference, and you tell where an object is with respect to that place. A point on the earth's surface is a good reference point that you are familiar with, so it's always a good idea to use that as a frame of reference. And by using a small enough part of the earth, we may act as if it were a flat plane.

The study of vectors enables us to discuss positions and motions in a plane where earlier we discussed motion along a straight line. As you can see, vectors enable you to discuss the familiar motion along a straight line as well as motion in a plane perpendicular to the surface of the earth, parallel to the surface of the earth, and in any plane between. Vectors free us from the straight-line world, enabling us to tackle the universe. And if you remember our opening discussion in this chapter as to why vectors were created in the first place and how they can have any point of reference, you'll realize that vectors work anywhere in the universe—well, the universe near the surface of the earth, at least, because that is our reference point.

Remember your friend on the couch? You can begin applying to him these new concepts of describing motion. If we let his initial *position* be your reference point, we can say his initial position, a vector quantity, is $\vec{X}_0$ and his final position is $\vec{X}_f$. When he moved from one end of the couch to the other, he had a change in position of $\Delta\vec{X} = \vec{X}_f - \vec{X}_0$.

> **def•i•ni•tion**
>
> **Position** is a vector used to locate a point from a reference point or frame of reference.
>
> **Displacement** is a vector and is defined as a change in position.

Did you notice that this is the difference in two vectors? And that means that $\Delta\vec{X}$ is a vector drawn from the head of position $\vec{X}_0$ to the head of position $\vec{X}_f$. Earlier we said that your friend experienced a change in position. We called that change $\Delta\vec{X}$, the distance traveled along the couch, because it was a scalar quantity. However, $\Delta\vec{X}$ is a vector quantity, because it has magnitude, which is the distance traveled, and direction. In physics, $\Delta\vec{X}$ is the vector called *displacement.*

Understanding Velocity

Let's go back to your friend again. (He must really be getting tired of sitting on the couch this long!) We used that simple situation to define average speed. In the same frame of mind, and using the same frame of reference as we did before (even though we didn't call it that), let's define a new concept, *velocity.*

To begin, we'll take the displacement of your friend and divide by the time required to make the displacement. $\frac{\Delta\vec{X}}{\Delta t} = \vec{V}$ is the equation resulting from that operation, and it states that the displacement divided by the time is equal to the average velocity. Velocity is a vector quantity, as you can see. It results from the division of a vector $\Delta\vec{X}$ by the scalar Δt, or, in this case, t, the time for the displacement to take place. The velocity vector has the same direction as $\Delta\vec{X}$ and has the speed for a magnitude.

Velocity is a vector quantity while speed is a scalar quantity. Velocity and speed are not the same things, and they are not interchangeable in general. If you refer back to Table 3.1, you can see just why those quantities were found under vector and scalar headings. Occasionally a discussion might involve objects in motion along a straight line in one direction;

> **def•i•ni•tion**
>
> **Velocity** is the rate of change of displacement. Velocity is a vector quantity that has the same direction as the displacement, and its magnitude is the speed.

then speed and velocity will have the same values. The instant the object changes direction, only velocity is defined. Velocity can only be used when an object moves off the straight line into a plane or changes direction and moves backward. For example, you might say that your car has a speed of 60 mi/hr, the reading from your speedometer. If you say that your car is traveling 60 mi/hr*N*, you've included both speed and direction, which is now the velocity of your car.

The velocity reading would be a combination of a reading from the speedometer and a reading from a compass. The important thing to remember is that speed and velocity are two different quantities. You will get plenty of practice with these ideas in later sections. Below are some vector problems. Some are easier than others. Try working them out the best you can. The solutions are in Appendix C.

Problems for the Budding Rocket Scientist

1. Does a car speedometer indicate speed or velocity? If your answer is speed, then what additional instrument would help you specify your velocity? If your answer is velocity, explain how.
2. A ship sails 60 miles in a direction 30° north of east, then 30 miles due east, then 40 miles 30° west of north. Where is the ship with respect to its starting point?
3. An excellent swimmer who can swim 1.7 m/s in still water tries to swim straight across a river. The current in the river causes the swimmer to move instead at an angle 32° from his intended direction with a speed of 2.0 m/s. What is the speed of the river current?

The Least You Need to Know

- Scalars are used to express size or distance, while vectors are used to define both size and direction.
- Velocity is a vector quantity. It describes both magnitude (size) and direction.
- The combination of size and direction traveled is called displacement and it is also a vector.
- The process of identifying the parts of a vector is called resolution.
- The study of vectors enables us to discuss positions and motion in a plane rather than in one direction only.

Chapter 4

Motion in Two Dimensions

In This Chapter

- Walk around until you're "displaced"
- Human cannonballs and boats
- Acceleration with a change in velocity
- Speed, velocity, and centripetal acceleration

So far, although we described position in two and three dimensions, we have been discussing motions in only one dimension and in one direction. In the real world, we are much more interested in the motion of objects in two dimensions and sometimes in three, too. You'd have a hard time getting to any of your destinations if all you could do was move in only one direction. This chapter introduces you to motion in two dimensions.

The Keys to the Kingdom

The key to understanding motion in two dimensions is to realize that an object's motion can be broken into two perpendicular components (as we did at the end of the last chapter), or components that are separated in direction by 90°. For example, motions in two dimensions can be separated into a component perpendicular to the ground (the y-direction) and one parallel to the ground (the x-direction).

You may already be familiar with the concept of locating a point or a line on a graph. The two-dimensional graph has a vertical line called the y-axis and a horizontal line called an x-axis. And if you can tell me the name of this system of graphing, you get to move to the head of the class. The system is called the *Cartesian coordinate system.* We won't be doing a lot of work with this system, but it's always good to know where the idea of an x- and y-axis comes from. We'll come back to x and y in just a little while.

def•i•ni•tion

The **Cartesian coordinate system** is named after the seventeenth-century French mathematician and philosopher René Descartes. His idea was to represent geometric points by numbers. In a plane, like this page, two numbers on a rectangular grid define a point in space. And given at least two points, you can then define a line.

Displacement Revisited

You were introduced to the concept of displacement at the end of Chapter 3. Because it's such an important part of motion, let's look at a real-world example: taking a walk. Suppose you take the following route around the neighborhood. Start at your front doorstep, walk 3.0 blocks north, then 6.0 blocks east, then 8.0 blocks south, 2.0 blocks west, and finally 1.0 block north. You have walked 20.0 blocks. Now you want to calculate your displacement from your front doorstep to the spot marking the end of your walk.

Do you remember that displacement is a vector quantity? If you do (or even if you don't), the individual displacements must be added together to find the displacement from your front door. Choose a convenient scale for constructing the magnitudes of displacements; perhaps 1 cm represents 1 block. Construct your own solution and compare your results with our solution given in Figure 4.1.

Johnnie's Alert

Preserve the magnitude and direction of all vectors when you perform any type of vector operation.

Applying what we have previously learned, your answer will have two significant figures as well as two parts—magnitude and direction. You may have already used the method that we're outlining here to solve some problems from the last chapter, but in case you haven't, the technique employs the use of a ruler and a protractor, and your solution involves a scale drawing. This technique is called a *graphic solution.*

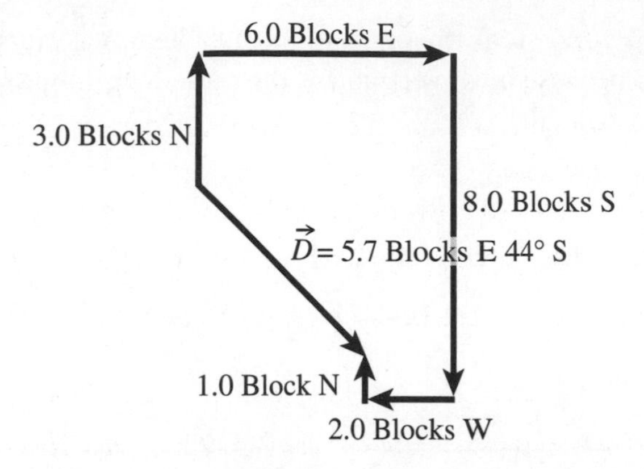

Figure 4.1

The solution to a displacement problem shows how different distances traveled relate to displacement.

We found a displacement of 5.7 blocks E44°S. When you measure the magnitude with the ruler, you find 5.7 cm, where 1 cm represents 1 block on your scale drawing. Drawing a horizontal line through the foot of the displacement vector enables you to measure the angle with a protractor. Another way to solve the problem is to add all north and south displacements together and then add all east and west displacements together. The result obtained turns out to be two mutually perpendicular components of the displacement vector. When we used this method, we found 4.0 blocks south and 4.0 blocks east for the two components. Can you do that as well and come up with the same values?

Another way that the magnitude of the displacement vector can be calculated is by using the *Pythagorean theorem*, where the magnitudes of the components make up the legs of a right triangle. The same right triangle will enable you to measure the appropriate angle with a protractor. Let's apply that theorem to this problem and see if we do indeed come up with the same solution.

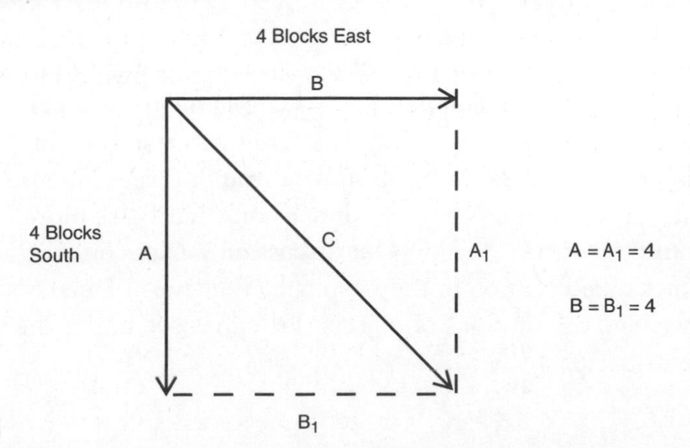

Figure 4.2

Application of the Pythagorean theorem to find the magnitude of a displacement vector.

When we apply the Pythagorean theorem to this problem, we can solve it the following way. Remember, because both vectors are the same length, we can use either vector to form our right triangle.

$$\vec{A}^2 + \vec{B}_1^2 = C^2 \text{ or } A_1^2 + B^2 = C^2$$

Substituting our values, we get the following:

$$4^2 + 4^2 = C^2, \text{ then } 16 + 16 = 32, \text{ and } \sqrt{32} \text{ or } 5.656 \text{ or } 5.7$$

def•i•ni•tion

A **graphic solution** is achieved by using drawing instruments to construct a scale drawing. A ruler provides the measurement of magnitude to scale and the protractor enables you to measure angles. The **Pythagorean theorem** is a theorem from plane geometry relating the legs and hypotenuse of a right triangle. The square of the hypotenuse is equal to the sum of the squares of the two legs. The legs are the mutually perpendicular sides and the hypotenuse is the side opposite the right angle in the right triangle.

If you can solve this problem using the two methods suggested here, you will not only have a check on your work but you have a head start on similar vector problems discussed later. As you can see, the displacement is not the distance traveled. It has magnitude and direction resulting from the addition of all of the displacements on your trip. Compare the magnitude of displacement with the distance traveled—20 blocks!

Motion in the x-y Plane

Earlier we briefly introduced you to the idea of using the x and y graphing system to understand motion in more than one dimension. As you know from experience, when objects move through space in the real world, they rarely move in one dimension. Real objects such as birds, airplanes, and balls move through three-dimensional (3D) space. It is possible to model motions in three dimensions, but that's more complicated than we want to deal with right now. Our discussion focuses on motion in two dimensions, sometimes called motion in the x-y plane. This type of analysis is particularly useful for describing the motions of objects through space under the influence of gravitational forces.

Let's think about how we can apply the x-y plane to the flight of a human cannonball. When our human cannonball is shot out of the cannon at some initial velocity, the only acceleration affecting her (ignoring air friction) is the impact or acceleration of gravity, and gravity acts in one direction: downward, perpendicular to the ground. (Actually, gravity acts toward the center of the earth, but locally this appears to be perpendicular to the ground.) If we ignore air currents, there is no force acting parallel to the ground. So if we divide the velocity vector of the human cannonball into two components, the x and y components, the problem becomes simpler to understand.

The x-component of her velocity (parallel to the ground) is unchanged during her flight. The y-component of her velocity changes gradually, starting with a positive (up) value, slowly changing to zero (as the acceleration of gravity slows her down), and then changing to a negative (down) value (as the acceleration of gravity speeds her up). The combination of these x- and y- direction motions produces a shape known as a *parabola*. Objects such as baseballs and human cannonballs that move through the air follow a parabolic trajectory. We take a look at these trajectories in more detail in Chapter 7. For now, let's just treat these motions as straight-line vectors.

Let's assume that the human cannonball is shot from her cannon at a 30° angle to Earth and that she leaves the cannon at a velocity of 5.0 m/s. What are her vertical and horizontal velocity components? Figure 4.3 shows a triangle that represents the vertical and horizontal components of her velocity. To answer the question we can use a graphic solution or the Pythagorean theorem.

def•i•ni•tion

A **parabola** is the shape defined in the x-y plane by the equation $y = x^2$. It looks like an upside-down bowl.

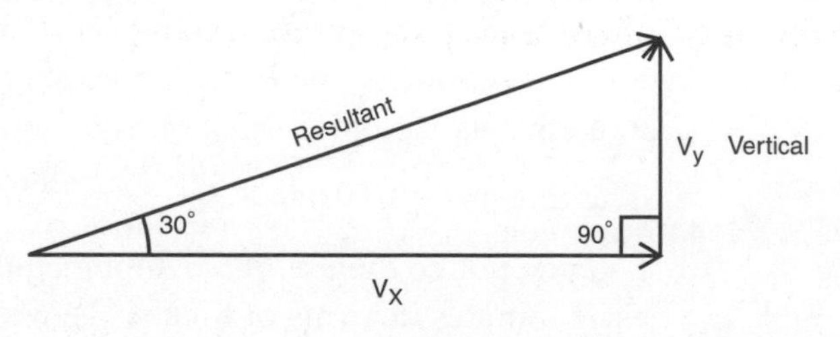

Figure 4.3

The vertical and horizontal components of the initial velocity of the human cannonball.

Because we have already covered the graphic solution method, you can get some practice setting up a scale drawing based on Figure 4.3. Just remember to make it as accurate as possible, so use a ruler and protractor. If you did it according to scale, you should come up with the values of $(v_x) = 4.3$ m/s, and $(v_y) = 2.5$ m/s. Then we can use the Pythagorean theorem to check our work.

$$v_x^2 + v_y^2 = c^2, \text{ or } (2.5)^2 + (4.3)^2 = 5.0^2,$$
$$6.25 + 18.49 = 24.74 \approx 5^2$$

Understanding the Velocity of a Boat

Underlying all methods of solving perpendicular vectors is the x-y plane, but it's useful to see other ways to solve problems and to think in as many different ways as possible.

Suppose you can row a boat in still water at 4.0 mi/hr and you want to travel across a river that is flowing from north to south at 5.0 mi/hr. If you maintain your velocity directly across the river for 10.0 minutes, you reach the other side. What is the velocity of the boat relative to the ground? How far down the river will the boat land? What is the minimum boat velocity (relative to the water) required to travel in a straight line across the river?

As usual, analyze the problem before attempting a solution. There are several questions to be answered, but all of them involve uniform motion. Approach the questions one at a time. List the information given in terms of the type of motion and the type of quantities involved.

The boat will move through the water at 4.0 mi/hr whether the water is moving or not; that is, $\vec{V}_{bw} = 4.0$ mi/hr E (the subscript *bw* represents our vector boat in water), assuming you start on the western side. The river has a uniform velocity relative to the ground of $\vec{V}_{wg} = 5.0$ mi/hr S (the subscript *wg* represents our vector, water to ground). The time can be tricky to interpret unless you note that we should express it in hours so that all information is in hours and miles. The time is $t = 10.0\,\text{min} \times \dfrac{1hr}{60\,\text{min}} = 0.167hr$.

Did you notice the use of a defining equation to write minutes in terms of hours? We want to find the following three vectors: $\vec{V}_{bg}$ (the subscript *bg* represents our vector, boat to ground), then $\vec{D}_{wg}$

Newton's Figs

If you did not interpret the problem in this section as described, don't worry. You will develop this type of critical thinking and focused reading by the time you complete this book.

(the letter D stands for distance in this case), for time t, and lastly, $\vec{V}_{bw}$ for the boat to travel east in a straight line.

Refer to the vector diagram in Figure 4.4 to visualize the problem and begin a solution. The diagram is a graphic solution, where we assigned 1 cm to represent 1.0 mi/hr.

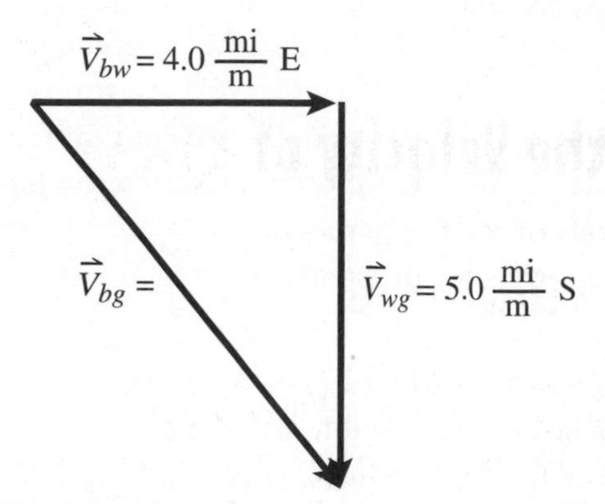

Figure 4.4

The velocity of a boat crossing a stream as an example of a component velocity solution.

Use any scale you choose to solve the problem. Note that we do not need a scale drawing to find $\vec{D}_{wg}$ because we know the time and velocity. Therefore, $\vec{D}_{wg} = \vec{V}_{wg} t$ = (5.0 mi/hr S)(0.167 hr) = 0.85 mi S. Does it surprise you that the boat lands 0.85 mi downstream from where it started? We've now answered one of the questions, but there is no reason to follow any particular order to answer all the questions.

Looking at Figure 4.4 again, we can see that $\vec{V}_{bg} = \vec{V}_{bw} + \vec{V}_{wg}$ and the graphic solution is 6.4 mi/hr S39°E.

The figure also shows that $\vec{V}_{bg} + (-\vec{V}_{wg}) = \vec{V}_{bw}$, where $\vec{V}_{bg} = 4.0 mi/hr E$ when the boat travels east in a straight line. You can construct a graphic solution or you can see by the symmetry of the problem that $\vec{V}_{bw} = 6.4 mi/hr\ N39°E$. That means that if you want to go straight across the river you must paddle like a stern-wheeler or get a motor to maintain the required speed through the water.

You can see that this problem involves an object moving through a medium (in this instance, water) while the medium is moving relative to the ground. You need to specify the frame of reference to clarify your solution to a motion problem. Because all motion is relative, you can see that it is essential to specify relative to what point or to what frame of reference. When you did that in this last solution, you arrived at an understandable answer by stating motion relative to the water or to the ground.

Uniform Circular Motion

The physical world is filled with many examples of circular motion: children on a merry-go-round, planets orbiting the sun, the propeller on an airplane, or even a yo-yo. At some time in your experience, you may have whirled some object about your head at a constant speed. A string tied to the object and the other end of the string held in your hand defined the circular path of the object. You probably even released the object at some point and watched it fly away in a line that was tangent to the circular path at the instant of release. If that was a long time ago, just keep all of the details in mind. Because a circle is a plane figure, and uniform motion in a circular path is motion in a plane, you use vectors to describe the motion.

Speed and Velocity in a Circular Path

Now that you have a pretty good idea of what circular motion is from the world around you, and perhaps some personal experience of it, let's take a look at how physics views it. The distinction between speed and velocity reveals that there is only one vector acceleration that results from the motion of the object in a circular path at a constant speed. The speed can be calculated in terms of the time required for the object to make one *complete revolution* or *circular path.*

Given that the speed is constant, the time required to make one complete revolution is called the *period of motion.*

def•i•ni•tion

One **complete revolution** or **circular path** is the circumference of the circle. The **period of motion** is the time required to complete one revolution.

Because the period is a very special time in physics, it is usually represented as T in seconds or some other unit of time, such as minutes or perhaps hours. The speed then would be the distance traveled in one revolution divided by the time to complete one revolution. The geometry of the circle enables you to calculate the distance traveled in one revolution by calculating the circumference of the circle. The speed is calculated to be $V = \frac{2\pi R}{T}$, where R is the radius of the circle and T is the period of the motion.

Have you noticed that mud on a spinning tire flies off tangent to the tire at the instant it leaves the tire and travels in a straight line? That is, the straight-line path is perpendicular to a radius of the tire. That means a 90° angle is formed between the radius of the tire and the straight-line path of the mud. Whenever any object is moving in a circular path at a constant speed, it has a velocity that is tangent to the circular path at every instant.

As you learned in Chapter 3, vector quantities are written in bold with an arrow over them and scalar quantities are represented in lighter-face italicized letters. In the following discussion, you'll see how that's put into practice. We use $\vec{A}$ to represent velocity and v to represent speed. A position vector $\vec{R}$, length R that is drawn from the center of the circle along a radius to a point on the circle indicates the position of the object on the circular path relative to the center of the circle. Two positions of an object moving in a circular path at a constant speed have been identified in Figure 4.5. The two positions can be anywhere on the path.

We chose $\vec{R}_0$ to be the position at the beginning of a time interval Δt or t. Either one may be used to represent a small time interval. $\vec{R}_f$ is the position of the object at the end of the time interval. The corresponding velocities of the object are represented by vectors tangent to the path at those positions. All this information is used along with Figure 4.5 to guide you through calculations of quantities implied by the physics of this situation.

Centripetal Acceleration

Let's suppose that you are given the speed of an object as v. That means that $v = v_0 = v_f$ because the speed is constant. Look at the diagram in Figure 4.5. Notice that d is a *chord*, and s is a corresponding *arc* of the circular path. If you recall from geometry, small angles the length of d are approximately equal to the length of s and for that reason you can feel free to substitute s for d when the need arises.

def•i•ni•tion

In mathematics an **arc** is a portion of a curved line, as of a circle. A **chord** is the straight line that intersects an arc or a curve.

The angles θ (the Greek letter theta) and ϕ (the Greek letter phi) are small angles, as small as you want to make them. So approximating the length of the chord with the length of the arc is legitimate. It will be easy for you to understand if we refer to the angles in terms of their right sides and left sides. In other words, if you imagine standing at the vertex of an angle and looking out along its sides, the side on your right is the right side and the one on your left is the left side.

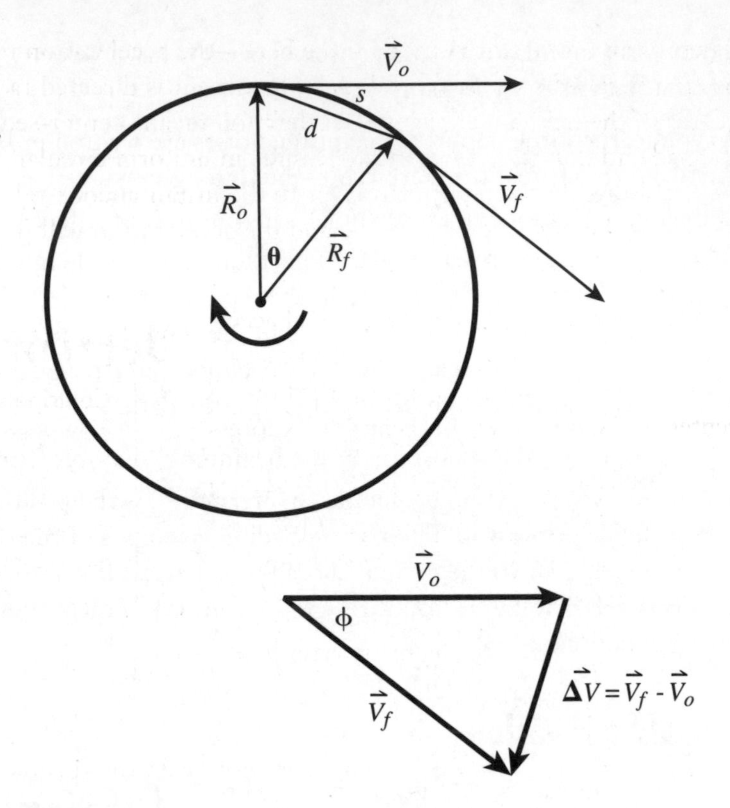

Figure 4.5

The object with uniform circular motion is located by the position vectors $\vec{R}_0$ and $\vec{R}_f$.

You can see that even though the speed is constant, the velocity changes at every instant. Why? Because at every moment the movement of the object on the circle is constantly changing direction, and any change in direction is a change in velocity. This change in velocity can be calculated as before. That is, construct the velocity vectors with their feet together so that you can find $\vec{V}_f - \vec{V}_0 = \Delta\vec{V}$. Be sure to remember that the magnitude of $\vec{V}_f$ equals the magnitude of $\vec{V}_0$, which also equals v. But the magnitude of $\Delta\vec{V}$ is ΔV (and remember that Δ, the Greek letter delta, means "change in"), which simply means that the length of the change in the velocity vector is equal to the change in velocity determined by the geometry of the triangle in the construction. Can you see that it is not Δv , because the speed v is constant? It does not change. Because there is a change in velocity, there must be acceleration. Although we could work out the proof for acceleration using the two triangles in the figure, we'll save you time and just tell you that the equation for finding acceleration is $a = \frac{v^2}{R}$.

There is an important point to make here—the acceleration for uniform circular motion is called *centripetal acceleration* because it is directed radially inward toward the center of the circle. Centripetal acceleration means center-seeking acceleration. It is the one and only acceleration that results in uniform circular motion. If the acceleration is always directed perpendicular to the instantaneous velocity vector, then the motion will end up being a circle, and the acceleration will point toward the center of that circle.

Any time we are discussing centripetal acceleration, it is important to understand that the direction is radially inward toward the center of the circle. At this point let's express centripetal acceleration completely by $\vec{a}_c = \frac{v^2}{R}$ toward the center of the circle. And because speed is equal to distance divided by time (something you already know), we can substitute $\frac{2\pi R}{T}$ for v in the expression $\vec{a}_c = \frac{v^2}{R}$

def•i•ni•tion

Centripetal acceleration is the center-seeking acceleration of a body with uniform circular motion. Furthermore, it is the only acceleration defined by uniform circular motion.

and get $\vec{a}_c = \frac{\left(\frac{2\pi R}{T}\right)^2}{R} = \frac{4\pi^2 R}{T^2}$. If you are not given the speed you can solve for it by simply rearranging our previous formula to $v^2 = \vec{a}_c R$ or $v = \sqrt{\vec{a}_c R}$. If you know any of the two variables in the equation, you can solve for the missing one by simply rearranging the formula in terms of what you're solving for. So, for example, to solve for period or time, we could use the formula, $T^2 = \frac{4\pi^2 R}{\vec{a}_c}$ or $T^2 = \frac{4\pi^2 R}{\vec{a}_c}$. Can you rearrange the formula to solve for the radius, R?

Physics Phun

1. Calculate the speed of an object moving in a circular path of radius 2.0 m if it has centripetal acceleration of 2.6 m/s².
2. Calculate the period of the motion of the object in (1).
3. Calculate the radius of the circular path of an object if it has centripetal acceleration of 2.0 m/s² and a speed of 3.0 m/s.

So What's the Angle?

Up until now we've used the term circular motion to describe circular movement in two dimensions. However, in physics we often refer to circular motion as *angular motion.* You'll understand why in just a little bit. There is really no difference between the two terms, but the way that you can solve angular motion problems doesn't involve vectors. In other words, we're going to discuss some ideas in angular motion that we haven't covered yet and that may be helpful in solving problems of a different kind.

def•i•ni•tion

Circular motion, or more generally **angular motion,** is measured in **radians,** which is an angular measure scaled to a circle.

What is the unit of distance in linear motion? We're hoping you said the meter. That's something that should be engrained in your mind by now. Well, angular motion is measured somewhat differently. Imagine that you are facing north. We'll assign a value of 0° to this direction. If you rotate your body so that you are now facing east, you have undergone angular motion. But what if we asked you how far you've gone? It would be hard to describe your movement in meters, so instead we would say that you have rotated through an angle of 90°. If we ask you to rotate until you're facing south, your angular motion would now be 180°. If you then face due west, you've now rotated 270°, and when you are again facing north, you have rotated through 360°. As you can see, angular motion is measured in degrees (°) or radians, called rad for short. But radians have no unit of measure. They are simply a special unit unto themselves that are used to measure a change in angular motion.

Newton's Figs

A radian is a unit used to measure angular position. There are 2π *(read 2 pi)* rad in a circle. And to give you an idea of how they relate to degrees, let's convert them back and forth. If there are 2π rad in 360°, then there are π rad in 180°. To convert from degrees to radians we simply multiply: Radians = Degrees × $(\pi/180)$. And to convert from radians to degrees we multiply: Degrees = Radians × $(180/\pi)$.

Another way to measure angular motion is in revolutions (one full rotation is a revolution). A rad is an angular measure scaled to the circle. It is the angle for which the arc length on a circle of radius r is equal to the radius of the circle. So in one complete revolution the arc length is the circumference of the circle. We can then say that there are 2π rad in one complete revolution, π rad in a half revolution and $\pi/2$ rad in a quarter revolution. If you look at Figure 4.6 you'll get a clearer idea of what this all means.

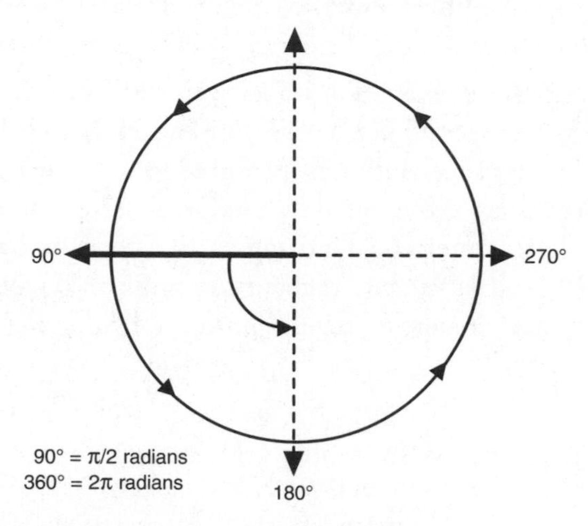

Figure 4.6

A circle showing the relationship between degrees and radians.

Crankshafts and Merry-Go-Rounds

Some cars have a gauge called a tachometer that measures the number of revolutions per minute (rev/min) of the engine's crankshaft. You'll notice that when your car is not moving the tachometer will read somewhere between 500 and 1,000 rev/min, known as the idle speed. Idle speed, you say? How can you have speed when you're not moving? The speed referred to is the angular speed, or the number of revolutions per minute that the engine's crankshaft is making. When you press your foot on the accelerator you feed more fuel to the engine, causing the crankshaft to rotate at a higher angular velocity—whether or not the car is moving. When you put your car into drive the transmission (automatic or manual) transfers the motion of the crankshaft to the wheels.

Angular velocity is simply a measure of the rate of change in angular position. So, for example, imagine that you're standing as before and you make one full revolution in 10 s. Your angular speed is 360° per 10 s, or 36° per second; you could also say 0.1 rev/s, or 0.2π rad/s. All of these expressions are equivalent, but the most commonly used units in problems are radians per second and revolutions per minute. And because radians are dimensionless units, the unit of angular velocity is simply 1/s.

Now, because there are angular velocities there must also be angular accelerations. But what does it mean to have an angular acceleration? Let's picture a stationary merry-go-round. Because it's not rotating, its angular velocity is zero. Okay, start pushing it so that it turns at a velocity of one revolution every 4 s, or 0.25 rev/s. The merry-go-round has been accelerated to this angular velocity. What happens if it continues to move at a constant angular velocity? Yes, you're right—its angular acceleration drops to zero.

As you know, accelerations are changes in velocity, so the units of angular acceleration are expressed in radians per second per second. And again, because radians are dimensionless units, angular acceleration is measured in units of $1/s^2$. In physics, angular position is most often represented by the Greek letter theta (θ), angular velocity by the Greek letter omega (ω), and angular acceleration by the Greek letter alpha (α). In the same way that velocity is change in position divided by change in time, angular velocity (ω) is change in angular position divided by change in time, or

$$\omega = \frac{\Delta\theta}{\Delta t}$$

and angular acceleration is change in angular velocity over change in time, or

$$\alpha = \frac{\Delta\omega}{\Delta t}$$

And one more way to think of angular velocity is as a linear velocity (v) divided by a radius (r), or

$$\omega = \frac{v}{r}$$

We hope this isn't all Greek to you. Just remember that all of this is simply another way to look at circular motion, along with another way to solve problems that involve angular motion, velocity, or acceleration. You'll get a chance to work a few of these types of problems at the end of this chapter—which also happens to be right now!

Problems for the Budding Rocket Scientist

1. A 0.3 kg weight is attached to a 1.5 m long string and is whirled around in a horizontal circle at a speed of 6 m/s. What is the centripetal acceleration of the weight?

2. A merry-go-round rotates once every 5 s. (a) How many degrees does it rotate through half a revolution? (b) How many radians? After 20 revolutions, (c) how many degrees has it rotated through? (d) How many radians?

3. The merry-go-round in the last problem started at rest. If it took 5 s to get it to a rotation rate of once every 5 s, what angular acceleration did the merry-go-round undergo?

The Least You Need to Know

- The key to understanding motion in two dimensions is to realize that an object's motion can be broken into two perpendicular components.
- The technique of drawing a scale diagram to find displacement is called a graphic solution.
- The distinction between speed and velocity reveals that there is only one vector acceleration that results from the motion of the object in a circular path at a constant speed.
- The acceleration for uniform circular motion is called centripetal acceleration because it is directed radially inward toward the center of the circle.
- Angular motion can be measured in terms of radians or revolutions.

Part 2

How and Why Things Move

In Part 1, we introduced you to some fundamental ideas about motion. You're now ready to tackle some of the reasons why things move, the laws that govern that motion, and how you can figure out how much you'll weigh on the moon. This section explains how satellites stay in orbit around the earth, why heavy and light objects all fall at the same rate of acceleration, and perhaps even shows why the Dark Force isn't all Darth Vader made it out to be. Along the way, you'll meet some more luminaries such as Galileo and Newton and see how important their contributions were to physics. So let's continue our descent down the rabbit hole and see if we can catch up to the white rabbit.

Chapter 5

Forces and Motion

In This Chapter

- Newton's laws of motion
- A practical application of a vector
- The parallelogram method
- Forces and uniform motion

In 1687, Sir Isaac Newton made available what is considered to be the greatest work in physics ever published. It was titled "The Mathematical Principles of Natural Philosophy," or *Principia* for short. What he did was to bring together all of the knowledge about physics that had been discovered so far, and expanded on it. He combined and synthesized ideas that would remain unchanged for more than 300 years, and even then the alterations that have been made to his theories are minimal. (Although Einstein did a good job of standing some on their head.) This chapter will be your introduction to his three laws of motion. We look at them in detail and see how and why they are so useful in our study of dynamics, or the world of inertia, force, and mass.

The Three Laws of Motion

To remove confusion from some of the terminology that was being tossed around in physics, Newton began his *Principia* by defining what he meant by inertia, mass, and force. He used these terms in his axioms of motion, which he presented as statements needing no proof. That is, they were accepted as universal principles that could only be tested and applied.

Newton inherited from Galileo and Descartes the essential idea that motion along a straight line with a constant speed was the natural state of any body, needing no further explanation. This is Newton's first law, the law of inertia. Stated in his own words:

> First law: Every body continues in its state of rest, or of uniform motion in a straight line, unless it is compelled to change that state by forces impressed upon it.

What would be the path of our solar system's planets if there were no force acting on them? The first law tells us that in the absence of any force, a body will continue to move in a straight line. So if there were no force on the planets, they would travel in straight lines, not in their elliptical orbits around the sun.

Newton, like Galileo before him (and we'll get to Galileo in a couple of chapters), realized that an object's *inertia* (or its resistance to acceleration) was somehow connected to its *mass*. He defined mass as the quantity of matter that rises conjointly from an object's density and size. The greater the mass of an object, the more difficult it is to get it moving if it's at rest, and also to prevent it from continuing in motion with a constant velocity. This idea led to Newton's second law:

> Second law: The change in motion is proportional to the force impressed; and is made in the direction of the straight line in which the force is impressed.

In other words, force is equal to the product of mass times acceleration.

Let's look at it in a little more detail. What Newton meant by "motion" involved not only a body's velocity, but also its mass. It is a quantity we call *momentum*, the product of mass m and velocity v. Stated as an equation, the second law is $F = \frac{d}{dt}(mv)$. The *d/dt* is a mathematical symbol meaning "the instantaneous rate of change of." When we understand that, the equation expresses Newton's second law almost as we would in an English sentence in which F is the subject and "=" is the verb. The mathematical sentence says, "Force is equal to the change of momentum of an object."

If the second law is applied to a body for which the mass m is a constant, then by a rule of differentiation (just a little calculus wizardry; don't worry if you can't follow, we're just trying to get to the formula that may be the most familiar to you) this means that $d(mv)/dt = m\, dv/dt$. And because we know that the instantaneous rate of change of velocity is acceleration, $a = dv/dt$, we can finally arrive at Newton's idea that acceleration is caused by forces, and the formula, $F = ma$.

def•i•ni•tion

Inertia is resistance to acceleration. **Mass** is the amount of substance or stuff that there is in an object. **Momentum** is the product of mass times acceleration. And **force** is a push or a pull.

Pushes and Pulls

To understand Newton's second law, we need to understand the concept of force. A *force* in physics is merely a push or a pull. If you hit me with your light saber, you've exerted a force. If I summon the Dark Side, I haven't exerted a force on you—well, at least not yet. I've been at rest. When you push on something, you can feel yourself exerting a force. When armed with that sensation, you can look around and find countless examples of things exerting forces on other things. Pushes, pulls, gravity, tension in a string, and friction are all examples of forces that enter Newton's second law. But these forces must originate outside the object whose motion we're trying to describe. In other words, only external forces acting on an object can change its motion.

The force F need not be just one force acting on the body. It is the vector sum of all external forces acting on the object. Even though vectors hadn't been invented yet, Newton knew that forces have both a magnitude and direction. Whenever we write $F = ma$, F symbolizes the vector sum of external forces acting on a body. The acceleration of an object is the result of the total force acting on it.

Action and Reaction

Newton needed one additional law to express what happens when several bodies interact with each other. His third law is:

> Third law: To every action there is always opposed an equal reaction: or, the mutual actions of two bodies upon each other are always equal, and directed to contrary parts.

When you push on anything—a door, a pencil—it pushes back on you with a force equal in magnitude but in the opposite direction. In other words, you can't touch without being touched. That's the essence of the third law—a law of interactions. Sometimes it's difficult to isolate the action-reaction pairs of forces in Newton's third law. As a guide, remember that they always act on different bodies, never on the same body. If you know one force—for example, you pull on a rope—you can find the reaction force by turning the sentence around: the rope pulls on you.

The Basics

Before we get to some vector problems involving force, we need to spend a little time discussing and examining the basic units of mass, momentum, and force. As you already know, the fundamental unit of mass in the metric (SI) system is the kilogram. Do you remember the unit of mass in the CGS system? It's the gram. This is old hat to you by now, right? So let's take a look at some of the new fundamental units.

Momentum, being a product of mass and velocity, is a derived quantity. In the metric system, the basic unit is kg m/s. It has been suggested that this unit be called a Descartes, after the French mathematician/philosopher, or a clout, which is a little more descriptive. In the British system, momentum comes in units of slug ft/s.

Through Newton's second law, force should have the same units as mass times acceleration. Therefore, the SI unit of force is the kg m/s^2, called a *newton* and abbreviated N. One newton is the force required to accelerate 1-kg mass at 1 m/s^2:

$$1N = 1kg\frac{m}{s^2}$$

def•i•ni•tion

A **newton** is the force required to accelerate a mass of 1 kilogram at a rate of one meter per second squared. A **dyne** is the force required to accelerate a mass of 1 gram at a rate of 1 centimeter per second squared. A **pound** is the force required to accelerate a mass of 1 slug 1 foot per second squared.

The unit of force in the CGS system is called the *dyne* and is the force that will accelerate a 1g mass at an acceleration of 1 cm/s^2. Using 1 kg = 10^3 g and 1 m/s^2 = 10^2 cm/s^2, can you show that 1 N = 10^5 dyne? (Not too hard, just substitute some values and convert the units.)

In the SI and CGS systems, mass, length, and time are the fundamental quantities. Force is a derived unit. In the British system, however, the standard quantities are force, length, and time. The unit of force in the British system is the *pound*, abbreviated lb, which is 1 slug ft/s^2. I've worked out the conversion units this time (I can't make you do all of the work!) and we get 1 lb = 4.45 N.

When an object is in free fall, gravity accelerates it downward at a constant acceleration g. (Which tells you that acceleration and gravity are the same, an insight of Einstein's.) Newton's second law tells us that the force must be $F = mg$, where the direction of the force is vertically downward, toward the center of the earth. This is what we mean by the weight of an object; weight is the force of gravity acting on an object (whether it's falling or not). Being a force, weight is a vector. If we call the magnitude of the vector W, then $W = mg$ near the surface of the earth.

In countries that still use the British system, people are often confused between kilograms and pounds. These units refer to different physical quantities. Yet labels list the weight of an item in pounds along with its mass in kilograms and don't specify that one is weight and the other is mass. Unlike the mass of a body, which is an intrinsic property of a body, the weight of a body depends on its location. If you know the mass of an object, you can find its weight if you know the acceleration of gravity at that location. Moving an object around on the surface of the earth doesn't change its weight very much, but moving it to the moon changes it considerably, without affecting its mass. We get into these ideas in more detail in the next chapter when we discuss gravitational force.

Newton's Figs

It's an important concept in physics to understand and differentiate between mass and weight. The two are often mistakenly interchanged. Mass is an intrinsic property of a body, while its weight depends on the acceleration of gravity.

The Vector Nature of Force

Let's review briefly what we've learned so far and extend those concepts. As you already know, the expression $F = ma$ is called Newton's second law. Another way to write this formula so that you can more easily identity force as a vector is $\vec{F} = m\vec{a}$, where $\vec{F}$ is a vector, m is a scalar, and $\vec{a}$ is a vector. The equation states that if a single force is applied to an object that is free to move, the object will accelerate in the direction of the applied force with a magnitude that is directly proportional to the magnitude of the force. In all cases, the single force and the acceleration it causes act in the same direction.

We usually think of a push or pull as a result of objects in contact. In some cases, such as that of a magnet, a planet, or an electric charge, there can be a push or pull even though the two objects may not be touching. We must add forces and subtract forces just like any other vector. It may be more convenient in some cases to add forces in a somewhat different way than you have done before to find the resultant force. Some forces do push or pull by being in contact with an object. If several forces are acting on the same point in a body, you may add them together two at a time by drawing a diagram that begins with the feet of their vectors together.

Let's apply that idea in the solution of this problem: an SUV is stuck on a muddy road. Two ropes are tied to the front bumper, and a man pulls on one of the ropes with a 2.50×10^2 N (where N stands for newtons, not the direction north) force that makes an angle of 30° with a line down the middle of the road. Another man pulls on the other rope with a force of 3.00×10^2 N at an angle of 50° with the same line. What is the total force exerted on the vehicle? Both ropes are pulled parallel to the ground. If you refer to Figure 5.1, which illustrates how this problem is laid out graphically, you'll see just how we can solve it.

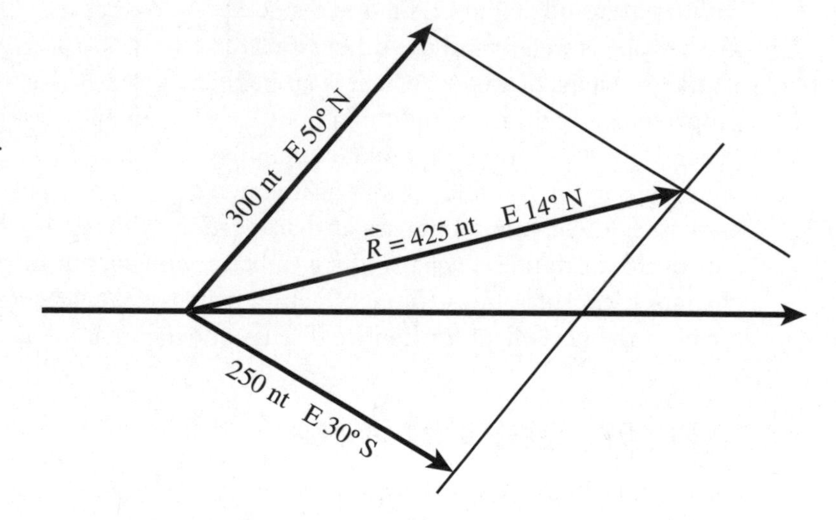

Figure 5.1

The vector diagram of forces illustrates the parallelogram method of adding two vectors.

The method of adding the forces in Figure 5.1 works only for the addition of two vectors at a time and is called the *parallelogram method.* (The method of adding vectors that we used in the last chapter is called the *closed polygon method*, and it works for any number of vectors. If you look back to Figure 4.1 in the previous chapter, it's easy to see why the method is called that.) Essentially, what you want to do is create the shape of a parallelogram from the two forces involved. A simplified version of the parallelogram method can be seen in Figure 5.2. As you can see, the resultant of two vectors acting at any angle may be represented by the diagonal of a parallelogram. The two vectors are drawn as the sides of the parallelogram and the resultant is its diagonal. The direction of the resultant is away from the origin of the two vectors.

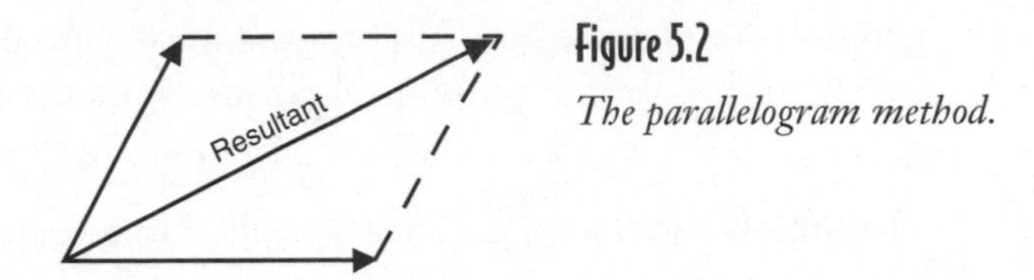

Figure 5.2

The parallelogram method.

By using a ruler and protractor as our tools for solving graphical problems (drawn to a scale of your choice), we come up with a resultant vector of $\vec{F} = 425nt$ (and in this particular answer we're using nt for newtons so that we don't confuse you with the N used to designate the direction north; in all other instances we'll use N for Newtons) at an angle of E14°N with the line down the middle of the road. We also calculated it to three significant figures even though the original angles did not indicate that much precision. According to Newton's second law, the vehicle will tend to accelerate at an angle of 14° with the middle of the road if these are the only two forces acting on the vehicle. That may be just enough to get the SUV off the high center and free to move.

def•i•ni•tion

The **closed polygon method** of adding vectors requires that any number of vectors may be added together by joining them foot to head until the sum is complete. The resultant vector is found by closing the polygon with a vector drawn from the foot of the first vector in the sum to the head of the last in that order. The **parallelogram method** of adding vectors requires that the feet of two vectors be located at the same point. A parallelogram is then constructed by drawing a line parallel to one of the vectors through the head of the other. A second line is constructed parallel to the second vector passing through the head of the first. The resultant vector is found by joining the feet with the opposite vertex of the parallelogram in that order.

Components of Forces

You often see people pulling or pushing an object to make it accelerate. You know that not all the force you apply to a rope or a stick attached to the object acts to accelerate the object parallel to the ground. Only the component of the force parallel to the ground acts to do the pulling or pushing along the ground.

Let's look at another example to illustrate what we're talking about. Suppose you apply 354 N of force to the handle of a lawn mower, and the handle makes an angle of 49° with the horizontal. How much of the force pushes parallel to the surface of the ground? Refer to Figure 5.3 for a graphic solution. (The subscript v represents vertical, the subscript h represents horizontal, and the subscript R represents resultant.)

Johnnie's Alert

The more methods you have for solving a problem, the better you'll get at solving them. It's always better to have more than one way to look at something. This helps you in many other areas of your life besides physics.

The graphic solution shows that only 231 N of the applied force acts to push the lawn mower parallel to the ground. Does the graphical solution look familiar to you? In Chapter 3, we showed you how to resolve a force into two mutually perpendicular components. Well, that's exactly what we did here, too. If this is unclear to you, try constructing the graphical solution yourself—practice this method of problem solving so that it becomes familiar to you. You can also use the Pythagorean theorem to check your answer and see whether you come up with the same values.

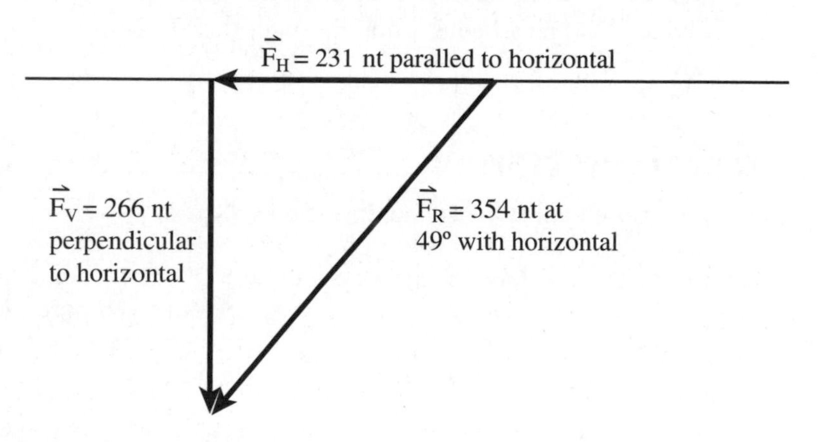

Figure 5.3

The diagram of the components of a force illustrates the resolution of a force into mutually perpendicular components.

Net Force and Uniform Motion

The forces we've discussed so far have included component forces and total or resultant forces. You know that the single resultant force can be used to replace many forces acting on an object. Newton's second law shows what a single force tends to do to an object. Taking all of these ideas into account, the question might be, what happens if we consider all forces acting on an object and arrive at one single force that can replace all of the forces acting on the object? That force is what we call the *net force*. The net force of an object can give us an indication of the type of motion an object has.

What this means is that an object can travel with uniform motion only when the net force on the object is zero. Otherwise the object would follow Newton's second law and accelerate in the direction of the net force. Remember Newton's first law? In some sense, the first law can be thought of as describing objects that are in *equilibrium.* In other words, when the forces acting on an object add to zero, then we say that the forces acting are in equilibrium. Another way to describe this is to say that the net force is zero. The net forces acting on the objects are zero, whether they are at rest or moving in a straight line at constant velocity. Why do they have to be moving in a straight line? Because if an object is moving in a curved path, its direction is changing, so its velocity is changing, and it is accelerating; and if it is accelerating, it is not in equilibrium. (But we know you knew that.)

def•i•ni•tion

When all the forces acting on an object add (as vectors) to zero, the forces acting are in **equilibrium.**

There are many situations in which we do not want systems to accelerate and in those cases the net force acting on a system must be zero. A building is a good example of this requirement. When a building or bridge is designed, it is important that the net force acting on it be zero. That result ensures that the building will stand and the bridge will not collapse. If you're sitting in a chair right now, the forces acting on you are in equilibrium. Gravity is pulling you down, and your chair is pushing you up. These forces are in exact balance, equal and opposite, and you sit there at rest reading these words. Whenever forces act on an object, no matter what their sources may be, the motion of the object may be explained by applying all of Newton's laws.

Let's end this chapter by examining a problem that brings together a lot of what we've been discussing. Suppose that you know there are only three forces acting on an object. The forces are 20.0 dynes $E15°N$, 30.0 dynes $W15°N$, and 40.0 dynes $S15°W$. What is the net force on the object? Figure 5.4 provides you with a graphic solution to this problem. The map directions are given so that you can construct this problem yourself. By doing so, you'll get a better understanding of how the solution was arrived at.

Two of the forces $\vec{F}_1$ and $\vec{F}_2$ were added using the parallelogram method, then that resultant was added to the third force $\vec{F}_3$ to get the net force. We used the parallelogram method to obtain our net force, too.

After drawing everything to scale and constructing our parallelograms, we obtain a final result of 37 dynes $W54°S$ for the net force. Did you come up with the same result, too? We hope so. And below are some problems for you to work on to put all that you've learned in this chapter into practice.

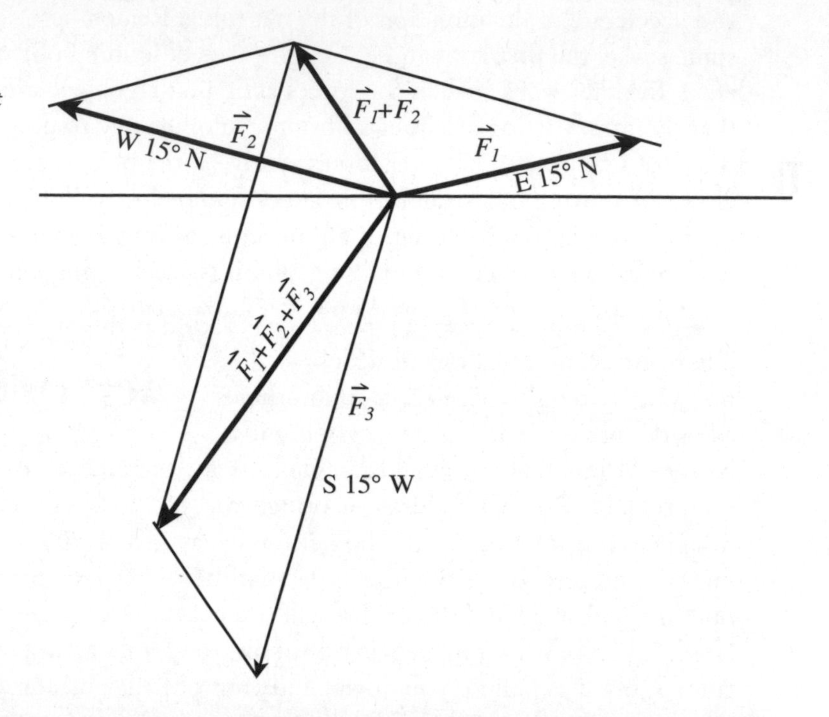

Figure 5.4

A graphic solution for the net force due to three different forces acting at a point uses the parallelogram method of adding vectors.

Physics Phun

1. How much net force must you apply to an object having a mass of 5.0 kg to cause it to accelerate at 6.0 m/s²?
2. A net force of 135 lb is applied to an object having a mass of 2.0 slugs. What is the acceleration?
3. Three forces act on a pumpkin that is placed on the frozen lake with the boy. The forces are 50.0 N North, 50.0 N ***E*60°*S*,** and 40.0 N ***W*30°*S*.** What is the net force on the pumpkin?

Problems for the Budding Rocket Scientist

1. A force acts on a 2-kg mass and gives it an acceleration of 3 m/s². What acceleration is produced by the same force when acting on a mass of (a) 1 kg? (b) 4 kg? (c) How large is the force?

2. A force of 70 N gives an object of unknown mass an acceleration of 20 ft/s^2. What is the object's mass?

3. A boy having a mass of 75 kg holds a bag of flour weighing 40 N. With what force does the floor push up on his feet?

The Least You Need to Know

- A force in physics is either a push or a pull. It lies at the core of Newton's second law.
- Momentum is a derived quantity because it is the product of mass and velocity.
- A single resultant force can be used to replace many forces acting on an object. It is called the net force.
- Whenever forces act on an object, no matter what their sources may be, the motion of the object may be explained by applying all of Newton's laws.

Chapter 6

Free Fall and Gravity

In This Chapter

- Galileo's inclined plane experiments
- Atmosphere, air friction, and terminal velocity
- The acceleration of gravity
- What really happens when an object falls
- Weight, gravity, and force

If you drop a lead weight and a piece of wood at the same time, which one do you think will hit the ground first? We tend to think that heavier objects fall faster than lighter ones. Galileo wanted to find out the answer to that question as well, so he performed his famous experiment of dropping two very differently weighted objects from the leaning tower of Pisa. Guess what he found out? The heavy and light objects hit the ground at the same time.

Demonstrating that all bodies fall at the same rate has become one of the classic experiments of physics. Innumerable students have witnessed a feather and a penny fall at the same rate in a large vacuum tube as a classroom demonstration. When astronaut David R. Scott of the Apollo 15 mission found himself on the airless surface of the moon, he could not resist repeating this classic experiment for the whole world to see. As he said then, he could not have got to where he was standing without Galileo's discovery.

Galileo is credited with establishing the scientific revolution that began in the fifteenth century because he established the idea that an observation or theory needs to be tested by experimentation. The validity of a theory can be determined only by collecting information from an experiment. So we're going to begin this chapter by taking a look at his experiments and the "laws" that he derived from these experiments.

Uniform Accelerated Motion

Armed with his new idea of experimentation, Galileo began to develop the elementary laws of dynamics. He wanted to know how a body moves. When you drop a stone, it falls faster and faster until it hits the ground. Galileo wanted to understand the mathematical principle that governs that accelerated motion. Objects fall too fast to be able to study them in detail, so he devised a method to slow down the fall. He built an inclined ramp, which is called an inclined plane in physics. The flatter he made the ramp the slower the ball rolled. He needed a method to accurately time the ball as it rolled the distance of the ramp. At first he counted beats while a musician played, but he later used a water clock that measured time by filling a container with water.

His results showed that the velocity at which the ball rolled down the plane was proportional to the amount of time that went by. More precisely, the distance traveled increased with the square of the time that passed. In the first unit of time, the ball rolled a certain distance, one unit. After 2 units of time, the ball rolled 4 units of distance. In 3 units of time, it rolled 9 units in distance, after 4 units of time, it rolled 16 units of distance, and so on. As he made the plane steeper, the ball rolled faster, but the motion always had the same property—that is, the total distance traveled was proportional to the square of the elapsed time.

That idea can clearly be expressed by the formula $D \propto T^2$. Galileo concluded that the law would hold if the plane were vertical—in other words, if the object were in free fall. Since the speed increased at a uniform rate, Galileo called this motion *uniform accelerated motion.* Take a look at the following table and you'll see how time and distance mathematically relate to each other.

Galileo's Inclined Plane Experiment

Time Elapsed	Distance Traveled
0	0
1	1
2	4
3	9
4	16
5	25
6 (and so on)	36 (and so on)

Free Fall and Friction

Galileo continued to carry out experiments with various weighted objects. He dropped them, rolled them down inclined planes, and basically learned everything there was to know about falling bodies. He showed by experimentation that objects appear to fall to earth with a constant acceleration. That value is $g = 9.8\frac{m}{s^2}$ in the MKS system, $980\frac{cm}{s^2}$ in the CGS system, and $32\frac{ft}{s^2}$ in the FPS system. All of this acceleration is due to the force of gravity. That force is so important that it is assigned a special letter, *g*. In other words, the acceleration of gravity near the surface of the earth is referred to as 1*g* or simply *g*. Larger accelerations can be referred to as 2*g*, 3*g*, and so on. If we want to sum all of this up in one simple statement, we could say that any object that is falling to the surface of the earth owing to the acceleration of gravity is in free fall and can be called a *freely falling body*.

def•i•ni•tion

Uniform accelerated motion is the movement of an object as it falls freely or down an inclined ramp. Mathematically, this motion is seen as the relationship of the distance traveled to the square of the time elapsed.

A **freely falling body** is an object that is influenced by the force of gravity alone.

Now let's suppose an object is falling near the surface of the earth. You know that gravity is pulling it downward. Another force opposes the motion of the object, the same force any falling object experiences: air friction. Air friction is caused by the medium of our atmosphere. Anything that moves through this medium experiences some form of slowing down caused by air friction, called drag. This is the reason why a feather and a rock fall at different velocities here on earth, where there is air friction to slow the feather down, and why they fall at the same velocity on the moon, where there is no atmosphere. Figure 6.1 helps to illustrate the two forces acting on an object, with mass *m*, falling near the surface of the earth.

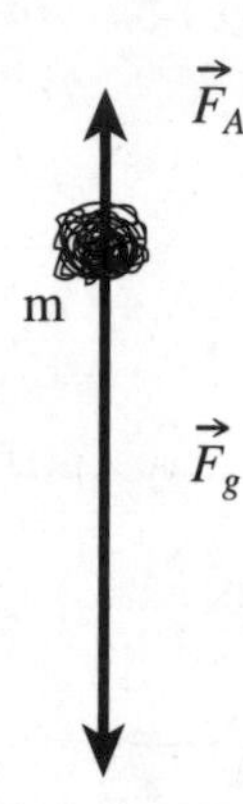

Figure 6.1

The motion of an object can be described using two forces on a falling object.

If you look at the vectors involved, you'll notice that the net force acting on the object is the sum of all the forces acting on the object. That means that the object in Figure 6.1 appears to have a net force acting downward. When we calculate the net force we find $\vec{F}_{net} = \vec{F}_A + \vec{F}_g$. We recognize this as a vector sum that is read as this: the net force is equal to the sum of the force of air friction and the force of gravity.

Terminal Velocity

When objects are dropped from a great height, they eventually stop accelerating and achieve what is called *terminal velocity*. This is the velocity at which the object will not accelerate any more. A leaf has a much smaller terminal velocity than a bowling ball. A feather has a smaller terminal velocity than a hammer, and a flat piece of paper has a smaller terminal velocity than the same piece of paper crumpled up.

def•i•ni•tion

Terminal velocity is the maximum velocity of a falling object in air. The air friction on a falling body will increase with velocity until the upward drag of air friction is equal in magnitude to the downward force of gravity on the body. At that point there is no net downward force, and its downward speed no longer changes.

Parachutes are a perfect example of the usefulness of lowering one's terminal velocity. A parachute adds almost nothing to a person's mass, but it tremendously increases his or her surface area. A typical parachute has an area that is close to 100 times that of a human, and from the preceding examples, we can see that the key to lowering terminal velocity is a higher ratio of surface area to mass. The piece of paper is the best example. The same mass object (a crumpled piece of paper) has a much lower terminal velocity when we allow it to have the greatest possible surface area (open and flat).

And just in case you're wondering why a relatively dense object reaches a terminal velocity, think about what is taking place in terms of Newton's laws. Did you make the connection that terminal velocity implies zero acceleration? If you did, that's great! But how can a body have zero acceleration? Well, we can apply Newton's first law to that situation and see that there is no net force acting on the body. That means that the force of gravity must be equal in magnitude but opposite in direction to the force of friction.

We can take this further by concluding that a falling body that is relatively dense compared to the amount of surface area exposed to the air will experience an increase in air friction with increased speed. When the force of friction reaches the magnitude of the force of gravity, the sum of the two will be zero, and the object will travel with a constant velocity called the terminal velocity. That is very cool!

Johnnie's Alert

Remember that the terms *velocity* and *speed* are interchangeable only when motion is along a straight line in one direction.

Describing the Motion of a Falling Object

What Galileo had measured was something called the acceleration of gravity, a concept we've briefly talked about. Let's look at it in a little more detail and see what really happens when an object falls.

Remember our letter g? That is the constant value of the acceleration of gravity only at the surface of the earth, and only for the particular mass and radius of our planet. However, based on these givens, we can determine a lot about the time it takes for an object to fall to earth if we know the value of acceleration. Let's assume that we are close enough to the surface of the earth that air resistance is not a significant factor. The general equation that we can use to determine the position of an object (x) after a time (t) that has a known initial position (x_0), initial velocity (v_0), and constant acceleration (a) is

$$x = x_0 + v_0 t + \frac{1}{2}at^2$$

So, for example, if an object is dropped from the height of 100 m and the only acceleration acting is that of gravity (9.8 m/s²), then we can say that the initial velocity (v_0) is 0 m/s, and the initial height (x_0) is 100 m. We must take a moment to consider direction. If height, measured upward from the ground, is positive, then the acceleration of gravity, pointing downward, must be negative, or $x = x_0 + v_0t + \frac{1}{2}at^2$. (You will often see the letter [*a*] for acceleration be replaced in an equation with the letter [–*g*], because the direction of the acceleration of gravity is always down or of a negative value.)

Using all of this information, we have

$$0 = 100m + 0m/s(t) + \frac{1}{2}(-9.8m/s^2)t^2$$

or

$$0 = 100m + 0m/s(t) + \frac{1}{2}(-9.8m/s^2)t^2$$

$$100m/(4.9m/s^2) = t^2$$

$$t = 4.5s$$

so the object takes 4.5 s to strike the ground.

Let's try another one together. We'll use the acceleration equation to solve a simple problem in free fall. Suppose that you are standing at the top of a three-story building, and you want to know how long it will take for the water balloon in your hand to hit the sidewalk below. Ignoring air friction, the only acceleration involved is the acceleration of gravity, and the height of the building is 12 m.

In the acceleration equation we use x = 12 m (when the balloon hits the sidewalk), x_0 = 0 m (we take the balloon's starting point at the top of the three-story building to be zero), v_0 = 0 m/s (the balloon starts from rest), and g = 9.8 m/s² (only gravity is acting on the balloon). When we insert these values into the equation and solve it for *t*, we determine that

$$x = x_0 + v_0t + \frac{1}{2}at^2$$

or

$$12m = 0m + 0m/s(t) + \frac{1}{2}(9.8m/s)t^2$$

or

$$24/9.8s^2 = t^2$$

$$t = 1.6s$$

Did you notice that in both of these problems the initial velocity (v_0) is zero? That's because the object was dropped in each problem.

If we assume an ideal condition where the object is not influenced by air friction, then the motion is uniformly accelerated motion. The acceleration is the acceleration caused by gravity, which is constant near the earth's surface.

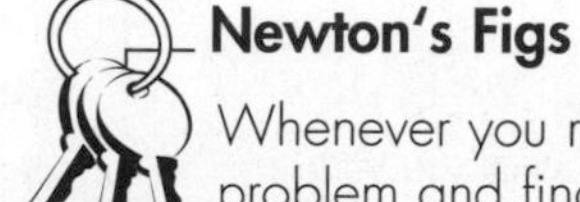

Newton's Figs

Whenever you read a physics problem and find that an object is dropped or is initially at rest then you are given that $v_0 = 0m/s$.

You can solve the preceding problem, like others that we've used in previous chapters, by simply plugging the values into the formula. In some cases you may have to solve the problem algebraically first to put it into terms of the correct variable that you're solving for, as we did with some equations at the end of Chapter 4. But most of the time the solution will be pretty straightforward.

Let's take a look at another problem, only this time we'll be solving for a different value as well as feet instead of meters.

Suppose an object is dropped from a point above the surface of the earth and falls for 3.0 s. How far does it travel?

$$\text{Solution: } x = x_0 + v_0 t + \frac{1}{2}at^2$$

$$x = 0 + (0)(3.0s) + \frac{32ft}{s^2} \times \frac{9s^2}{2} = 144ft$$

In this problem, the height or initial position is unknown, so $x_0 = 0$. There is also no initial velocity, so $v_0 t = (0)(3.0s) = 0$. The only part of the equation that we have to plug in values for is $\frac{1}{2}at^2$. We know the values for *(a)* and the time *(t)*, so we just put those numbers in and solve for *x*, the position of the object, or the distance it traveled.

More on Weight and Mass

In Chapter 5, we briefly discussed some of the differences between weight and mass. We're now going to review some of those ideas and see how weight and mass relate specifically to gravity.

As we already know, *weight* is a force caused by a gravitational field, like the earth's field, and is therefore a vector quantity. Mass is a scalar (see Chapter 3 for more information on vectors and scalars). Weight is a derived quantity and mass is a fundamental quantity. Both are measures of matter, but they are entirely different measures.

> **def•i•ni•tion**
>
> **Weight** is the force of gravity on any object. It is a vector quantity with direction radially inward toward the center of the earth. The direction as described locally is down.

Newton's second law shows us that mass and weight are proportional but not equivalent. Using that law we can state the relationship symbolically as $W = m_g g$. The statement reads as this: weight is equivalent to the product of gravitational mass and the acceleration due to gravity.

You can see that mass and weight are proportional but not equal, and the constant of proportionality has the value of the acceleration due to gravity. The value is also called the earth's gravitational field constant. It has units of N/kg. One kilogram has a weight of 9.8 N here on the earth. That is the amount of force with which the earth pulls on a kilogram. Of course, that is the force with which the kilogram pulls on the earth also. Why doesn't the earth accelerate toward the kilogram as much as the kilogram accelerates toward the earth? That's right—because the earth is much more massive. Can you see that if you were to multiply mass in kg times 9.8 N/kg, you would get an answer in newtons? If you are given the gravitational mass of an object, you are able to calculate the weight of the object. Also, if you are given the weight of an object, you can calculate its gravitational mass.

Weight is a derived quantity, and it's always a good idea to check the units of any new quantity when it is introduced. Again, using Newton's second law, we can find that $W = m_g g$, which means that in the MKS system W is expressed in $\frac{kg \cdot m}{s^2}$ or N. In the CGS system, W is expressed in the units $\frac{g \cdot cm}{s^2}$ or dynes. In the FPS system, W is expressed in the units $\frac{slug \cdot ft}{s^2}$ or lb.

If you're having trouble solving a physics problem, double-check the units used in your answer. For instance, if you're calculating the weight of an object and you find that the units are $\frac{slug \cdot ft}{s^2}$, then you know that you have either made a mistake in algebra, unit analysis, or substitution. It may be that you're using the wrong idea or maybe even solving the wrong problem. The wrong units suggest an error, so it's always important to go back and check all of your work.

Physics Phun

1. Your physics teacher has a mass of 5.0 slugs. How much does he weigh?
2. The athletic trainer weighs 490 N. What is her mass?

Understanding the Gravitational Field

Have you noticed that we have emphasized that the force of gravity discussed so far has been a force "near" the surface of the earth? The reason is that the force on a 1-kilogram mass near the surface of the earth is about 9.80 N. It is correct to say about 9.80 N because the weight of a 1-kilogram mass at the north pole is nearly 9.83 N, and it weighs about 9.78 N at the equator. As you can see, the force of gravity on a unit mass varies slightly depending on where on the earth's surface it is measured. Even though the magnitude of the force of gravity on a unit mass varies a bit depending on the location on the surface of the earth, the direction remains radially inward toward the earth's center.

It doesn't matter where you measure the force of gravity, on the earth's surface or above it; the direction is always radially inward toward the center of the earth. Above the surface of the earth, the force on a unit mass varies from point to point as you move away from the earth's surface.

No matter where you place a unit mass above the surface of the earth, the force of gravity pulls on the mass in a direction back toward the earth. To help you to visualize what is going on here, imagine that the surface of the earth is just one of many spherical surfaces having the center of the earth as their common center. If you measure the magnitude of the force of gravity on a unit mass at each point on any one surface, you'll find that the measurements are the same. Even though that is correct, you may find a different reading for each surface. In fact, the measurement of the force of gravity on a unit mass becomes smaller as you move away from the center of the earth. The situation just outlined is what physicists call a *gravitational field.*

def•i•ni•tion

The earth's **gravitational field** is that region of space where the force of gravity acts on a unit mass at all locations on or above the surface of the earth.

Later we will calculate the magnitude of the earth's gravitational field at any altitude above the surface of the earth. It is important at this time to be able to calculate the magnitude of the gravitational field at any point given the force of gravity on a unit mass. So let's use Newton's second law to calculate the force of gravity.

By Newton's second law: $\vec{F}_g = m_g \vec{g}$

Dividing both sides of the previous equation by m: $\vec{g} = \dfrac{\vec{F}_g}{m_g}$

If you use a unit mass in the last equation, it's possible to calculate the force of gravity on a unit mass that is the magnitude of the gravitational field. The vector represents the gravitational field at any point. Remember that the direction of that vector is radially inward toward the center of the circle as indicated in Figure 6.2 for a couple of altitudes above the surface of the earth.

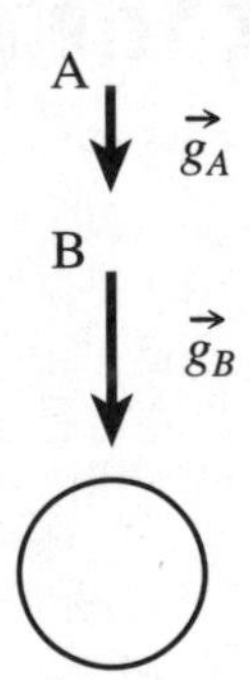

Figure 6.2

The gravitational field vectors represent the gravitational field for two different altitudes.

Imagine concentric spherical surfaces with the vectors $\vec{g}_A$ and $\vec{g}_B$ at every point on the surface, and you'll get an idea about the gravitational field of the earth for two different altitudes. The units of the gravitational field in each system are: $\dfrac{N}{kg}$, $\dfrac{dynes}{g}$ or $\dfrac{lb}{slug}$.

Even though we've used only two altitudes, it's important to know that any mass placed anywhere in the region of space surrounding the earth will be affected by the gravitational field.

Understanding Force Fields

The idea of a force field is important to understanding several different ideas in physics. For example, there are electric fields and magnetic fields. You're probably familiar with both. The source of the force as well as the unit particle that mediates them is different from gravity but the basic concept of field is the same. If you've ever been caught out in a thunderstorm and felt the hairs on your body stand up, you knew that you were in an intense electric field.

Usually there are two or more bodies involved in the definition of a field—two charges or two magnetic poles, for example. In the thunderstorm example, the earth is one body defining your thunderstorm, and a charged cloud above the earth is the other. You are a particle affected by the electric field. Later you will find that the earth is not the only body involved in the definition of the gravitational field of the earth. You can think of all the people on earth as being little particles in the gravitational field of the earth.

Problems for the Budding Rocket Scientist

1. How far will a braking car travel while stopping? If the car goes from 10 m/s to a full stop, then the change in velocity is 10 m/s. If it decelerates at a rate of 2.5 m/s^2, it will take 4 s to stop. How far has the car traveled?
2. Shortly after leaping from an airplane, a 91.8-kg man has an upward force of 225 N exerted on him by the air. What is the resultant force on the man?
3. To measure the mass of a box, we push it along a smooth surface, exerting a net horizontal force of 150 lbs. The acceleration is observed to be 3.0 m/s^2. What is the mass of the box?

The Least You Need to Know

- Galileo is credited with establishing the scientific revolution that began in the fifteenth century because he established the idea that an observation or theory needs to be tested by experimentation.
- Any object that is falling to the surface of the earth owing to the acceleration of gravity is in free fall and can be called a freely falling body.
- Terminal velocity is the maximum velocity that a falling object will not accelerate beyond.

- It doesn't matter where you measure the force of gravity, on the earth's surface or above it. The direction is always radially inward toward the center of the earth.
- Weight is equivalent to the product of gravitational mass and the acceleration due to gravity.

Chapter 7

Projectile Motion

In This Chapter

- Galileo's understanding of projectile motion
- Horizontal motion of objects thrown near the earth
- Vertical motion of objects near the earth
- Simultaneous vertical and horizontal motion
- The flight of a dense object

What goes up must come down. That's essentially the motion of just about any projectile that's launched or thrown through the air. Our old friend Galileo was the first person to accurately describe this motion mathematically. So in this chapter we're going to continue our examination of objects moving in two dimensions.

Projectiles combine both horizontal uniform motion and vertical uniform accelerated motion. This combination has been called complex motion, composite motion, and combined motion. We'll refer to it as combined motion. Each motion can be analyzed separately and then combined to show the final result. Remember the human cannonball we discussed back in Chapter 4? If she was impressed with your understanding of motion in the x-y plane, wait until she sees how adept you are at solving any type of projectile motion. She'll hire you on the spot as her gunnery mate.

Early Ideas of Projectiles and Falling Objects

Before we get down to the nuts and bolts of projectile motion, it might be a good idea to revisit some earlier ideas about how and why objects fall to the earth the way they do.

For almost 1,800 years, the prevailing ideas in physics and astronomy were based on or could be traced back to the theories of Aristotle. Before the discovery that the earth was not at the center of the universe, these were some of the popular beliefs:

- Objects fall to the earth because they are naturally attracted to the center of the universe, which is earth. If the sun were the center of the universe, objects would fall toward the sun. (No one knew anything about gravity yet!)
- The earth does not move. When a person is moving—for example, on a horse—he or she can feel a breeze. If the earth were moving, we could feel a continuous wind. Also, if the earth were moving, an apple falling from a tree wouldn't fall straight down. It would land somewhere off to the side, because the tree and the ground beneath the falling apple would have moved in the time it took for the apple to hit the ground.

If you think about it, these ideas seem pretty reasonable. It would take Copernicus (the guy who explained that the sun, not the earth, was at the center of the universe), Kepler (who developed the laws of planetary motion), Galileo (who through experimentation discovered the mathematics governing the movement of pendulums, falling bodies, and projectile motion), and Newton (who put it all together and then some) to change the way people understood physics and astronomy.

Galileo used the knowledge that he gained from his experiments to refute the argument against the motion of the earth. He argued that the reason an apple doesn't land some distance away from the tree was because the apple maintained a forward motion imparted by the movement of the tree on the moving earth, and therefore doesn't get left behind.

The experiment that he used to arrive at this knowledge went something like this. If you shoot a cannon with the muzzle parallel to the ground, the cannonball will travel horizontally until gravity pulls it to the ground. If a friend drops a cannonball from the same height as the cannon muzzle, the cannonball will hit the ground at exactly the same time as the one shot from the cannon. The one shot from the cannon just covers a larger horizontal distance before it lands. The ball shot from the cannon has a large horizontal velocity that enables it to travel a greater distance, but that velocity doesn't keep it in the air any longer than if it were just dropped. That's an amazing discovery! The shape of the trajectory of the cannonball is a term you where introduced to back in Chapter 4. Do you remember the name? Yes, it's a parabola.

Newton's Figs

Socrates, Plato, and Aristotle were three Greek philosophers who strongly influenced the development of western civilization. Of the three, Aristotle's theories concerning the structure of the universe and his ideas about physics dominated Western thought for over 1,800 years. From roughly 300 B.C.E. to the birth of the scientific revolution in the fifteenth century, few, if any, questioned the theories of Aristotle. It was the advent of scientific experimentation that eventually changed our understanding of the physical laws governing our world.

Instantaneous Velocity

Now that we have some background information about the early efforts to understand projectile motion, let's take a look at how we can analyze the mechanics behind this motion by breaking it down into its two basic parts, horizontal and vertical motion.

Because horizontal is the simplest to describe, that's a good place to start. Under the conditions we're going to discuss, air friction does not play a significant part in our description of motion. That's why we say the motion is ideal, so that you'll be aware that there is no net force on the object acting in the horizontal direction while it is traveling horizontally in the air.

Let's suppose we launch an object horizontally with an initial horizontal velocity. In order to distinguish this velocity from any other we may encounter in this discussion, we'll label its magnitude v_{0H}. And as we've already mentioned, because this is the motion of an ideal object, the force of air friction is negligible or zero. Therefore, the motion of this object horizontally is uniform with constant velocity magnitude, v_{0H}.

If we analyze our horizontal movement using the x-y graphing techniques that we've made reference to in previous chapters, we can call the horizontal displacement x. The magnitude of the horizontal displacement can then be calculated as $x = v_{0H}t$, where t is the time the object is in motion horizontally. In Figure 7.1, you can see that the path of the object is plotted with the two vectors representing its horizontal velocity as the object moves along the path. The horizontal velocity vector tangent to the path shows the velocity at the instant of launch. As you learned in Chapter 4, the motion of an object in a circular path at a constant speed has the velocity vector tangent to the path at every point in the path. Here is a similar situation. Only now we have an example of an object falling to earth in a parabolic trajectory. The velocity in both of these types of motion is called the *instantaneous velocity*. What this essentially means is that soon we'll be able to find the velocity of a moving body (and in this particular chapter, a projectile) at any instant of time.

def•i•ni•tion

Instantaneous velocity is the velocity of an object at any instant of its motion. The instantaneous velocity vector is tangent to the path of motion of the object at every point along the path.

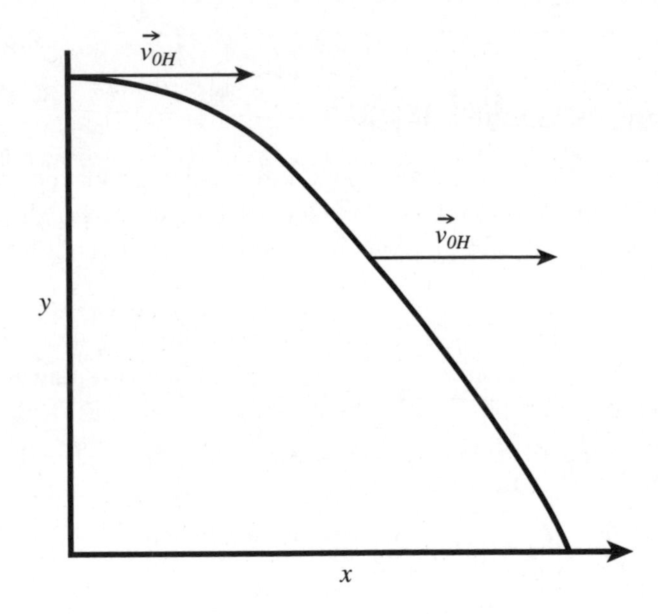

Figure 7.1

The path of an object near the surface of the earth launched horizontally.

The horizontal velocity in this case is the instantaneous velocity only at the time of launch. At all other points along the path of the object in Figure 7.1, it is the horizontal component of the instantaneous velocity. Usually when we refer to the velocity of an object in motion, we are referring to the instantaneous velocity unless we specify a component of the velocity or the average velocity.

Another important point to mention here is that when we discuss projectiles (or for that matter most falling or moving objects), we're referring to ones that are solid or dense, not light and hollow like a ping-pong ball. Tennis balls, ping-pong balls, and similar objects that would be significantly affected by air friction are much harder to analyze. It's always easiest to understand the movement of objects under ideal conditions, and then progress to include factors that make calculations more challenging.

What Happens When an Object Is Thrown Vertically Upward?

That's a review and brief analysis of dense objects dropping and traveling horizontally in the atmosphere of the earth. And we know that an ideal object launched horizontally will travel at a constant speed if we ignore the effects of gravity. The force of gravity will accelerate an object dropped from a point above the surface, and ideally it will travel with uniformly accelerated motion.

Keeping all of that information in mind, let's next consider the motion of an object under ideal conditions when it is launched straight up. Before we can describe the motion we must identify physical conditions at the object's launch time, upon returning to the earth, and for the total amount of time it is actually in the air.

A good way to identify an important piece of information or two is by thinking about the initial conditions of the problem, and what they imply. Keep in mind that the object must have an initial velocity upon launch. Not only that, but the initial velocity is vertical only and directed upward. If the object is launched, what is implied?

Newton's Figs

It's a good idea to read textbook problems with a pencil and paper at hand. Try to focus on what the problem states, and then write down what is given or implied and what you're going to find.

For one thing, if the object goes up and comes back down, somewhere up there it slows down, stops, turns around, and returns to the earth. It stops for only an instant. How long of a time is an instant? That's a good question. An instant is shorter than the smallest time that you can think of. However, the object must stop if it comes back down. We know that it has a net force on it; otherwise, it would continue to travel in a straight line forever.

If it stops up there somewhere, what can we say about the distance from the surface of the earth? What is the acceleration at that point? These are all good questions to ask yourself, and all are implied in the setup of the problem. The more you can think critically about these types of problems, the better you'll get at solving them.

Johnnie's Alert

Discussing motion along a vertical straight line involves vector quantities. However, the convention that is used to indicate the direction is a plus sign for motion upward and a minus sign for motion downward. That means the vector will be properly represented with the magnitude of the vector and the appropriate algebraic sign.

So now that we've formulated some basic questions concerning how events might unfold when an object is shot into the air, let's start working on solving those questions with algebraic formulas.

To begin with, we can represent the initial velocity of launch by v_{0V}, which means the initial vertical velocity or the vertical velocity when time is zero or when we start measuring time. The net force is the force of gravity, so the acceleration is $-g$. (Remember, acceleration is often represented this way.) The point where the object stops is a very important piece of information that we can label $v_{fV=0}$; this represents the trip upward. We can then represent the height above the surface of the earth as y. That means $y = 0$ initially (because the object hasn't left the ground) when $t = 0$. The formula we can use is one that we introduced you to in Chapter 6, with some slight modifications. This was the original formula:

$$x = x_0 + v_0 t + \frac{1}{2}at^2$$

In our updated version, we replace x with y, eliminate x_0 as our initial position, replace $v_0 t$ with $v_{0V}t$, and substitute $-g$ for a. Study the formula below, and you'll see that it's very close to the formula above.

$$y = v_{0v}t - \frac{1}{2}gt^2$$

This formula gives us the height of an object at any time t because we know the initial velocity and the acceleration of gravity. Armed with this formula you have the ability to solve the problems in this Physics Phun sidebar.

Physics Phun

1. A stone is thrown straight up with an initial velocity of $5.0 \times 10 \frac{mi}{hr}$. How high is it in 1.0 s? In 2.0 s? In 3.0 s? What is its maximum height above the surface of the earth?
2. A steel ball is thrown upward with an initial velocity of $1.00 \times 10^2 \frac{km}{hr}$. What is its maximum height? How high is it in 3.0 s?

Two Motions: Uniformly Accelerated (Vertical) and Uniform (Horizontal)

We've now discussed both horizontal and vertical motion, but independently of each other. Now we're going to combine them so that we can analyze, first, projectiles that are launched from a horizontal direction (such as an object shoved off of a table) and, second, objects that are shot upward, like a cannon or rocket.

Let's begin with a simple example of combined motion. Suppose we drop a stone from a bridge and three seconds later you see a splash. What is the height of the bridge? According to what we know already we can use the formula $y = -\frac{1}{2}gt^2$. Remember that when we drop an object, its initial velocity is zero, so we don't need $v_0 t$ at the beginning of the equation. Plugging the given values into our formulas, we get $y = -\frac{1}{2}(32\frac{ft}{s^2})(3s^2) = -100\,ft$.

But what happens to the stone if instead of dropping it we throw it out horizontally, with a velocity of 10 ft/s? Because you've done similar things many times in your life, experience tells you that the stone will describe a curved trajectory and fall some distance away from the base of the bridge. To draw the trajectory of the stone in this example, we must consider the stone as having two independent motions: 1) horizontal motion with the constant velocity that was imparted to it by your arm at the moment of release; and 2) vertical motion of free fall with the velocity that increases proportionally to time. (Remember Galileo's experiments with the inclined plane? $D \alpha T^2$)

The result of adding these two motions is shown in Figure 7.2. On the horizontal axis we plot equal stretches corresponding to the distances traveled by the stone during the first second, the second second, and so on. On the vertical axis we plot the distances, which increase as the squares of the numbers in accordance with the law of free fall. Actual positions of the stone are shown by the small circles that lie on the curve you know as a parabola. If we throw the stone with twice the velocity, it will cover in its horizontal motion a distance that is twice as large, while its vertical motion remains the same. As a result, the stone will fall twice as far from the base of the bridge, but its time of flight through the air will be the same.

A Graphical Solution to Projectile Motion

The analysis we described above refers to a projectile that was launched from a horizontal position. The other type of projectile we want to discuss is one that is shot from the ground up into the air. Earlier we came up with some questions about what

def•i•ni•tion

The **time of flight** of a projectile is the total time it is in the air. The **maximum height** of a projectile is the projectile's greatest distance above the earth or the level of launch. The **range** of a projectile is the maximum distance it travels horizontally. It is measured from the point of launch to the point of return to the same level.

would happen to an object thrown up into the air. Well, now's our chance to apply what we've learned so far and extend those ideas into solving a problem. So let's say that a projectile is launched into the atmosphere with an initial velocity v_0. We are to find the *time of flight*, *maximum height*, and *range* of the projectile. We'll let T represent the time of flight, Y the maximum height, and X the range. All of these quantities are labeled in Figure 7.3. And all of the quantities needed to solve the problem are labeled in that same figure.

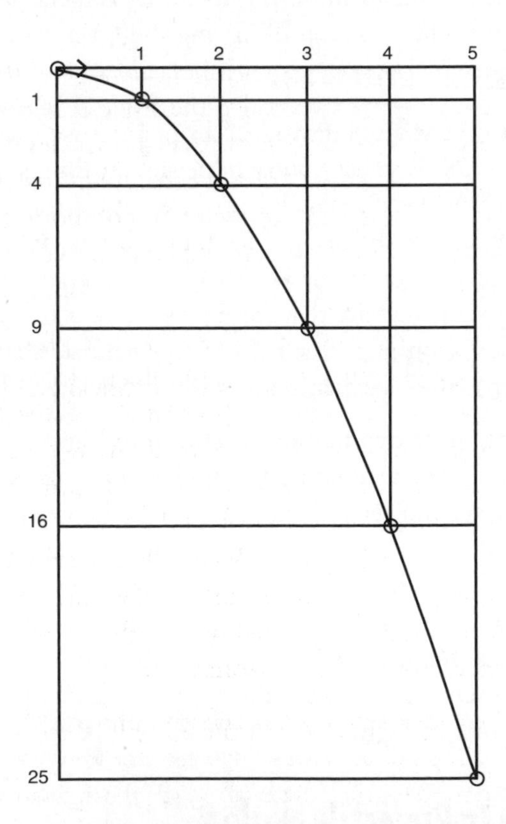

Figure 7.2

The addition of uniform motion in a horizontal direction and accelerated uniform motion in a vertical direction. The resulting curve is a parabola.

Essentially what we're providing you with in this example is a graphical solution that you can use to solve this type of problem. There are no numeric values given, only variables and symbols that represent where you would put the values in. In going through this construction, we'll also provide you with information that will explain what the variables represent and the role that they play in solving this type of problem.

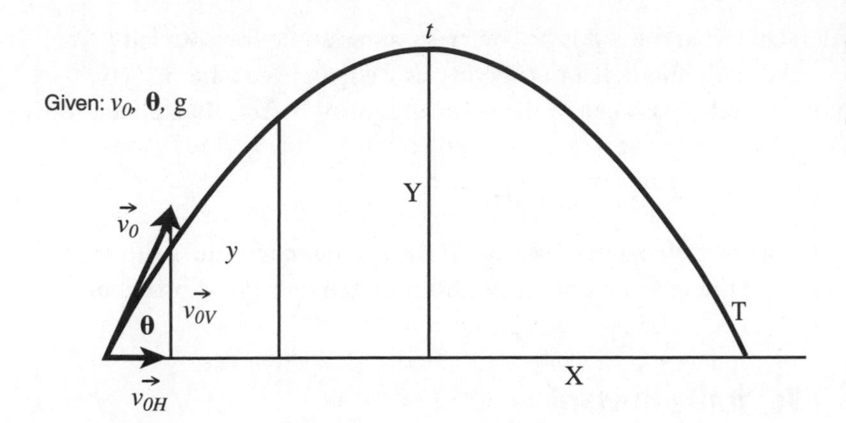

Figure 7.3

The path of a projectile launched at or near the earth's surface.

The two most important variables and symbols are $\vec{v}_0$ and the angle θ. The vector $\vec{v}_0$ not only gives you the magnitude, but also the angle θ, which is the angle at which the projectile was launched. We also know that the acceleration is $-g$ because the object is in motion above the surface of the earth. Our analysis begins with the realization that the projectile is undergoing two types of motion simultaneously: uniform motion horizontally and uniformly accelerated motion vertically. That gives us a plan to follow as a solution.

Describing the horizontal motion requires that we are given the initial horizontal velocity. We can deduce the initial horizontal velocity from the initial velocity by using an idea developed in the section "Components of Vectors" in Chapter 3. We must resolve the initial velocity into two mutually perpendicular components, one vertical and one horizontal. We did that for you in Figure 7.3. Now, we know that the initial velocity is the instantaneous velocity at the point of launch, so it is tangent to the path of the projectile at that point. We constructed the angle θ that is given as the direction of v_0 along with the projections of the initial velocity on the horizontal and vertical axes. That was accomplished by dropping a perpendicular from the head of the initial velocity vector to the horizontal.

The geometry of the constructed right triangle indicates the magnitude of the horizontal component of the initial velocity. Remember how we constructed graphical solutions using the parallelogram method? Well, that's what we did in this example. We labeled the initial horizontal component v_{0H} and the initial vertical component v_{0V}. Did you remember that these are vectors acting along straight lines, so direction is indicated with a plus or minus sign?

The best way to understand what we did is to construct an example for yourself. Choose any scale you like, with an angle of trajectory as well, and see whether you can construct a graphical solution similar to the one in Figure 7.3. You'll just need to do the following two steps:

- Construct the given angle θ with a protractor.
- Construct the vectors to scale so that you can measure the magnitudes of the components as part of the information given in the statement of the problem.

Launching a Projectile Vertically Upward

A graphical solution is definitely one way to solve a projectile problem, but isn't necessarily the quickest or most efficient. An algebraic solution is a much quicker and perhaps easier method. So let's analyze our drawing in Figure 7.3 in a little more detail and come up with some formulas that we can use to solve this type of problem using algebra.

One of the first things we can notice about the trajectory of the projectile is that it takes the same amount of time to reach its maximum height as it does for it to return to the earth.

The little t at the top of the path in Figure 7.3 represents two things to us, the maximum height of the projectile and the halfway point of the time it takes to travel from its launching point to its return to earth. Knowing that little bit of information we can say that the total time $T = 2t$.

Because we're using two different types of letters to represent time in our analysis of the problem, it might be good to use just one more letter to represent any time the projectile is in flight. So let's assign the letter τ for that purpose. (Normally you would only need to use one letter to represent the amount of time that something is in the air, but since we've needed to break this problem down into parts, we've used more letters for time than you would ever need.) We know that $T = 2t$. The horizontal motion can be described with the expression $x = v_{0H}\tau$, that is, the distance traveled horizontally is the horizontal component of the initial velocity multiplied by the time. That means that the distance traveled horizontally for half

Newton's Figs

A dense object launched near the surface of the earth at a low initial velocity will undergo two types of motion simultaneously, and they are independent of each other. Vertically the motion is uniformly accelerated because of the acceleration due to gravity, and horizontally it is uniform motion.

the trip is given by $x = v_{0H}t$, and for the whole trip the range is $X = v_{0H}T = v_{0H}(2t) = 2v_{0H}t$!

Our problem is two thirds solved. We have a formula for 1) the time of flight and 2) the range. All we need is a formula for the maximum height and we'll be ready to solve most projectile problems.

Our third equation is one that you're already familiar with, so let's write that one down and we'll be ready to move on: $y = v_{0V}\tau - \frac{1}{2}gt^2$.

The maximum height is the height when $\tau = t$ so $Y = v_{0V}t - \frac{1}{2}gt^2$.

A summary of our solution so far includes:

1. $T = 2t$: The time of flight is twice the time to reach maximum height. You will calculate the time to reach maximum height later.
2. $X = 2v_{0H}t$: The range is twice the product of the horizontal component of initial velocity and the time required to reach maximum height.
3. $y = v_{0v}t - \frac{1}{2}gt^2$: The maximum height is the product of the vertical component of the initial velocity and the time required to reach maximum height less one half the product of the acceleration due to gravity and the square of the time required to reach maximum height.

We're sure glad that we can use algebraic symbols and don't always have to express these ideas in words! However, our problem will be solved only when we have expressed all three of the parts of the analysis in terms of just the information given: v_{0V}, θ, and $-g$. We'll need to find the time required to reach maximum height in terms of one or more of the quantities given.

There is a piece of information that is implied as we found from an earlier analysis, and that is the final vertical velocity at maximum height; the projectile stops moving vertically at that point. We know that v_{0V}, $-g$, and v_{fV} all = 0. And by using the definition of acceleration to solve for the time required to reach maximum height, we get:

$-g = \dfrac{v_{fV} - v_{0V}}{t}$ The definition of acceleration

$t = \dfrac{v_{0V}}{g}$ Multiplying and simplifying

Now we can solve the problem by substituting the value for the time required to reach maximum height into all three of our partial solutions.

1. $T = 2t$, so $T = 2\left(\frac{v_{0V}}{g}\right) = \frac{2v_{0V}}{g}$.
2. $X = 2v_{0H}t$, and $X = 2v_{0H}\left(\frac{v_{0V}}{g}\right) = \frac{2v_{0H}v_{0V}}{g}$.
3. $Y = v_{0V}t - \frac{1}{2}gt^2$, therefore

$$Y = v_{0V}\left(\frac{2v0V}{g}\right) - \frac{g}{2}\left(\frac{v_{0V}^2}{g^2}\right) = \frac{2v_{0V}^2}{g} - \frac{g}{2}\frac{v_{0V}^2}{g^2} = \frac{2v_{0V}^2}{g} - \frac{v_{0V}^2}{2g} = \frac{v_{0V}^2}{2g}.$$

There we have it—a complete algebraic solution for the time of flight, maximum height, and the range in terms of only the information given. Let's work through a couple of problems together and then you'll get a chance to work on some on your own.

A Baseball from Center Field

Baseballs are fun objects to analyze. We've all played with them at some time in our lives. Think of a baseball as a good approximation of an ideal projectile. Let's apply what we've learned to the flight of a baseball.

Suppose a batter hits a 145-g baseball into deep center field and the ball is caught. A runner at third base tags up and runs for home plate. The center fielder throws the ball to the catcher. It arrives 4.0 seconds after it was thrown, and the catcher catches the ball at the same height from which it was thrown. We can find out what maximum height the ball reaches, its initial vertical component of velocity, its net force at maximum height, and its acceleration at maximum height.

Let's begin the solution with the time of 4.0 seconds. That's called the time of flight in our analysis of projectile motion, so $T = 4.0s$ and $t = 2.0s$, where t is the time required to reach maximum height. We solve for t algebraically and find $t = \frac{v_{0V}}{g}$, so $v_{0V} = gt$, and $v_{0V} = \left(980\frac{cm}{s^2}\right)(2.0s) = 2.0 \times 10^3 \frac{cm}{s}$ observing two significant figures. The maximum height for a projectile was calculated to be $Y = \frac{v_{0V}^2}{2g}$, so

$$Y = \frac{\left(2.0\times10^{3}\frac{cm}{s}\right)}{2\left(980\frac{cm}{s}\right)} = 2.0\times10^{3}cm = 2.0\times10m.$$

Remember that the maximum height is measured from the level of launch. At maximum height the acceleration of the ball is the acceleration due to gravity and is the same for all objects near the surface of the earth. The vertical velocity is zero at that point, but not the acceleration. Finally, the net force on the ball at maximum height is the same as for any other point in the flight because the baseball is treated here as an ideal projectile. That means by Newton's second law,

$$W = mg = 145g\times980\frac{cm}{s^{2}} = 1.42\times10^{5}\,dynes.$$

Will you ever look at a baseball the same way again?

The Hang Time of a Punt

Now let's take a look at football. Have you ever heard the announcer discuss the hang time of a punt? The hang time of the football is the same as the time of flight of a projectile. Suppose the hang time for a punt of a football with a mass of 0.028 slugs is 4.5 seconds. How many feet does it go in the air above the punter's toe? What is the weight of the football at maximum height? What is its initial vertical velocity?

You know that the weight of the football is the same everywhere near the earth's surface. Newton's second law enables you to calculate the weight as $W = mg = 0.028slugs\left(32\frac{ft}{s^{2}}\right) = 0.90lb$. Because the time to reach maximum height is given by $t = \frac{v_{0V}}{g}$, then $v_{0V} = gt = 32\frac{ft}{s^{2}}\times2.25s = 72\frac{ft}{s}$. The maximum height you find to be given by $Y = \frac{v_{0V}^{\;2}}{2g}$, then $Y = \frac{\left(72\frac{ft}{s}\right)^{2}}{2\times32\frac{ft}{s^{2}}} = 81ft$.

As you can see, we used the algebraic solutions we developed in this chapter. Most of these are common enough equations that you'll find in most physics books, perhaps with slight variations to the variables and symbols. But you should be able to recognize them all as functional equations that will work in almost all problems dealing with projectiles. In the next chapter, we'll cover the last of Newton's laws, that of universal gravitation.

Physics Phun

1. An Idaho potato is launched from a spud gun at an angle of $(3.00 \times 10)^\circ$ with a muzzle velocity of $5.00 \times 10\frac{m}{s}$. What is the time of flight, maximum height, and range of the potato?
2. A steel ball of mass 114 g is launched from ground level with an initial velocity of $615\frac{m}{s}$ at an angle of $(4.50 \times 10)^\circ$. What is the time required to reach maximum height?

   What is the acceleration of the ball at the point of maximum height? What is the velocity of the ball at the point of maximum height? What is the net force on the ball at the point of maximum height? How long does it take the ball to descend from the maximum height? Calculate the time of flight, the maximum height, and the range of the ball.

Problems for the Budding Rocket Scientist

1. A marble with velocity of 20 cm/s rolls off the edge of a table 80 cm high. How long does it take to drop to the floor? How far, horizontally, from the table edge does the marble strike the floor?
2. Another marble traveling at 100 cm/s rolls off the edge of a level table. If it hits the floor 30 cm away from the spot directly below the edge of the table, how high is the table?
3. How fast must a ball be rolled along a 70 cm high table so that when it rolls off the edge it will strike the floor at a distance of 70 cm from the point directly below the table edge?

The Least You Need to Know

- For almost 1,800 years, the prevailing ideas in physics and astronomy were based on or could be traced back to the theories of Aristotle.
- Instantaneous velocity is the velocity of an object at any instant of its motion.
- A dense object launched near the surface of the earth at a low initial velocity will undergo two types of motion simultaneously, and they are independent of each other.
- At maximum height the acceleration of an object is the acceleration due to gravity and is the same for all objects near the surface of the earth.

Chapter 8

Newton's Universal Law of Gravitation

In This Chapter

- Gravity, mass, force, and distance
- Harmonic motion
- Systems in equilibrium
- How an Earth satellite stays in orbit
- Use of a satellite to measure mass

The force of attraction between two objects is proportional to the product of their masses and inversely proportional to the square of the distance between them. When you first read Newton's definition of the universal law of gravitation, it sounds more like love than the defining equation of how the force of gravity operates.

Whereas all of the material we've studied so far is important to understanding how bodies move, the examination of gravitational force reveals how bodies interact with each other. It is a powerful predictive tool that explains the shape of the orbits of the planets in our solar system and can even be used to find unknown celestial bodies. We begin our quest to understand the nature of gravity by looking at how it relates to some ideas of motion that we've already studied.

How Is Force Related to Mass and Distance?

In Chapter 7 we discussed the path of a projectile in terms of the horizontal component and the vertical component of the instantaneous velocity. We drew a diagram of the path of the projectile and we called the shape of the path a parabola. Let's consider the two equations we used to describe the motion of the projectile. We used $x = v_{0H}\tau$ and $y = v_{0V}\tau - \frac{1}{2}gt^2$. These two equations are called parametric equations with the parameter τ.

If we solve for the parameter in either equation and substitute the results in the other equation, we can eliminate the parameter. Solving for the parameter in the first equation, we get $\tau = \frac{x}{v_{0H}}$. Now let's substitute for τ in the other equation, and we get this:

$$y = v_{0v}\left(\frac{x}{v_{0H}}\right) - \frac{1}{2}g\left(\frac{x}{v_{0H}}\right)^2 = \frac{v_{0v}}{v_{0H}}x - \frac{1}{2}g\left(\frac{x}{v_{0H}}\right)^2 = \frac{v_{0v}}{v_{0H}}x - \frac{g}{2v_{0H}^{\ 2}}x^2$$

This last equation is the equation of a parabola. And although this is a great formula to use for solving the path a projectile can take, there are other ways to do this, too. Let's examine the components of the force of gravity on a projectile.

The force of gravity is the net force on the projectile we discussed in Chapter 7. The path of the projectile is a parabola, as shown in Figure 8.1.

The force of gravity is shown acting on the object at one point along the path. The components of the force of gravity perpendicular to instantaneous velocity are represented by $\vec{F}_{g\perp}$, and the components of the force of gravity parallel to the instantaneous velocity are represented by $\vec{F}_{g\parallel}$. We know by Newton's first law that an object tends to continue moving in a straight line unless acted upon by a net force.

The object traveling the parabolic path would continue traveling tangent to the path at any point if it did not have $\vec{F}_{g\perp}$ deflecting it out of the straight-line path at every instant that it is in the air. The component $\vec{F}_{g\parallel}$, on the other hand, acts parallel to the instantaneous velocity, causing the object to speed up along the path. Therefore, the force of gravity on the object has one component deflecting the object into a curved path and another component speeding the object up along the path, creating the nice

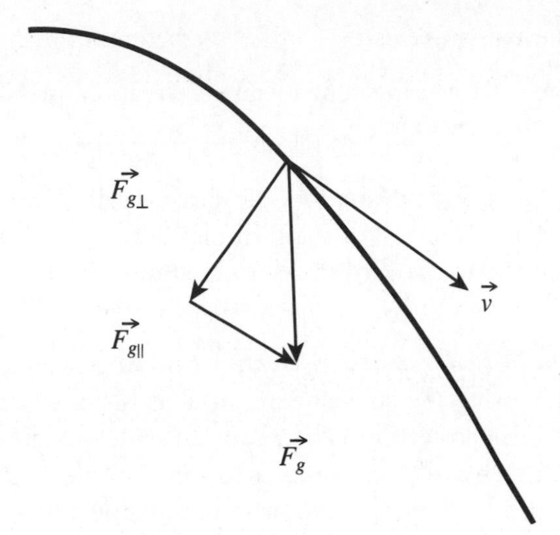

Figure 8.1

The components of the force of gravity on a projectile influence the motion of the projectile.

smooth parabolic *trajectory* in Figure 8.1.

Our next step is to examine in more detail an idea we introduced back in Chapter 4, centripetal acceleration. If that concept is a bit rusty right now, it might be a good idea to go back and review that material before moving on.

We had two equations we could use to represent centripetal acceleration. The first, $a_c = \frac{v^2}{R}$, represents the magnitude of the centripetal acceleration of an object traveling in a circular path at a constant speed. The second way to express the same idea is with the vector equation $\vec{a}_c = \frac{4\pi^2\vec{R}}{T^2}$, where $\vec{R}$ is the positive position vector from the center of the circle to the object at any time, and T is the period or the time required for the object to complete one revolution.

The centripetal acceleration is the only acceleration the object experiences when it is traveling in a circular path at a constant speed. Unlike the projectile in the parabolic path, there is no increase in the speed. The centripetal acceleration is always perpendicular to the instantaneous velocity, and that means it is always pointed directly toward the center of the circular path. And as we already know by Newton's second law, the only way you can have acceleration is by a net force acting

def•i•ni•tion

The **trajectory** is the arced or curved path of a projectile. **Centripetal force** is the center-seeking force on an object at every instant that deflects the object into a circular path at a constant speed.

in the direction of the acceleration.

The magnitude of the force causing centripetal acceleration, by Newton's second law, is given by $f = ma = \frac{mv^2}{R} = \frac{m4\pi^2 R}{T^2}$, and is always directed radially inward toward the center of the circular path. For that reason, the force is called *centripetal force*. It is the only force causing acceleration. (All of the formulas in the preceding equations are equivalent. We simply replaced a with the two equations we discussed in the earlier paragraph.)

Newton's laws hold in *inertial frames of reference*, or in other words, frames of reference either at rest or moving at a constant speed. This is similar to our notion of an ideal condition or an ideal projectile. When you're first learning how forces interact with one another, it's much easier and simpler to solve problems that don't involve lots of other factors such as friction, wind, or rain, to name a few.

Simple Harmonic Motion

If you've ever watched a pendulum swing back and forth, you were observing *periodic* or *simple harmonic motion*. Another example is the vibration of a string on a musical instrument. It can also be related to the period an object takes to complete one cycle in a circular path. So let's see how that idea works in regard to centripetal force.

Imagine that an object of mass m moves with a constant speed around a vertical circular path of radius R, as shown in Figure 8.2. As m moves around the circle, the projection of m moves back and forth along the diameter, AB, of the circle. When m starts at A and makes one complete trip around the circle, the projection of m starts at A, then travels to B and back to A. All of this movement takes place in the period

def•i•ni•tion

An **inertial frame of reference** is a frame of reference that is at rest or moving at a constant speed. For our purposes, the earth is an inertial frame of reference. **Simple harmonic motion** is the to-and-fro motion caused by a restoring force that is directly proportional to the magnitude of the displacement and has a direction opposite the displacement. An example is the motion exhibited by a vibrating string or a simple pendulum. The **period** of simple harmonic motion is the time for the object in motion to complete one cycle. The path from A to B and back to A is one cycle for the projection of m, as in Figure 8.2.

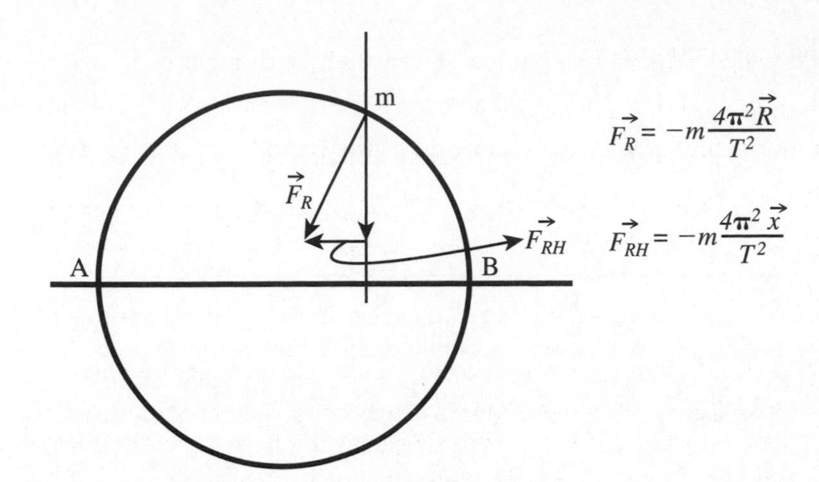

Figure 8.2

The horizontal component of centripetal force models the restoring force for simple harmonic motion.

of motion of *m*, the time to make one complete revolution. The period of the motion of the projection of *m* is exactly the same as the period of the motion of *m*.

The force on *m* is the centripetal force labeled $\vec{F}_{RH}$ in Figure 8.2. You can think of the horizontal component of the centripetal force labeled $\vec{F}_{RH}$ in that diagram as the force acting on the projection of *m*. That is, this is a model related to uniform circular motion that can be used to explain another type of motion and its causes. Notice that as m moves counterclockwise from its first position in that diagram, the direction of $\vec{F}_{RH}$ is to the left until *m* gets to a point directly above the center of the circle. At that instant, $\vec{F}_{RH} = 0$. As *m* continues counterclockwise, the direction of $\vec{F}_{RH}$ is now toward the right. It keeps that direction until *m* is at a point directly beneath the center of the circle, where $\vec{F}_{RH}$ becomes zero again.

The direction of $\vec{F}_{RH}$ is toward the left after that point, and remains pointing left through the starting point and until *m* is at a point directly above the center. At that point $\vec{F}_{RH}$ becomes zero again and the motion repeats. Not only does the direction of $\vec{F}_{RH}$ change as *m* moves around the circle, but its magnitude changes, too. When *m* is directly above the center, $\vec{F}_{RH} = 0$; it then increases in magnitude until *m* is at A, where $\vec{F}_{RH}$ is maximum. As *m* continues to move counterclockwise, $\vec{F}_{RH}$ decreases in magnitude and becomes zero when *m* is directly below the center of the circle. It then increases in magnitude and reaches maximum magnitude when *m* is at B.

The motion of the projection is said to be *periodic* with period T. That is, $\vec{F}_{RH}$ goes from zero to maximum to zero to maximum then back to zero while *m* is making one complete revolution. The complete trip of *m* in one revolution and the corresponding trip of its projection is called a cycle. The units of T can then be thought of as $\frac{s}{cycle}$ (seconds per cycle).

But you're probably more familiar with the reciprocal of the period that is called the

def•i•ni•tion

A **cycle** is one complete trip for an object moving in a circular path at a constant speed as well as the corresponding trip of its projection on the diameter shown in Figure 8.2. The time for one cycle is called the period T, which means that you can think of the units of T as $\frac{s}{cycle}$. The **frequency** of the motion is the reciprocal of the period and is given by $f = \frac{1}{T}$; it has units of $\frac{cycles}{s}$ or just $\frac{1}{s} = s^{-1}$ because a cycle is not a unit of measurement. Another common unit to measure frequency is the hertz, where $1 Hz = 1 s^{-1}$.

frequency. That is, the frequency is given by $f = \frac{1}{T}$ and the units of frequency can be thought of as $\frac{cycles}{s}$ (cycles per second).

Just as the centripetal force is directed opposite the radius vector of the circle, the horizontal component of the centripetal force is directed opposite the displacement of the projection along the diameter of the circle. The displacement is labeled x and is measured from the center of the circle to the projection.

Did you notice that $\vec{F}_{RH}$ is maximum when the displacement is maximum and $\vec{F}_{RH} = 0$ when the displacement is zero? While m travels at a constant speed, the projection changes velocity. The velocity of the projection is a maximum when the displacement is zero; in fact, at that instant its velocity is the same as the instantaneous velocity of m. The velocity of the projection is zero when the displacement has a maximum value.

That was a lot of discussion to explain the periodic rotation of an object in a circular path. If you had a problem following all of that, we can simplify the movement into a pendulum swing and describe it in terms of gravity.

So let's suppose that a mass is attached to the end of a string and set in motion. The mass comes to rest at the top of the swing, then falls back in the other direction, gains speed, reaches maximum velocity at the bottom of its swing, rises on the other side, and again comes to rest at the top of its swing. Then the process repeats itself. When we examine the energy of motion in upcoming chapters, we'll be able to show how this entire process reflects the balance between gravitational energy and kinetic energy (a term that you'll know very well, very soon).

Equilibrium and Simple Harmonic Motion in a Spring

Another example of simple harmonic motion is that of a spring. If you hang a mass from a spring, the mass will stretch the spring a certain amount and may even continue to move up and down, but it will eventually come to rest. (If you've ever jumped off a platform with a bungee cord or have seen someone else jump, that's a perfect example of what we're talking about.) When the mass does finally come to rest—or in terms of physics, when the pull of the spring upward on the mass is equal to the pull of the force of gravity downward on the mass—the system, spring, and mass are said to be in *equilibrium.*

> **def•i•ni•tion**
>
> A system is in **equilibrium** when the sum of the forces is zero. If one force is used to balance the effects of two or more other forces, that force is called the equilibrant and is equal in magnitude and opposite in direction to the resultant of the other forces.

Before we analyze the harmonic motion of the spring, let's talk about the concept of equilibrium a bit. To begin with, there are two types of equilibrium, stable and unstable. Pendulums, roller coasters, and guitar strings are examples of systems with points of stable equilibrium. These are instances in which something set into motion continues in motion for a while, but as time goes on it eventually comes to a rest. (The types of energy involved that slow it down and bring it to a stop will be discussed in Chapters 10 and 11.) Examples of unstable equilibrium include a pencil balanced on its point, a row of dominoes, and a house of cards. All these systems have excess potential energy (a term we discuss in detail in Chapter 11), which can easily be turned into other forms and eventually will be.

Ideas of equilibrium and stability can be applied to areas besides physics. One example is an ecological system. Simple ecological systems, such as a small jungle, tend to be stable. When there are enough little animals around for the big animals eat, the system is in equilibrium. If the predators eat too many prey, however, then there aren't enough little animals and the big animals start dying. The presence of fewer predators allows the little animals to recover; then the big animals recover, too.

Let's go back to our analysis of the spring. If we displace the mass up or down from the equilibrium position and release it, the spring will undergo simple harmonic motion caused by a force acting to restore the vibrating mass back to the equilibrium position. That force, called the restoring force, is directly proportional to the magnitude of the displacement and is directed opposite the displacement. The necessary condition for simple harmonic motion is that a restoring force exists that meets the conditions stated symbolically as $\vec{F}_R = -k\vec{x}$, where k is the constant of proportionality and $\vec{x}$ is the displacement from the equilibrium position. The minus sign, as usual, indicates that $\vec{F}_R$ has a direction opposite that of $\vec{F}_{RH}$.

So what's a constant of proportionality? In mathematics, two related quantities x and y are called proportional if there exists a functional relationship with a constant, nonzero number k such that $y = k$ times x. For example, if you travel at a constant speed, then the distance you cover and the time you spend are proportional, the proportional constant being the speed. Get it? Similarly, the amount of force acting on a certain object from the gravity of the earth at sea level is proportional to the object's mass.

We can perhaps understand all of this better by putting together all of the pieces of information we have so far. So let's begin by stating that $\vec{F}_R = \vec{F}_{RH}$. This means that the restoring force causing simple harmonic motion is exactly the same as the horizontal component of the centripetal force that causes the motion of the projection along the diameter in Figure 8.2. There's a lot of information that we can get about simple harmonic motion from this formula.

$\vec{F}_R = \vec{F}_{RH}$. The restoring force is equal to the horizontal component of the centripetal force.

$-k\vec{x} = -m\dfrac{4\pi^2\vec{x}}{T^2}$ Substituting definitions of both.

$k = \dfrac{m4\pi^2}{T^2}$ Because $-\vec{x}$ is the same in both cases, and the constant k is the same in both cases.

$T^2 = \dfrac{m4\pi^2}{k}$ Multiplying both sides of the equation by $\dfrac{T^2}{k}$.

$T = 2\pi\sqrt{\dfrac{m}{k}}$ Regrouping and finding the principal square root of each member of the equation.

After working through the algebra we arrive at the equation that shows us that the period of simple harmonic motion (T) is calculated using the mass of the object and the constant of proportionality for the displaced object. It reveals that the period is directly related to the mass and inversely related to the constant of proportionality.

And if we apply that understanding to the spring, for a given spring, the larger the mass, the longer the period of vibration. Also, given a certain mass, the stiffer the spring, the shorter the period of vibration. You may have already seen a description of this sometime in your life, especially if you've ever installed a spring on a screen door.

A Simple Pendulum

Do you know what a *simple pendulum* is? It's called simple because the mass of the string, or whatever the pendulum bob is attached to, has a negligible mass in relation to the mass of the bob. When set in motion using small angles, the simple pendulum undergoes simple harmonic motion. Small angles here are angles less than 10°, because if you use a large angle, say 90°, and release the pendulum, it will tend to drop straight down and bang around until there is a smooth swing. You want a nice smooth swing from the beginning in order to have simple harmonic motion.

Let's look at the simple pendulum in Figure 8.3 and see how the derivations of the period of simple harmonic motion applies to this familiar object undergoing simple harmonic motion. We know from our previous example that it is simple harmonic motion because there is a restoring force directly proportional to the displacement from the equilibrium position and directed opposite the displacement.

def•i•ni•tion

A **simple pendulum** is a physical object made up of a mass suspended by a string, rope, or cable from a fixed support. It's called a simple pendulum because the string, rope, or cable has a negligible amount of mass compared to the mass of the object being supported.

Our pendulum is made up of a string of length l supporting a mass m. The vertical position of the pendulum is the equilibrium position. The force of gravity on the mass m is vertically downward, and the string pulls upward with a tension equal to the weight of the mass. (We didn't draw the force of gravity in that position because we wanted to avoid a confusing diagram.) The pendulum is displaced to the right by angle θ, which should be less than 10°. We had to draw the angle large enough so that you could see the geometrical relationships in the diagram.

The force of gravity $\vec{F}_g$ is drawn at the second position of the pendulum and is resolved into two components. One component is parallel to the string (keeping the

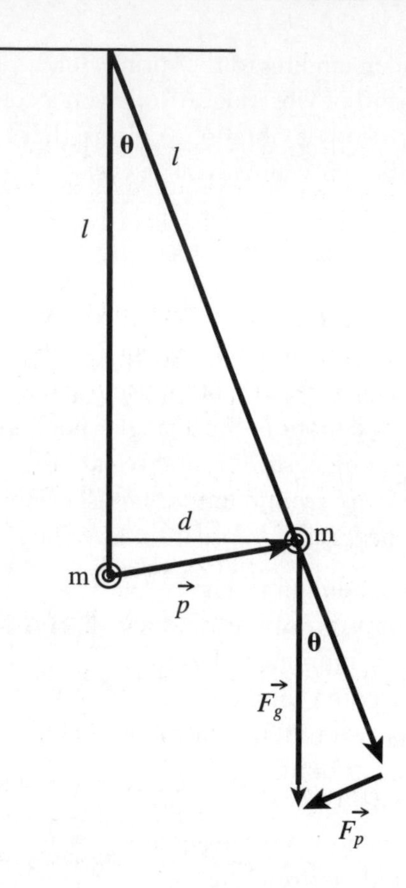

Figure 8.3

A simple pendulum is an application of simple harmonic motion.

tension in the string) and the other component, labeled $\vec{F}_p$, is perpendicular to the string. The acute angles in the two right triangles are labeled the same, θ, to indicate that they are equal. Do you remember from geometry that two angles are equal if their sides are parallel right side to right side and left side to left side? And we can also add that two right triangles are similar if an acute angle of one is equal to an acute angle of the other.

With the understanding of the relationship between the two triangles, we can say that:

$$\frac{\vec{F}_p}{d} = \frac{\vec{F}_g}{l}$$ Corresponding parts of similar triangles are proportional.

$$\vec{F}_p = \frac{d}{l}\vec{F}_g$$ Multiplying both equations by *d*.

$F_p = \frac{d}{l}mg$ By substitution.

Now, to put that all into English: as m swings to and fro, its path is a small arc of a circle. For small angles, the chord p is about equal in length to the corresponding arc, thus the chord p is about equal to the semi-chord d. The vector $\vec{p}$ is the displacement of m from the equilibrium position.

Therefore, $\vec{F}_g = -\frac{mg}{l}\vec{p}$, where we substituted p for d, and because $\vec{p}$ and $\vec{F}_p$ are in opposite directions we use a minus sign. This shows that there is a restoring force that is directly proportional to the displacement and acts in a direction opposite the displacement.

To take this a little further, that means that $k = \frac{mg}{l}$, and in our equation for simple harmonic motion, $T = 2\pi\sqrt{\frac{m}{k}}$, we can substitute for k and get $T = 2\pi\sqrt{\frac{ml}{mg}} = 2\pi\sqrt{\frac{l}{g}}$.

Essentially what that shows us is that the simple pendulum meets the criteria for simple harmonic motion and has a period of $T = 2\pi\sqrt{\frac{l}{g}}$. Among other things, the simple pendulum can be used to measure the acceleration due to gravity.

Do you know who discovered the mathematics that described the swing of the pendulum? He's someone who also was the first to use time to measure the speed of falling bodies. Yes, it was Galileo. At the age of 17, Galileo was studying medicine at the

Physics Phun

1. Calculate the period of a simple pendulum having a length of 25 cm. The pendulum is near the surface of the earth.
2. A simple pendulum having a length of 55.8 cm is used to measure the acceleration due to gravity. What is the acceleration due to gravity if the period of the pendulum is 1.50 s?
3. How much centripetal force is required to keep a 2.0-g object moving in a circular path of radius 111 cm with a period of 0.61 s?
4. What is the frequency of the pendulum in problem 2?
5. Discuss the acceleration of an object that is undergoing simple harmonic motion. What can you say about its acceleration when it has maximum velocity?

University of Padua. The story goes that one Sunday, while attending Mass, he noticed a large chandelier swinging overhead. As it swung back and forth, each swing getting shorter and shorter, he wondered if the time of each swing also shortened. He didn't have a stopwatch, so he measured the time by counting his pulse. To his surprise, he found that even though the swings got shorter, it took the same amount of time for the chandelier to complete each swing. This discovery intrigued him so much that he switched majors and began to study mathematics.

The Motion of Heavenly Bodies

There is a long history of the efforts of men to explain the motion of heavenly bodies. Many models were developed to explain the strange motion of the planets. Although all ancient cultures had some explanation of how the universe operated, the lineage of astronomical thought in the West culminates with the sun-centered system of Copernicus. It is from this point that modern astronomy begins.

Tycho Brahe replaced the system devised by Copernicus with his own geometric system, in which the sun goes around the earth and the other planets go around the sun. He was a skilled experimenter who made careful observations and recorded data that accurately identified the positions of planets and fixed stars.

Johannes Kepler worked in Tycho's laboratory during the last year or so of Tycho's life. Kepler had great respect for the work of Tycho and his records that filled volumes of books. Kepler felt that there must be an easier way to describe the motion of heavenly bodies than to record their positions in tables on page after page of laboratory records. Kepler was not a skilled experimenter like Tycho, but he had a strong mathematics background. Mathematics was not one of Tycho's strengths. Kepler had that special scientific sense that all of the volumes of Tycho's observations could be replaced by a simpler idea. Simplicity is a goal of all scientists when they are seeking an explanation for physical phenomena.

Kepler took the work of Tycho Brahe and developed a mathematical model to replace all of the observations Tycho had so carefully recorded. But Kepler had so much respect for the work of Tycho Brahe that he questioned one of his own mathematical models when it disagreed with an observation by Tycho by a fraction of a degree of arc. He abandoned that model based on eccentric circles for the paths of the planets because he didn't think that Tycho Brahe could make such a mistake in his observations. Eventually his hard work came to fruition when he developed a mathematical model that not only agreed with the work of Tycho but also made predictions that could be verified by observation. His model is summarized in the statement of

Kepler's three laws, listed here:

1. The planets travel in elliptical paths, with the sun at one focus.
2. A line joining the sun and a planet sweeps out equal areas in equal times.
3. The ratio of the cube of the mean radius to the square of the period of revolution is the same for all planets. A simple statement of Kepler's third law is $k = \frac{R^3}{T^2}$.

Gravitational Force Between Any Two Bodies

Kepler's laws describe the motions of the planets. Newton used the work of Kepler to discover the causes of the motions of the planets. He used the fact that even though the paths of the planets were elliptical they are so nearly circular that circular paths were used to calculate the force causing the motion. He used his own idea that

$$F_{sp} = m_p a_p = m_p \frac{4\pi^2 R}{T^2}.$$

What does this equation say? Basically it states that the force of attraction of the sun for a planet is equal to the product of the mass of the planet and the centripetal acceleration. Another way of stating the same thing is that the attraction of the sun for a planet is equal to the mass of the planet divided by the square of the period of revolution of the planet around the sun times the mean radius of the path of the planet times a constant. Newton thought that there would be an inverse square relationship for the distance between the planet and the sun. He used Kepler's third law to express the force in terms of R and m only.

That is, since $k = \frac{R^3}{T^2}$, then $T^2 = \frac{R^3}{k}$, and the force can be expressed as

$F_{sp} = m_p \frac{4\pi^2 R}{\frac{R^3}{k}} = m_p \frac{4\pi^2 k}{R^2}$. Newton used the same idea to calculate the force of attraction of the earth for the moon and arrived at a similar expression, except the constant k was associated with the earth—that is, $F_{em} = m_m \frac{4\pi^2 k_e}{R^2}$. He felt that there must be some property of the earth involved in the constant k_e so the force of the sun on a planet must have a similar relationship, that is, $F_{sp} = m_p \frac{4\pi^2 k_s}{R^2}$.

Satisfied that he had found that the force was directly proportional to the mass of the planet and inversely proportional to the square of the distance to the planet, he wondered what the constants $4\pi^2 k_e$ *and* $4\pi^2 k_e$ and similar expressions for other planets meant. He inferred that it must involve some property of the body about which the object is orbiting. He thought that property was the mass of the body, and stated $4\pi^2 k_s \propto m_s$ so $4\pi^2 k_s = Gm$ and similarly $4\pi^2 k_e = Gm$. Then the force can be stated $F_{se} = \frac{m_e G m_s}{R^2}$, and by his third law if the sun attracts the earth, then the earth must attract the sun, with the force $F_{es} = \frac{m_s G m_e}{R^2}$. When these are simplified, they are one and the same; that is, $F_{es} = F_{se} = \frac{G m_e m_s}{R^2}$ and the directions are opposite.

He then generalized this result to state his universal law of gravitation: any two particles in the universe attract each other with a force that is directly proportional to the product of their masses and inversely proportional to the distance between their centers of mass. Another scientist by the name of Cavendish determined an accurate value of G experimentally, and recently more accurate values of G have been determined as technology for making such measurements has improved. The value of G is 6.67×10^{-11}, and the units were assigned in the same way we assigned units before for the constant in Newton's second law. The universal gravitational constant, G, is

Physics Phun

1. Calculate the force of attraction, gravitation only, between Mary, whose mass is 51.3 kg, and Jim, who has a mass of 117 kg, if they are sitting in desks that are 1.25 m apart.
2. Assume that the earth is a perfect sphere with a radius of 6.28×10^6m. How much centripetal force acts on a 101-kg man if he stands at the equator? Remember that the earth revolves around its axis every 24 hours. If the man stands on a bathroom scale and finds his weight at the equator and at the north pole, will the scale read the same in both places? Explain your answer.

$6.67 \times 10^{-11} \frac{N \cdot m^2}{kg^2}$. Newton's universal law of gravitation is stated symbolically as

$$F = \frac{Gm_1m_2}{R^2}.$$

No one really knows how Newton arrived at his universal law of gravitation. But we've tried to give you an idea of how he might have reasoned through this work. Kepler started with Brahe's exhaustive records of observations and replaced them with a simple set of mathematical laws. Newton built on the work of Galileo and Kepler to develop his ideas. Newton's universal law of gravitation explains the motions of planets as well as causes for the motions. It had one difficulty with an explanation for the behavior of the orbit of Mercury. It took Einstein's theory of gravity to explain that little detail.

The Acceleration of Satellites of the Earth

We've found that centripetal force varies inversely with the distance and varies directly as mass. The restoring force for simple harmonic motion varies directly as the distance. What do you expect for the force of gravity at some distance from the earth? A good way to find an answer is to consider an object not near the surface of the earth but at a considerable known distance from the earth. Canada launched a satellite, Alouette I, into an orbit that was nearly circular. Any satellite must be given a velocity that is large enough to keep it in orbit. If the orbit is nearly circular, the force keeping the satellite in orbit is centripetal force. The centripetal force must be the force of gravitational attraction of the earth for the satellite at that altitude. We can use what we know about centripetal force to discuss the motion of a satellite orbiting the earth as well as the cause of that motion.

The satellite has an altitude above the surface of the earth of $h = 1.01 \times 10^6$, and the radius of the earth is $R = 6.38 \times 10^6 m$. The mass of the satellite at that altitude is the same as its mass on Earth. That means that the acceleration due to gravity must be different at that altitude than it is near the surface of the earth. By the time you complete this chapter, you will be able to calculate the acceleration at that altitude, but for now let's state that the acceleration due to gravity at $h = 1.01 \times 10^6$ above the surface of the earth is $g_a = 7.30 m/s^2$. What is the velocity that the satellite must have to maintain that altitude? Use the analysis suggested here to calculate a solution. Remember, this is just one approach to a solution.

$f_c = f_{ga}$ The centripetal force is equal to the force of gravity at that altitude.

$m\frac{v^2}{R_a} = mg_a$ Substituting definitions of the centripetal force at that altitude is equal to the force of gravity at that altitude.

$\frac{v^2}{R_a} = g_a$ Dividing both equations by m, which is the same at both positions.

$v^2 = R_a g_a$ Multiplying both equations by R_a.

$v = \sqrt{(R+h)g_a}$ Solving for v and substituting for R_a.

$v = \sqrt{\left[(6.38\times10^6 m + 1.01\times10^6 m)(7.30m/s^2)\right]}$ Substituting numeric values.

$v = 7.34 \times 10^3 m/s$

Let's make the answer a little more understandable by expressing the velocity in miles/hr. You need the defining equation 1 mi = 1610 m to make the calculation:

$$v = (7.34 \times 10^3 m/s)(1/1610 mi/m)(60 s/min)(60 min/hr) = 16{,}400 mi/hr$$

Another interesting detail that we can gain from this example is a comparison of the force of gravity at 1.01 × 106*m* with the force of gravity at the surface of the earth: $\frac{f_{ga}}{f_{g0}} = \frac{mg_a}{mg_0} = \frac{g_a}{g_0}$.

The ratio of the force of gravity at the altitude of the satellite to the force of gravity at the surface of the earth ($h = 0$) is equal to the ratio of the acceleration due to gravity at the altitude to the acceleration due to gravity at the surface of the earth. Using values in this example, we can express it as $\frac{f_{ga}}{f_{g0}} = \frac{7.30m/s^2}{9.80m/s^2} = 0.745$. That is, about 74.5 percent of the force of gravity at the surface of the earth is the force of gravity on the satellite at $1.01 \times 10^6 m$ above the surface of the earth. The force of gravity appears to vary inversely with the distance, but where does mass enter this discussion? There were two masses involved here, the mass of the earth and the mass of the satellite, and we used the mass of the satellite alone.

Let's take our discussion further and calculate the weight of a body at any altitude above the earth. Newton's universal law of gravitation will enable you to do that. Let m_e be the mass of the earth and m_b be the mass of a body, and let R_e be the radius of the earth and R_b the distance from the center of the earth to the body above the

Johnnie's Alert

Always include the radius of the earth as part of the distance to a satellite when it is orbiting at a given altitude. It is easy to forget that the acceleration or weight at the altitude depends on the distance made up of the altitude plus the radius of the earth. The distance in Newton's equation is the distance from the center of a mass to the center of the other mass.

earth. An easy way to compare two things is to find the ratio of the two:

$$\frac{W_e}{W_h} = \frac{\dfrac{Gm_e m_b}{R_e^2}}{\dfrac{Gm_e m_b}{R_h^2}} = \frac{\dfrac{1}{R_e^2}}{\dfrac{1}{R_h^2}} = \left(\frac{R_h}{R_e}\right)^2 \text{ and that means } W_h = \left(\frac{R_e}{R_h}\right)^2 W_e.$$

Suppose that a 160-lb man at the surface of the earth is placed at an altitude 4000 mi above the earth. What will be his weight at that altitude? The radius of the earth is about 4000 mi. Substituting numeric values in the algebraic solution to this problem, we get $W_h = \left(\frac{4000mi}{4000mi + 4000mi}\right) \times 160lb = \left(\frac{1}{4}\right) \times 160lb = 40lb$. If you're wondering why we have two identical values in the denominator, remember that R_h is the radius from the center of the earth plus the distance from the surface of the earth to the orbiting body.

Suppose you know that a satellite is at a specific altitude above the earth, and for some reason you need to know its acceleration. You can use Newton's second law and the universal law of gravitation to calculate the acceleration due to gravity at any altitude.

$$W_{sh} = f_{gsh}$$

$$m_s g_{sh} = \frac{Gm_s m_e}{R_h^2}$$

$$g_{sh} = \frac{Gm_e}{R_h^2}$$

The algebraic solution states: the acceleration due to gravity at any altitude is equal to the universal gravitational constant times the mass of the earth divided by the square of the distance between the center of the earth and the satellite. Calculate the acceleration due to gravity at the altitude used earlier in the example, $h = 1.01 \times 10^6 m$. The mass of the earth is $5.98 \times 10^{24}\ kg$.

$$g_{sh} = \frac{\left(6.67 \times 10^{-11} \frac{N \cdot m^2}{kg^2}\right) \times \left(5.98 \times 10^{24} kg\right)}{(1.01 \times 10^6 m + 6.38 \times 10^6 m)} = 7.3 \frac{N}{kg} = 7.3 \frac{\frac{kg \cdot m}{s^2}}{kg} = 7.3 m / s^2$$

We just calculated the value we gave you to work with earlier. Any time you need the acceleration due to gravity at any altitude, you can calculate it.

Calculate the centripetal acceleration of a natural satellite of the earth. The moon is about $3.8 \times 10^8 m$ from the earth.

$$a_c = g_{mh} = \frac{\left(6.67 \times 10^{-11} \frac{N \cdot m^2}{kg^2}\right) \times \left(5.98 \times 10^{24} kg\right)}{(3.8 \times 10^8 m)^2} = 2.8 \times 10^{-3} m / s^2$$

The mass of the moon is $7.34 \times 10^{22} kg$. What is the centripetal force needed to keep the moon in orbit?

$$f_c = f_g = m_m g_{mh} = (2.8 \times 10^{22} kg)(2.8 \times 10^{-3} m/s^2) = 7.84 \times 10^{19} N$$

Measuring the Mass of the Earth

Have you ever wondered how they measured the mass of the earth? Well, here's your chance to see if you've understood the concepts and formulas that we've introduced you to in this chapter. It might be a bit challenging, but not too hard. We'll even provide the major steps to the solution. All you need to do is figure why we did what we did, and make the final calculations.

Scientists study the mass of distant heavenly bodies using the same technique that is outlined in this section. The first bit of information you need is the period of the satellite used in our earlier example. Alouette I completes an orbit every 105.4 minutes; that is, the period of the satellite is 105.4 minutes. We also know that a centripetal force equal to the force of gravity keeps the satellite in orbit. You can use that information along with ideas developed in this chapter to construct a solution to the calculation of the mass of the earth.

$$f_c = f_g$$

$$m_s \frac{4\pi^2 R}{T^2} = G \frac{m_e m_s}{R^2}$$

Physics Phun

1. Calculate the acceleration due to gravity at an altitude of $5 \times 10^5 m$ above the earth.
2. Calculate the weight of a girl at an altitude of $6.6 \times 10^5 m$ if her mass is 55 kg.
3. Calculate the mass of the earth if an orbiting satellite has a period of 100 min and an altitude of $7.5 \times 10^5 m$.

$$\frac{4\pi^2 R}{T^2} = G\frac{m_e}{R^2}$$

$$m_e = \frac{4\pi^2 R^3}{GT^2}$$

$$m_e = \frac{4\pi^2(6.38\times10^6 m + 1.01\times10^6 m)^3}{\left(6.67\times10^{-11}\frac{N\cdot m^2}{kg^2}\right)\left(105.4\,\text{min}\times 60\frac{s}{\text{min}}\right)}$$

$$m_e = 5.97 \times 10^{24} kg$$

How did you do? Did you understand how we moved from one formula to the next? The equations are all ones that we've used in the chapter; the final calculation gave us the mass of the earth. In the next few chapters, we examine different forms of energy and how they relate to work.

Problems for the Budding Rocket Scientist

1. Use Newton's law of universal gravitation to determine the attractive force between you and the earth. Earth has a radius of $6.38 \times 10^6 m$ and a mass of $5.97 \times 10^{24} kg$. If your mass is 80 kg, what is the force of gravity exerted on you?
2. Knowing that the mass of Mars is one tenth the mass of the earth and that its radius is half that of earth, what is the acceleration due to gravity on the surface of Mars?
3. Calculate the force of attraction between two 90-kg spheres of metal spaced so that their centers are 40 cm apart.
4. A spring is stretched 4 cm when a mass of 50 g is hung on it. If a total of 150 g is hung on the spring and the mass is started in a vertical oscillation, what will the period of the oscillation be?
5. Find the length (in meters) of a pendulum that has a period of 2.4 s.

Part 3

Work and Energy

The entire universe is made of just two primary constituents: matter and energy. In the first two parts of this book, we examined some aspects of how motion is related to matter and energy. We now delve deeper into this relationship and define how movement expresses itself as work, power, and different forms of energy. We'll also discover another law that is fundamental to physics, the law of conservation of energy. The total amount of energy in the universe is, has been, and always will be the same as it is right now. The phrase *conserve energy* will take on a whole new meaning for you.

Chapter 9

Work and Power

In This Chapter

- What exactly is work?
- Calculating work
- Simple machines
- Work and watts

Power is a concept that is important in the everyday world. Some people crave it, others express it, and still others avoid it. But is that the kind of power we're talking about in physics? Well, it depends on how we define it. The same could be said for the word *work.* One definition that we can use is to say that power is the rate at which work is done. Doing a large amount of work in a small amount of time requires a lot of power. Having said that, are we any closer to the way in which physicists understand these two words? Read on and we'll discover together just how interconnected work and power really are.

Work in Motion

How do you use the word *work*? You might say, "Work was rough today," or "Who knew that relationships could be so much work?" Like many words, *work* has a different meaning in physics than it does in the everyday world.

However, what if you walked into your apartment, collapsed on the couch, and said, "It sure took a lot of work to get that dresser into the truck!" You would mean that you had to exert a force to move something. But before we explore that idea in detail, let's briefly step back a little and see how other scientists have understood the concept of work.

A person sitting quietly can suddenly move his arm and lift a cup or fold a newspaper (any number of work-related activities) and do work where no work previously seemed to exist. From this example, one could suppose that work or something equivalent to it could be stored in the human body and that this work-store could be called upon at need and converted into visible, palpable work. It seems as though this work-store is particularly associated with life, because living things seem to be filled with this capacity to do work, whereas dead things for the most part lay quiescent and don't work. (Think of all the animals that have been used to move, haul, or pull carts, wagons, etc.)

The German philosopher and scientist Gottfried Wilhelm Leibnitz (1646–1716), the first to get a clear notion of work in the physicist's sense, called this work-store *vis viva* (Latin for "living force"). However, it is clearly wrong to suppose that work is stored only in living things; the wind can drive ships and running water can turn millstones, and in both cases force is being exerted through a distance. Work, then, was obviously stored in inanimate objects, too. In 1807, the English physician Thomas Young (1773–1829) proposed the term *energy* for this work-store. This term gradually became popular and is now applied to any phenomenon capable of conversion into work. And as you know, there are many varieties of such phenomena and many forms of energy.

The first form of energy to be clearly recognized as such was that of motion itself. Work involved motion (since an object had to be moved through a distance), so it wasn't surprising that motion could do work. It was moving air, or wind, that drove a ship, and not still air; moving water that could turn a millstone, not still water. It was not air or water that contained energy, but the motion of the air or water. In fact, anything that moved contained energy, for if the moving object, whatever it was, collided with another, it could transfer its momentum to the second object and set its mass into motion; it would therefore be doing work upon it, for a mass would have moved through a distance under the urging of a force.

def•i•ni•tion

In physics, when forces act over some distance, we say that they do **work**. **Energy** is often defined as the ability to do work. If something has a large amount of energy, it has the ability to do a lot of work.

The energy associated with motion is called kinetic energy, a term introduced by the English physicist

Lord Kelvin (1824–1907) in 1856. The word *kinetic* is from a Greek word meaning "motion." And we devote a whole chapter (Chapter 10) to examining that form of energy.

Work and Force

In Chapter 5, we discussed the nature of forces and how forces result in accelerations that can be described by Newton's second law ($F = ma$). In physics problems, when forces act over some distance, we say that they do work. Expressed mathematically, the work done by a force is written as follows:

Work = Force × Distance

or

$$W = fd$$

Another way to look at it is that it takes twice as much work to lift twice the weight to the same height, or that it takes three times as much work to lift a given object three times as high. This all indicates how work is related to both force and distance.

From previous chapters, we know that distance is measured in meters (m), and forces are measured in newtons (N); so the unit of work is a Newton-meter ($N \cdot m$), which is sometimes called a joule (J). The joule is a derived unit, because it is derived from the fundamental units we defined earlier. (And by now we're positive you know what they are!) A joule (named after the English physicist James Prescott Joule) is also expressed as $1kg \cdot m^2s^2$. One important fact to remember is that the only force that matters in calculating work is the force exerted in the direction of motion. Let's examine that fact in a little more detail.

We already know that work involves displacement (or the distance traveled) and force and that work is therefore a scalar quantity. And we also know that we can calculate work by multiplying the magnitude of the applied force in the direction of the displacement of an object by the magnitude of the displacement. However, there are examples where the force is not applied in the direction of the displacement. So we need to know that if a force is applied to an object, and the object is displaced in a direction different from the direction of the force, then we must use the component of the applied force that acts in the direction of the displacement, Δx. This is shown in Figure 9.1 and 9.2.

The work done to move the object along a horizontal surface in Figure 9.1 may be calculated simply as $W_k = F_A \Delta x$ (and this is simply a variation on the basic formula of $W = fd$). This is an example where the force and displacement are in the same direction, so the calculation is straightforward.

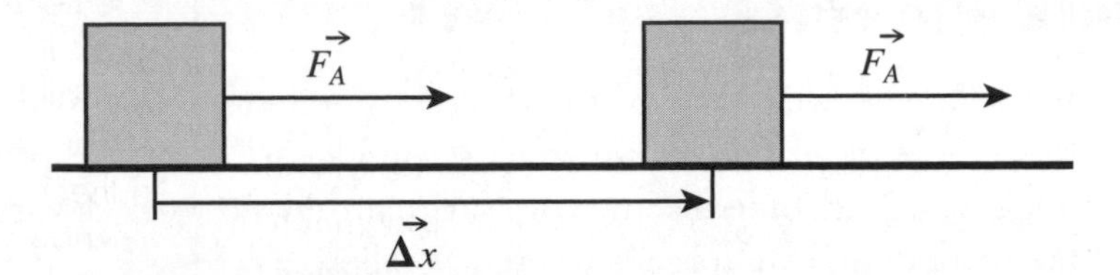

Figure 9.1

The object displaced by a horizontal force is a simple case for the calculation of work.

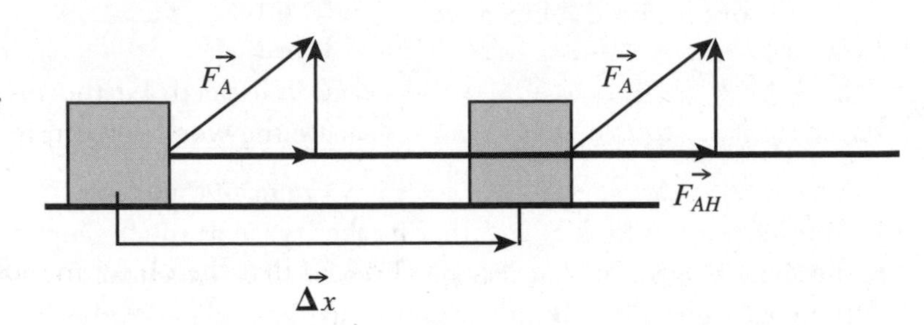

Figure 9.2

The object displaced by a force at an angle with the horizontal is a more complicated case for the calculation of work.

The example diagrammed in Figure 9.2 is a little more complicated because as you can see, the object does not move in the direction of the applied force $\vec{F}_A$. What that means is that we must resolve the force into components perpendicular and parallel to the displacement and use the component parallel to the displacement to calculate the amount of work done. The horizontal component of the applied force in Figure 9.2 is labeled $\vec{F}_{AH}$. The work done in that case is $W_k = F_{AH} \Delta x$. And all this really says is that the work can be calculated by multiplying the magnitude of the component of the applied force parallel to the displacement by the magnitude of the displacement.

There are problems that will require us to recognize the units of work in any one of the three forms. So far we showed you the unit in the MKS system, the joule. The unit of work in the CGS system turns out to be (dynes)(cm) or *dyne · cm*. The unit of work in the CGS system is called the *erg* (as in "Errrgg, that's a tough problem!"), which is the root of the word *energy*. The relationships among all of these units can be expressed as $1erg = 1dyne \cdot cm = 1\frac{g \cdot cm^2}{s^2}$.

Our last unit of work in the FPS system, weird as always, is the *ft · lb*. There is no other unit of work in the FPS than 1. The units of work expressed in fundamental units are $1ft \cdot lb = \frac{1slug \cdot ft^2}{s^2}$. And now that you have a basic idea of what work is all about, as well as the way it is expressed in the three different systems, let's "work" through some problems together.

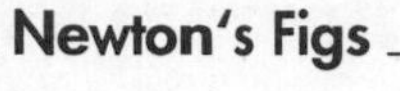

Newton's Figs

Even if some of the units of measurement in the three fundamental systems are easier to remember than others, it's a good idea to become familiar with all three. The problems that you solve in physics vary widely in these systems of measurement, so knowing them all will come in handy.

Example 1: Calculating Work Using a Horizontal Force

Suppose that a force of 54 N is required to roll a 981 N safe 11 m across the floor. How much work is done? The way the problem is worded, it appears that the force is applied in the direction the safe is moving and that 54 N is just enough force to move the safe at a constant speed. Furthermore, the safe stops when the force is removed. So we can calculate the work done in the following way:

$$W_k = F_A \Delta x = 54N \times 11m = 590N \cdot m \text{ or } 590 \text{ joules}$$

Unless stated otherwise, it is understood that the object moved has a constant speed.

Example 2: Calculation of Work Using a Component of the Force

Let's try another one. A force of 68 lb must be applied to the box in Figure 9.2 to move it across the surface in the diagram. The angle at which the force is applied is 42° with the horizontal and the distance the object moves is 12 ft. How much work is done? The analysis of the situation is the same as in Example 1, but we must determine the component of the force that is in the direction of the displacement in order to solve the problem. The best way to do this is by drawing the force to scale and measuring the angle with a protractor. Construct the projection of the force on a horizontal line to find the horizontal component of the applied force. We've had some good practice by now drawing graphical solutions, so it would be a great idea for you to solve this problem along with us to see how we arrived at the answer. In doing our drawing we came up with the horizontal component of the applied force as 51 lb. The solution is this:

$$W_k = F_{AH} \Delta x = 51lb \times 12ft = 610ft \cdot lb$$

Did you notice that the answer is given to two significant figures because the values given in the problem were given to two significant figures?

Example 3: Calculating Work Using Lifting Force

Okay, let's do one more and then move on. A woman lifts a 5.3-kg box from the floor of her garage and places it on a shelf 2.5 m above the floor. How much work did she do? This problem is a little different because the object is moved vertically. We know that if the box is caused to move upward at a constant speed near the surface of the earth, the force applied must be the weight of the box. By reviewing Newton's laws, we may figure out how to calculate the weight of the box. The analysis of the problem into its parts enables us to enjoy a quick solution to the problem.

$$W_k = F_A \Delta y = mg\Delta y = (5.3kg)(9.8m/s^2)(2.5m) = 130\ \frac{kg \cdot m}{s^2} \times m = 130N \cdot m = 130\ joules$$

These examples are good models for solving simple problems involving the calculation of work done. There's one other idea that should be included to make your collection of tools for solving work problems more complete. We told you earlier that it is assumed that the object moves with a constant speed unless you're told otherwise. The way we can be sure that the object moves with a constant speed is to make sure there is no net force. That will be true if the force you apply is equal in magnitude but opposite in direction to any other force on the object. You did that in the solution to Example 3. The force applied upward to the box was equal to the weight that acts downward.

In Example 1, can you guess what the other force was that kept the object from accelerating? That's right, friction was acting in the direction opposite to the direction of the applied force. The force of friction was equal in magnitude, but opposite in direction, to the applied force. Now look at Example 2. The force of friction would not be in the direction of the applied force. Did you realize that it would act in a direction opposite to the horizontal component of the applied force? If you did that's great! Forces acting on objects when work is done are not always constant. Graphical techniques or calculus are needed to handle those situations, but handling cases when forces are constant will prepare you to handle the other situations when they arise.

Any time we move something, there will be an opposing force, the force of friction. In many cases, it is negligible, and we don't have to account for it. In some problems, we'll have to find the force of friction. Usually the force that we'll have to apply to get the object to move at a constant speed is equal in magnitude to the force of friction. There is a force of moving friction and a force of starting friction. We already know that it is harder to get something to start moving than it is to keep it moving. The moving friction is the force we should concern ourselves with even though we're well aware that it may be a little less than the force we applied initially.

The force we apply at first has to get the object up to speed. Therefore, the object is accelerating. We are talking about the time after that initial acceleration when the object is moving at a constant velocity. The force of friction depends on a lot of things, such as the types of surfaces in contact, whether the surfaces are rough or smooth, whether the surfaces are bare and dry or lubricated in some way, and so on. So we should be aware of how those things can change the force of friction, but for now let's concentrate on the bare and dry surfaces that cause an opposing force.

The two surfaces that we have to consider have a characteristic coefficient of friction. Usually the coefficient of friction is either given in a problem if it is needed, or enough information is given for you to calculate it. Sometimes a reference to a table will be given for us to find the coefficient of friction when we need it. When we know the coefficient of friction for two surfaces, the force of friction is calculated as $\Im = \mu N$, and in words it means that the force of friction is equal to the product of the coefficient of friction and the normal force. The normal force means the force perpendicular to the surfaces.

Example 4: Calculating Work Using the Force of Friction

Okay, we lied; we're going to work one problem, and then you'll be on your own to work the Physics Phun problems following this example. Here's the problem. How much work is done to roll a metal safe, mass 121 kg, a distance of 20.5 m across a level floor? The coefficient of friction is 0.049. We want to move the safe at a constant speed and that means the force we apply must be equal in magnitude to the force of friction. In this case, the normal force is the weight of the safe, and we already know how to calculate that.

$$W_k = F_A \Delta x = \Im \Delta x = \mu N \Delta x = \mu W \Delta x = \mu m g \Delta x$$

The algebraic solution states that the work done is equal to the coefficient of friction times the weight of the safe times the distance moved.

$$W_k = (0.049)(121\,kg)(9.8\,m/s^2)(20.5\,m) = 1200\ joules$$

Are you ready to go for it all by yourself? Give it a try!

Physics Phun

1. A man dangling by his hands on the monkey bar flexes his biceps and raises his body until his head is well above the monkey bar. His mass is 5.22 slugs and his body is raised 61.1 cm. How much work did he do? Remember to express given information in the same system of measurement. In this case it is probably easier to change centimeters to feet and calculate the work done in *ft • lb*. Hint: 2.54 cm = 1 in, g = 32 ft/s2.
2. The coefficient of friction between a packing crate and the floor is 0.234. If the packing crate weighs 372 lb, how much work must be done to move it 21.5 ft across a level floor at a constant speed?
3. A woman pulls a sled across a level field of snow by applying a force of 68 N to a rope attached to the sled. The rope makes an angle of 38° with the snow. How much work does she do pulling the sled across the field a distance of 52 m?

Simple Machines

There are six simple machines that enable us to make our work easier. Sometimes we combine them to make complex machines using two or more of them. Simple machines can be broken into two groups. The first group is made up of the lever, the wheel and axle, and the inclined plane. The second group includes the pulley, the screw, and the wedge. We'll discuss one member of each family because all other members in the family have similar characteristics. But first let's discuss general characteristics common to both groups.

Johnnie's Alert

Although there are six simple machines, the three in the second group (the screw, the wedge, and the pulley) are really extensions of the three in the first group (the lever, the inclined plane, and the wheel and axle). The screw is an inclined plane wound around an axis; the wedge consists of two inclined planes set back to back; and the pulley combines the characteristics of both the lever and the wheel and axle.

In an ideal world, the simple machines would work with no loss in the work you put in compared to the work you get out. Any time we use a simple machine, we know that friction opposes any motion that may be involved. Another way of stating the ideal situation is to compare the work we get out of a machine with the work we put into it. As we already know, a ratio of two quantities is a good way to compare them. Ideally $\frac{W_{k\ out}}{W_{k\ in}} = 1$, but in our world, we know $\frac{W_{k\ out}}{W_{k\ in}} < 1$. That means the work we get out in the real world is less than the work we put in. We can express the inequality by introducing *efficiency*.

The efficiency is expressed as: efficiency $= \frac{W_{k\ out}}{W_{k\ in}} \times 100\%$. We can see that efficiency is expressed in such a way that we know the efficiency of any machine will never be 100 percent. The electrical transformer is about 98 percent efficient, and that is about the upper limit of efficiency.

def•i•ni•tion

The **efficiency** of a machine is a fraction greater than zero and less than one that expresses what part of input work the output work amounts to. The efficiency is usually stated as a percent obtained by multiplying the fraction by 100 percent.

Actual mechanical advantage (AMA) is the actual gain in force that a user realizes from a machine because it includes the effects of friction. **Ideal mechanical advantage** (IMA) is the theoretical gain a user expects from a machine. It is always more than the AMA because it does not include the effects of friction.

Because a machine is never perfect, something must be sacrificed when a machine is used. Distance or speed is gained by exerting considerable force. Riding a bicycle is a good example of sacrificing force to gain speed. Heavy weights may be lifted but only at the expense of distance and/or speed. Your auto mechanic can lift the motor from a car with a pulley system by applying a small effort force, but large amounts of chain links must pass through the pulley to lift the motor a small distance. Two terms are used to avoid explaining the limitations of a machine when it is employed. One term is *actual mechanical advantage*, and is represented by the letters AMA. The AMA is the real gain a user realizes when using a machine. The AMA is calculated as a ratio of forces as follows: $AMA = \frac{F_R}{F_e}$. Similarly, the *ideal mechanical advantage*, IMA, is the expected advantage if you did not have any friction.

The effort force, F_e, is the force applied to the machine. The resistance force, F_R, is the force the machine applies to an object.

The distance, S_e, is the distance through which the effort force moves. The distance, S_R, is the distance through which the resistance force moves.

The ideal mechanical advantage is calculated as $IMA = \frac{S_e}{S_R}$. Either or both of these ideas may be used to solve problems involving machines. Both can be used to discuss efficiency because they have their origin in the calculation of work. We can write the ratio of output work and input work in terms of forces and distances as follows:

$\frac{W_{k\,out}}{W_{k\,in}} = \frac{F_R S_R}{F_e S_e} = \frac{\frac{F_R}{F_e}}{\frac{S_e}{S_R}} = \frac{AMA}{IMA}$. Can you see how these new terms are very closely associated with our earlier discussion at the beginning of this section about friction and efficiency? In fact, efficiency $= \frac{AMA}{IMA} = \times 100\%$.

With these general properties of all simple machines in mind, we now investigate one member from each of the two groups of simple machines.

The Inclined Plane

Let's begin with the inclined plane. This is a simple machine that is in wide use today and has many applications. The inclined plane makes it easier for us to lift a heavy object from one level to another with force less than the object's actual weight because it is some form of a ramp. It may be in one of many forms such as stairs or streets. Take a look at Figure 9.3 as we discuss the details of an inclined plane. The inclined plane has length l and height h.

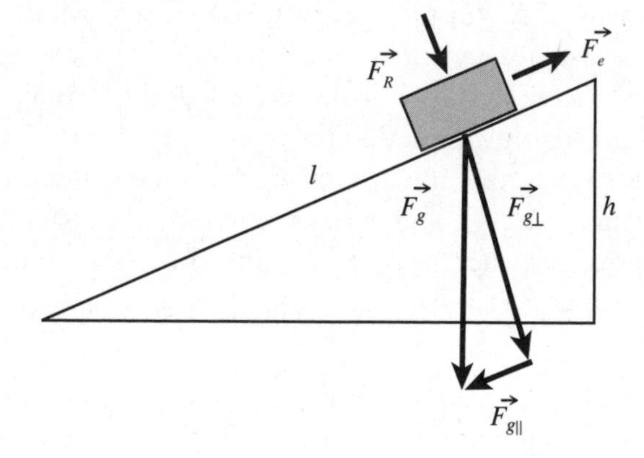

Figure 9.3

The inclined plane is one example of a simple machine.

The inclined plane can be used to raise the object vertically a distance h by applying a force parallel to the plane for a distance l rather than lifting the object straight up. The magnitude of the weight of the object is F_g, which has been resolved into components parallel to l and perpendicular to l, $F_{g\parallel}$ *and* $F_{g\perp}$, respectively.

The weight of the object or the resistance force is F_R. If we want to pull the object up the plane at a constant speed, then we must apply the effort force F_e parallel to l and up the plane in the opposite direction of $F_{g\parallel}$. The magnitude of F_e is the sum of $F_{g\parallel}$

and the force of friction. That is, $F_e = F_{g\|} + \Im = F_{g\|} + \mu F_{g\perp}$. The length of the inclined plane is the distance S_e and the height is the distance the resistance force moves S_R. We will determine the values of $F_{g\perp}$ and $F_{g\|}$ when we construct the force triangle and resolve the weight into two mutually perpendicular components. We'll then be able to calculate AMA = $\frac{F_R}{F_e} = \frac{F_g}{F_{g//} + \mu F_{g\perp}}$ and IMA = $\frac{S_e}{S_R} = \frac{l}{h}$. We can solve any inclined plane problem with the information outlined here. So why don't we do exactly that.

Problem:

An inclined plane is 8.00 ft long and 2.00 ft high. How much force must be applied parallel to the plane to pull an object weighing 2,020 lb up the inclined plane at a constant speed if the coefficient of friction is 0.049? What is the efficiency of this machine? A construction of the inclined plane to scale and resolving the weight into two mutually perpendicular components yields the following information: $F_{g\|}$ = 506 lb, $F_{g\perp}$ = 1,960 lb, S_e = 8.00, $S_R = 2.00ft$, $F_f = \mu N = (0.49)(1960lb) = 96lb$.

The effort force $F_e = F_{g\|} + F_f$ = 506 lb + 96 lb = 602 lb.

Solution:

The efficiency of this machine can be calculated as follows:

$$\frac{AMA}{IMA} \times 100\% = \frac{\frac{2020lb}{602lb}}{\frac{8.00\,ft}{2.00\,ft}} \times 100\% = \frac{3.36}{4.00} \times 100\% = .84 \times 100\% = 84\%$$

This machine is 84 percent efficient, and that means we get back 84 percent of the work we put into the machine. That's pretty good efficiency, considering it would take a lot more work to lift the weight straight up. Of course, in that case we might use a pulley to lift the weight directly off the ground and vertically into the air. Any time we can use a simple machine to make our work easier, the better off we are.

The Lever

The lever is a member of the other group of simple machines that deserves our attention. An understanding of the lever reveals the basic principles of all other members of that group. The lever, like the inclined plane, has associated with it the ideas represented by the symbols F_R, F_e, S_R, S_e, IMA, and AMA.

Newton's Figs

One force that doesn't seem to be the result of direct contact of one body upon another is gravitation. Gravitation seemingly exerts a force from a distance and produces motion without involving direct contact between bodies. Such "action at a distance" troubled Newton and many physicists after him. And since no one really knew the cause of gravity, it was included in the list of mechanical forces since it behaved in relatively the same way as other mechanical forces. So the study of the motions of the heavenly bodies that result from and are controlled by gravitational forces is called celestial mechanics.

The effort force, the force applied by the user of the machine, is the same for the lever as it was for the inclined plane. The resisting force is the same, too. The force the machine works on is often the weight that is being lifted. In order to understand S_e and S_R, which play the same role for the lever as the inclined plane, we need to realize that S_e and S_R are arcs along circles. Furthermore, $S_e \alpha l_e$ and $S_R \alpha l_R$ are labeled for us in Figure 9.4.

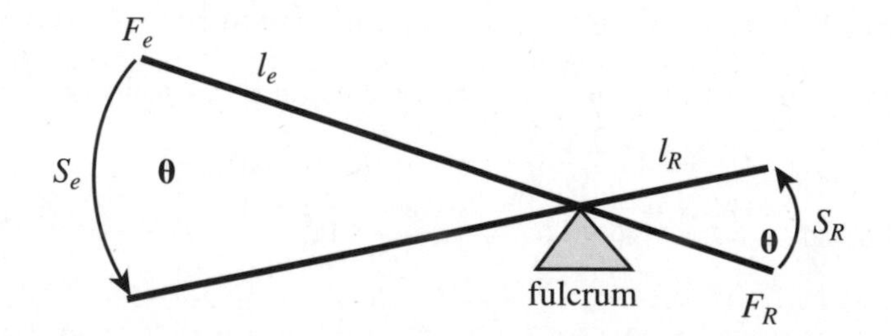

Figure 9.4

A lever as a simple machine showing necessary parts.

The output part is on the right side of the diagram and the input on the left. A lever pivots about a point called the fulcrum. There are three different types of levers. The one used in the diagram lifts a large force, or weight, with a smaller effort force. We could do it in reverse. That would be the case if you wanted to gain distance instead of force. (We'll save that for another problem later.) In the diagram, the effort force is applied to the left end that moves through a circular arc labeled S_e, which represents the effort distance. The resistance force is applied to the end of the lever on the right in the diagram and that end moves along the circular arc labeled S_R. The distance moved by the effort force is calculated in terms of l_e. Similarly, the resistance force moves a distance that is expressed in terms of l_R.

$$IMA = \frac{S_e}{S_R} = \frac{l_e}{l_R}$$

The lever shown in the diagram is used to multiply force. This same lever can be used to multiply distance if the effort force is applied at the other end. An example of a lever used to multiply distance is the use of scissors. Some levers have the fulcrum at one end and the effort force is applied at the other end, such as an oar for a boat where the blade is the fulcrum. The same arrangement can be used where the resistance force is on the end opposite the fulcrum and the effort force is applied between the fulcrum and the resistance force.

The lever, like any other simple machine, will multiply force or multiply distance. Even though we know that a machine cannot do both in our world because of friction, we can often assume friction is negligible in the solution of many problems unless enough information is given to prove otherwise. But friction is necessary. It would be impossible to roll a wheel or to take a step walking without it. Like many things in our world, it can be considered useful at times and a hindrance at others. Let's work a problem together so that we're sure we understand the concepts and the math.

Problem:

A board 8.3 m long is used as a lever with the fulcrum between the resistance force and the effort force. The fulcrum is placed 3.3 m from the resistance force. (a) Calculate the IMA of the simple machine. (b) What is the minimum effort force that should be applied to lift the weight of a 55-kg mass?

Solution:

We know that the IMA is the ratio of resistance and effort arms. That is, (a) $IMA = \frac{l_e}{l_R} = \frac{5.0m}{3.3m} = 1.5$. Because we don't have enough information to determine the efficiency of the machine, we can figure out the force needed if we assume that it is 100 percent. That would be the minimum force needed. If there were friction, it would just mean that the actual force would have to be more than that. That means, (b) $AMA = \frac{F_R}{F_e}$ and $F_e = \frac{F_R}{AMA} = \frac{(55kg)(9.8m/s^2)}{1.5} = 360N$. What if we had multiplied by 1.5 instead of dividing? We would have given an answer that would say that it takes more force to lift the object than it weighs. This can't be right if the fulcrum is nearer to the object than the force being applied by the lifter. Thinking about the problem to see if our answer makes sense can really help us when we work problems.

Physics Phun

1. A crate weighs 555 N and is pulled upward along a 31 m inclined plane at a constant speed to raise it 2.3 m above the starting point. The force required to pull it along the plane is 55.5 N. Calculate (a) the work output, (b) the work input, (c) the efficiency of the machine.
2. A holiday nut is placed 2.54 cm from the hinge of a nutcracker. If you exert a force of 51 N at a point 15.2 cm from the hinge, what resistance force does the nut exert?

Work, Power, and Time

The definition of work as the product of a force and the distance through which it acts says nothing about the time it takes to act. We often find it preferable to accomplish a particular amount of work in a short time rather than in a long period of time and are therefore interested in the rate at which work is done. This rate is spoken of as *power*. We finally have a working definition of power that we spoke about in the opening paragraph of this chapter. And as you can see, it's a bit different from the way it is often used in economics, politics, or psychology. In these occasions, power refers to social interaction; in physics, it's a little more straightforward.

def•i•ni•tion

Power is defined as the work done per unit of time or as the rate at which work is done. A common unit of power is the **watt**.

As mentioned previously, power is the work done per unit of time. Power depends on force applied in the direction of displacement, the displacement, and time. The definition of power in symbols is $P = \frac{W_k}{t}$. As we always do, let's consider the units of power in all three systems of measurement.

In the CGS system, the units of power from the definition are $P = \frac{ergs}{s}$. In the MKS system, the units of power are $P = 1\frac{joule}{s} = 1watt$

In the FPS system, the units of power are: $P = \frac{ft \cdot lb}{s}$.

A very common unit of power that fits into neither system was originated by the Scottish engineer James Watt (1736–1819). He had improved the steam engine and made it practical toward the end of the eighteenth century, and he was anxious to know how its rate of work pumping water out of coal mines compared with the rate of work of the horses previously used to operate the pumps. In order to define horsepower, he tested horses to see how much weight they could lift through what distance and in what time. He concluded that a strong horse could lift 150 pounds through a height of 220 feet in 1 minute, so one horsepower was equal to 150 × 220, or 33,000 foot-pounds/minute. (Which is 550 foot-pounds per second.) That means that if you have 300 horses under the hood of your car, you have a car that will do 165,000 ft-lb of work every second. That's a lot of power!

This inconvenient unit (horsepower) is equal to 745.2 joules/sec, or 7,452,000,000 ergs/sec. As previously mentioned, a joule/sec is defined as a watt and was named so in honor of James Watt. We can also say then that one horsepower is equal to 745.2 watts. The watt is the most commonly used measurement in electricity, and horsepower is the most common unit used in mechanical engineering as well as cars.

Power is also what our utility company sells us in the form of electricity. However, the thing we pay our power company for is work, not power. You pay the electric company based on your kilowatt-hours. These are units of work and not power. Take a look at the units. We're multiplying a unit for power times a unit for time. Now look at the equation that defines power. If we multiply both sides by time, we get $Pt = W$. So kilowatt-hours are units of work.

Let's suppose we calculate the work required for a 100-watt bulb to provide light in our room. We can say that every second the amount of work is 100 watts of work. In an hour, the bulb requires:

$$(100 watts)(3600 s) = 3.6 \times 10^5 watt - s = 1 \times 10^2 watt - hr = 1 \times 10^{-1} kilowatt - hr \text{ or } 0.1 kw - hr$$

Let's do one final problem together to close this chapter before moving on to Chapter 10:

A man's mass is 82.0 kg. If he walks up a flight of stairs 3.0 m high in 9.0 s, what power has he used?

Solution:

$M = 82.0$ kg, $t = 9.0$ s, $h = 3.0$ m, $P = ?$

$$P = \frac{W_k}{t} = \frac{(82.0kg)(9.8m/s^2)(3.0m)}{9.0s} = 270 watts$$

(Remember that the units of power in MKS are joules/sec or watts.)

Problems for the Budding Rocket Scientist

1. A student is pushing a box of her belongings across a wooden floor. A constant force of 200 N is required, and she pushes the box over a distance of 10 m. How much work does the student do?
2. A 4-kg object is slowly lifted 1.5 m. (a) How much work is done against gravity? (b) Repeat for a scenario in which the object is lowered instead of lifted.
3. In the course of a 30-minute workout session you lift a total of 10,000 kg, an average of 50 cm. (a) How much work do you do? (b) How much power was required for your workout?

The Least You Need to Know

- The only force that matters in calculating work is the force exerted in the direction of motion.
- The way to be sure that an object moves with a constant speed is to make sure there is no net force.
- The force of friction is equal to the product of the coefficient of friction and the normal force. The normal force means the force perpendicular to the surfaces.
- Any device that transfers a force from the point where it is applied to another point where it is used is called a machine.
- A machine not only transfers a force, it can often be used to multiply that force, too.

Chapter 10

Kinetic Energy

In This Chapter

- Describing kinetic energy
- How does work enter the scheme?
- Raindrops and kinetic energy
- Following kinetic energy
- Kinetic energy and air pressure

As the name implies, the kinetic energy of an object is related to its motion. Moving objects have energy—that is, they have the ability to do work. Why don't you let a baseball hit you in the face? Because of the significant work it would do on your nose! And the faster the baseball is moving, the more work it can do, the more energy it has. The more massive the baseball, the more work it can do, and the more energy it has, too. So while you're remembering to duck or catching the baseball in a glove, you're getting the chance to experience kinetic energy firsthand.

What Has Kinetic Energy?

As we've already mentioned, the energy associated with motion is called *kinetic energy.* Below is a list of some common examples of how kinetic energy can manifest in our everyday world.

Johnnie's Alert

Now is a good time to review the description of motion characterized by a constant speed as well as motion characterized by constant acceleration. The discussion of kinetic energy will require knowledge of the description of motion of an object.

- A hurricane hitting the coast, crashing giant waves into barriers; trees bending in the wind and roofs of homes flying away like leaves
- A towering punt with an unbelievable hang time; a receiver running under the nearly 1-pound football, catching it, and running
- A medicine ball passed from person to person
- A swerving unguided missile in the form of a giant, free-rolling truck wheel

The notion of energy is one of the few elements of mechanics not handed down to us from Isaac Newton. Mechanics is the branch of physics that deals with motions and forces. The branch of mechanics that specifically deals with motion is called dynamics; the branch that deals with motions in equilibrium is called statics. Archimedes was the first great name in the history of statics (he was the Greek thinker who invented the screw and said that with the right lever he could move the world). And Galileo was the great name in the history of dynamics.

def•i•ni•tion

Kinetic energy is the energy an object has when it is in motion.

Work and Kinetic Energy

Exactly how much kinetic energy is contained in a body moving at a certain velocity? In this context, we'll use v for initial velocity and v' for the final velocity. Because we are discussing motion and force in a straight line in the same direction, we won't bother with vector notation.

The motion is uniformly accelerated motion, so we have three ideas to guide us in our analysis, along with Newton's second law to help us with the force due to gravity. Let's solve algebraically in terms of the quantities involved.

$\frac{v+v'}{2}=\bar{v}$, $\Delta x=\bar{v}t$, $a=\frac{v'-v}{t}$ This is information we know for describing uniformly accelerated motion.

$$\Delta x = \bar{v}t$$

$$\Delta x = \bar{v}\left(\frac{v' - v}{a}\right)$$

$$\Delta x = \left(\frac{v + v'}{2}\right)\left(\frac{v' - v}{a}\right)$$

$$\Delta x = \frac{v'^2 - v^2}{2a}$$

This is a description of the motion in terms of things we could observe: initial velocity, final velocity at its destination, acceleration, and a distance traveled before arriving at a safe destination.

Now the task is to find a way to include the force and look for a relationship with work: $a\Delta x = \frac{v'^2 - v^2}{2}$

Now what? If we only had mass … hmm, the giant wheel from our examples above has mass (m), so let's multiply both sides of the last equation by *m:*

$$ma\Delta x = \frac{mv'^2 - mv^2}{2}$$

That will do it! Remember that mass times acceleration is equal to the net force. Did you notice that the term on the left of the equation represents work? That is,

$$F\Delta x = \frac{mv'^2}{2} - \frac{mv^2}{2}.$$

The quantity $\frac{1}{2}mv'^2$ is the kinetic energy. The last equation in our derivation states that the net work done is equal to the change in kinetic energy. In symbols, $W_k = \Delta KE$, where $KE = \frac{1}{2}mv'^2$.

Another way to state this equation is to say that the kinetic energy of an object is equal to half its mass times its velocity squared. From this formula it is apparent that the more massive an object, and the faster it is moving, the more kinetic energy it possesses. So a slow-moving truck and a fast-moving bullet may have an equivalent amount of kinetic energy, even though their masses are very different. The units of kinetic energy are determined by taking the product of

Newton's Figs

Many problems state given information explicitly while other information is implied. When an object is dropped from a point above the surface of the earth, it's implied that the initial velocity is zero, and the acceleration is the acceleration due to gravity.

the units for mass (kg) and velocity squared (m^2/s^2). So the units of *KE* are (not surprisingly) $kg \cdot m^2/s^2$, or joules. Let's work through a couple of problems together and look at all of these aspects of kinetic energy in more detail.

Example 1: Motion on an Inclined Plane

A crate, mass 2.00 kg, slides from rest down a frictionless inclined plane. The inclined plane is 4.00 m long and makes an angle of 15° with the horizontal.

(a) How large is the force that accelerates the crate down the plane?

(b) How fast is the crate traveling when it reaches the bottom of the plane?

(c) How much time does it take the crate to slide down the plane?

Newton's Figs

Resolving a given vector into components is often implied in the statement of a problem. You will decide whether or not finding the components is necessary as well as which components are needed. A large percentage of cases will involve components of a vector that are mutually perpendicular.

Whenever you're given just a word problem without a picture, it's a good idea to draw one. Visual information can really help a lot in solving problems. We did that and included it as Figure 10.1. Let's refer to that diagram as we consider the solution. Part of the solution is constructing the diagram correctly and labeling each part possible. We drew a horizontal line to represent the base of the inclined plane and constructed a 15° angle at one end. We then extended the one side of the angle to represent 4.00 *m*; we can use any scale we choose.

We then constructed the weight of the crate, F_g, and resolved it into components parallel and perpendicular to the plane (the side l is referred to as the plane in the context of problems dealing with inclined planes). We labeled each part and are now ready to construct the solution.

$m = 200kg \qquad v = 0$

$l = 4.00m \qquad \Im = 0$

$\theta = 15.0°$

(a) $F_{g\parallel} = ?$

(b) $v' = ?$

(c) $t = ?$

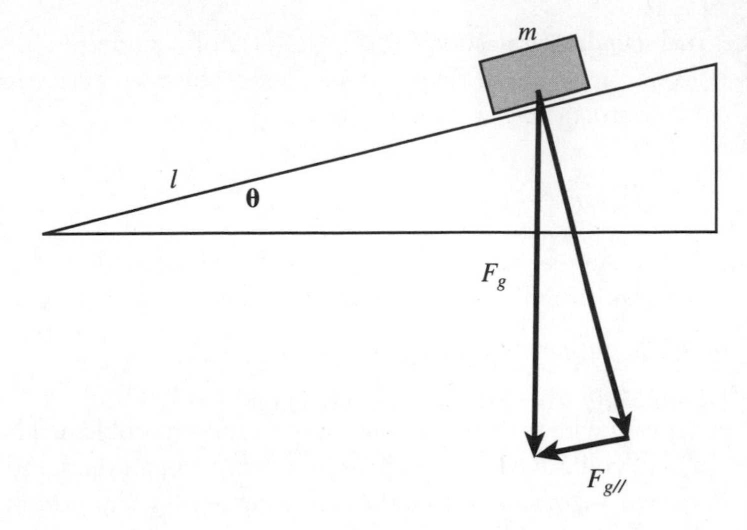

Figure 10.1

The inclined plane for Example 1 includes important labels.

(a) $F_g = mg = 2.00kg \times 9.8m/s^2 = 19.6N$. We can use this value of the weight to construct a vector to represent $5.07N$. When you construct the force triangle as we have in Figure 10.1, the geometry of the triangle will dictate the length of $F_{g\|}$ so that when you measure it using the same scale as you used to construct F_g, you will find that $F_{g\|} = 5.07N$.

(b) When we analyze the next part, we know that the motion down the plane is uniformly accelerated motion, and we can solve for acceleration in terms of the mass and the net force, the component of the force of gravity parallel to the plane. However, to find the final velocity we need the time to travel the length of the plane. There must be another way. Fortunately, there is, and we have just developed the relationship between kinetic energy and work. Let's try this idea:

$$W_g = \Delta KE$$

$$F\Delta x = \frac{1}{2}mv'^2, \text{ since } v = 0$$

$$F_{g\|}\Delta x = \frac{1}{2}mv'^2$$

$$v' = \sqrt{\frac{2F_{g//}\Delta x}{m}}$$

$$v' = \sqrt{\frac{2(5.07N)(4.00m)}{2.00kg}} = \sqrt{20.3m^2/s^2} = 4.5m/s$$

(c) When we have the final speed, we can solve for t.

$$a_{\parallel} = \frac{v' - v}{t}$$

$$\frac{F_{g\parallel}}{m} = \frac{v'}{t} \quad \text{Because } v = 0 \text{ and } \Im = 0.$$

$$t = \frac{mv'}{F_{g\parallel}}$$

$$t = \frac{(2.00kg)(4.50m/s)}{5.07N} = 1.78s$$

It's often good to think about the ideas that you'll be using to solve a problem. That way it's much more clear, and you can take yourself step by step through the solution. In the problem we just worked together, we used the following ideas: (a) uniformly accelerated motion, (b) Newton's second law of motion, (c) work, (d) kinetic energy, and (e) the relationship between work and kinetic energy. Let's try another problem.

Johnnie's Alert

It's always a good idea to check your understanding of the calculation of work as well as the units of work in each system. Having this information readily at hand will often make the solution of a problem obvious. Read Example 2 carefully before planning a strategy for a solution.

Example 2: Kinetic Energy of an Object Moving in a Circular Path

An object of mass 0.500 kg is moving in a circular path of radius 1.00 m at a constant speed with a frequency of $0.500s^{-1}$.

(a) What is the centripetal force?

(b) How much work is done?

(c) What is the kinetic energy of the object?

The first part of this problem is fairly straightforward because it just requires an expression for centripetal force. Because we are given the frequency, the mass, and the radius, the centripetal acceleration can be calculated, then multiplied by mass; we may use $F_c = \frac{m4\pi^2 R}{T^2} = m4\pi^2 Rf^2 = (0.500kg)(4\pi^2)(1.00m)(0.500s^{-1})^2 = 4.93N$ as a solution to (a).

Do you remember this problem from a previous chapter? And do you remember that frequency $f = \frac{1}{T}$? Because there is no displacement in the direction of the centripetal force (radially inward toward the center), no work is done by the applied force! To solve (c), we calculate the KE as

$$KE = \frac{1}{2}mv^2 = \frac{1}{2}(0.500kg)\left[(2\pi)(1.00m)(0.500s^{-1})\right]^2 = 2.47\ \frac{kg \cdot m^2}{s^2} = 2.47N \cdot m = 2.47\ joules$$

We're hoping that you observed the rules for scientific notation and significant figures in both examples. Those kinds of operations should be pretty familiar to you and almost second nature, or maybe even third. And although all of this is fresh in your mind, now would be a good time to work a few problems.

Newton's Figs

When you read a physics problem, note the units of measurement of each item of given information. If all of the information is not in one system of measurement, choose a system and use unit analysis to convert all measurements to your choice of CGS, MKS, or FPS.

Physics Phun

1. What is the kinetic energy of a baseball that weighs 5.1 oz and is thrown with a speed of 93 mi/hr?
2. A car weighing 2208 lb accelerates on a level road from 40.0 mi/hr to 60.0 mi/hr in 11.5 s. (a) Calculate the change in kinetic energy. (b) Calculate the force that produced the acceleration. (c) How much power, in horsepower, was developed?
3. An electron is moving at a constant speed of 1.00×10^9 *cm/s*. Calculate its kinetic energy if the mass of the electron is $9.11 \times 10^{-28} g$.

Kinetic Energy of Falling Objects

Most of us have heard about hailstones falling from the sky, making terrible dents in cars; some of us may have seen this firsthand. We know that hailstones have a tremendous amount of kinetic energy. So let's discuss the kinetic energy of falling objects briefly, and then later we'll take the idea up again in Chapter 11 when we discuss potential energy.

We've already learned that anytime something is in motion it has kinetic energy. If we drop an object from some point above the surface of the earth, we know that it is accelerated by the force of gravity and continues to gain speed until it reaches terminal velocity, hits the ground, or dents a car. The fact that it is dropped means that the initial speed is zero. Therefore, the kinetic energy of any falling object that was dropped is given by $KE = \frac{1}{2}mv'^2$.

Newton's Figs

Motion along a straight line in one direction can be described by using the terms velocity and speed interchangeably. In addition, make sure you can express work in fundamental units as well as joules, ergs, and *ft · lb*.

Now let's suppose that one of the hailstones had a mass of 4.00×10^{-4} slugs and is dropped from a cloud $5.00 \times 10^{2} ft$ above the earth. Using our knowledge of falling objects to calculate the speed of the object when it hits the earth, we can work this out in the following way:

$-h = 5.00 \times 10^{2} ft$ The distance the object falls.

$a = -g$ The acceleration of gravity $v = 0$, because the object is dropped.

$$-h = \bar{v}t = \left(\frac{v'+v}{2}\right)\left(\frac{v'-v}{2}\right)$$

$$-h = \frac{v'^2 - v^2}{-2g}$$

$$2gh = v'^2$$

$$v' = \pm\sqrt{2gh}$$

$$v' = \pm\sqrt{2(32.0\,ft/s^2)(5.00 \times 10^2\,ft)}$$

$$v' = -179\,ft/s$$

And just as a friendly reminder, the minus sign means it has a downward velocity because it is falling.

Now that we have the velocity, we can plug that into our formula and calculate the kinetic energy of the falling object, and we get

$$KE = \frac{1}{2}(4.00 \times 10^{-4}\,slugs)(-179\,ft/s^2) = 6.41\,ft \cdot lb$$

Remember the stone that we dropped back in Chapter 7 from the bridge and saw a splash 3.00 s after we dropped it? We found the height of the bridge in that problem. Now let's do the same experiment, but this time calculate the kinetic energy just as the stone hits the water if it has a mass of 5.00 g.

$$-h = \frac{v'^2 - v}{-2g}$$ From the last solution.

$-h = vt - \frac{1}{2}gt^2$ Solving for $-h$ in terms of t, v, and $-g$.

$-h = -\frac{1}{2}gt^2$ Because $v = 0$.

$-\frac{1}{2}gt^2 = \frac{v'^2 - v}{-2g}$ Substituting the last equation into the first.

$g^2t^2 = v'^2$ Because $v = 0$.

$KE = \frac{1}{2}mg^2t^2$

$$KE = \frac{1}{2}(5.00g)(980cm/s^2)(3.00s)^2 = 2.16 \times 10^7 \frac{g \cdot cm^2}{s^2} = 2.16 \times 10^7 ergs$$

Physics Phun

1. Suppose a bowling ball of mass 6.82 kg is dropped from a point 103 m above the earth. (a) Calculate the speed of the bowling ball just before it touches the earth. Remember, it's a good idea to list all of the information you are given and what you are to find. Then, if you need to, solve the problem algebraically before attempting to figure it all out. (b) Calculate the kinetic energy of the ball just before it touches the earth. (c) What would you expect to happen to the kinetic energy after the ball touches the earth?
2. A 2,000-lb car traveling at 60 mi/hr came to a safe, sudden stop. (a) What was the change in kinetic energy of the car? (b) How much work was done on the car? (c) What do you think caused the change in kinetic energy?

Where Does Kinetic Energy Go?

What do you think happens to the kinetic energy of an object that is in motion when it stops? A more complete answer will be explained after we discuss potential energy in Chapter 11, but we can take a look at part of that question now.

Remember the hailstones? What did they do when they hit the car? They dented the car and bounced off. And what do you suppose the bowling ball in the last set of Physics Phun problems did after it hit the ground? It probably made a dent in the earth and maybe bounced a tiny bit. Something happened to the kinetic energy.

The thing that happened in each of those situations is what we might have expected: work was done on the object to stop it. However, that is just part of the explanation.

If we felt the dent made by the hailstone or bowling ball, it would probably feel warm. If that's hard to imagine, consider pounding a nail into a piece of wood with a hammer. If you felt the nail, there would definitely be a temperature increase generated from the loss of kinetic energy of the hammer, and work was done to drive the nail into the wood. The increase in temperature indicates that there is thermal energy in the nail as a result of the transfer of kinetic energy.

So part of the story is that some of the kinetic energy goes into the distortion of the wood and some into heat. You've also probably seen, and heard, the operation of a pile driver. A huge engine lifts the large hammer and drops it on the pile and work is done driving the steel girder into the ground. If you hear it pounding, you know that some of the kinetic energy is changed to sound (a topic covered in Chapter 16).

The Kinetic Energy of a Projectile

It is interesting to consider kinetic energy in the context of a problem we discussed earlier. Back in Chapter 7, Figure 7.3, we considered the path of a projectile. Let's suppose that $\vec{v}_0 = 30m/s$ at an angle of 40.0° with the horizontal and its mass is 114 g. Calculate the kinetic energy of the object initially, at maximum height, and just before it touches the ground. Here again, we will be able to detect any losses of kinetic energy, if there are any, but we must wait until Chapters 11 and 12 to get a more satisfying answer to the problem.

Johnnie's Alert

Pick a solution to a physics problem that you understand and one that ties a lot of ideas together for you to use as a model for your study time. The path of a projectile is such a problem.

We can begin a solution for the problem of the projectile by realizing that everything necessary to construct a solution is not stated explicitly. We need to resolve the initial velocity into vertical and horizontal components. We do that by constructing a line tangent to the path initially at an angle of 40° with the horizontal. We know that the trajectory changes with different angles associated with the initial velocity. The horizontal component is 23.0 m/s and the vertical component is 19.3 m/s.

We've learned that the projectile involves two types of motion simultaneously. That is why we know that the acceleration of the projectile at maximum height is 9.80 m/s2 and the vertical velocity is zero. However, it still has its horizontal velocity. Let's use that velocity to find the kinetic energy.

The kinetic energy at maximum height is then

$$KE = \frac{1}{2}(0.114kg)(23.0m/s)^2 = 30.2\,joules.$$

The kinetic energy initially is $KE = \frac{1}{2}(0.114kg)(30.0m/s)^2 = 51.3\,joules$. That means there was a loss of kinetic energy from the initial launch to the point where the projectile reaches maximum height. Where did it go? Just before the projectile touches the ground, the instantaneous velocity is –30.0 m/s because it is going down. The kinetic energy at that time is $KE = \frac{1}{2}(0.114kg)(-30.0m/s)^2 = 51.3\,joules$. The kinetic energy has returned! If you don't quite understand where the kinetic energy went and then why it returned, it will all become clear after we go through the next couple of chapters together on potential energy and the conservation of energy. Until then, being able to think through the problem up until this point shows good comprehension on your part.

Bernoulli's Principle

Because kinetic energy has a direct relationship with motion, now is a good time to introduce you to some ideas that involve the movement of air and water. What we're specifically talking about are air pressure and fluid pressure. Both gases and liquids move or flow from one place to another. When they are under pressure, certain formulas can describe the characteristics that liquids and gases have. In Chapter 14, we go into fluid dynamics in much more detail, but because we're examining kinetic energy in this chapter, a brief look at how kinetic energy relates to the movement of gases and liquids is a good idea.

For right now, all we need to know is a few basic things about air pressure. For instance, when we drink liquid through a straw, we lower the air pressure inside the straw by sucking on it, creating a very small partial vacuum. This creates a significant difference between the air pressure inside the straw and the atmospheric pressure outside of it. The atmospheric pressure on the liquid outside the straw causes the liquid to rise in the tube to your mouth.

Newton's Figs

The figures in this book are drawings that were made with simple geometric drawing instruments or on the computer. Making a drawing and labeling it correctly is an important part of many physics problems. Practice making sketches and diagrams for yourself to help you construct solutions to problems.

Another case that you're probably familiar with is the operation of a vacuum cleaner. The vacuum cleaner removes air from above the rug, and the atmospheric pressure under the rug pushes the dirt and dust up into the vacuum cleaner. These are oversimplifications, but we wanted to review some general information so that you would have some simple examples that could be used to build new ideas from.

We've learned that kinetic energy is directly proportional to the square of the velocity. That means that, all other things being equal, the kinetic energy quadruples as the velocity doubles. We can see that the kinetic energy increases rapidly with increases in velocity.

Bernoulli's principle deals with fluids that have a nice smooth flow for equal volumes. The nice smooth flow is called a streamline flow. Cars and other vehicles are built with the streamline shape to allow fluids (which is how air behaves when it moves) to move around them with the streamline flow. Figure 10.2 shows the cross section of an airplane wing built so that air can move around it in a streamline flow. We've marked the path of some air particles that pass under the bottom of the wing in nearly a straight line. There are a couple of paths of particles that flow over the top. We exaggerated those paths to emphasize that both paths are traveled by air particles in the same amount of time. That means that the particles on top of the wing are traveling faster than those along the bottom.

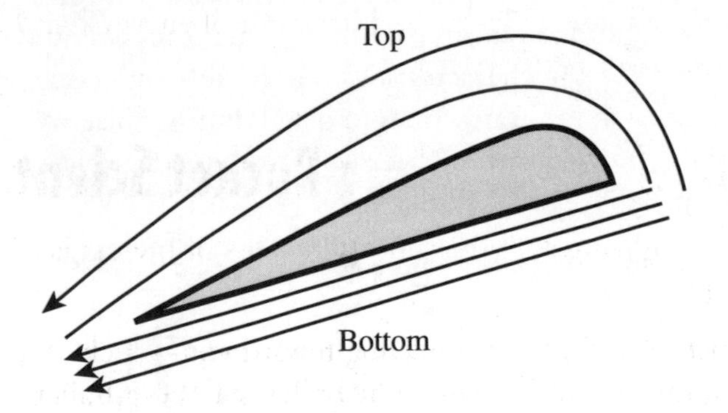

Figure 10.2

The diagram of air particles moving around the cross section of an airplane wing illustrates Bernoulli's principle.

Bernoulli's principle states: $\frac{KE}{V} + P =$ constant. Stated in words, the kinetic energy per unit volume plus the pressure is a constant. That means that the air particles going over the top of the wing in Figure 10.2 are traveling much faster than those under the bottom and the kinetic energy per unit of volume of air must be larger than on the bottom.

If $\frac{KE}{V} + P =$ constant, and $\frac{KE}{V}$ is larger, then p must be smaller if the sum of the two are to always be the same constant. Therefore, the pressure on top of the airplane wing must be less than on the bottom. The result is that the atmospheric pressure will tend to push upward on the bottom of the plane toward the region of lower pressure. (The particular design of the wing is called an airfoil.) It is much like the atmospheric pressure pushing liquid into the straw, where you have reduced the pressure inside the straw by removing the air.

Bernoulli's principle applies to liquids as well as gases such as the air. The most dramatic effects that we observe are the effects like those just discussed. If you can do anything to change the pressure of the streamline flow of a fluid by changing the speed of some of its particles, you will observe the effects just described. We revisit this idea when we discuss pressure in more detail in the section "How Does a Baseball Curve?" in Chapter 14.

We'll also be able to see that the difference in pressure above and below the wing of an airplane provides the lift necessary to keep the plane in the air along with the thrust of the motor. But that, too, comes later. The important thing for us to understand here is that knowledge of kinetic energy is important to have before we can discuss concepts like Bernoulli's principle. As with most of the way this book is laid out, each chapter builds on the knowledge gained in preceding chapters. Our next chapter introduces us to ideas found in another form of energy that is closely related to kinetic energy: potential energy.

Problems for the Budding Rocket Scientist

1. A car is moving at 100 km/h. If the mass of the car is 950 kg, what is its kinetic energy?

2. A truck and a bullet are moving toward you—duck! The truck is a 5,000-kg truck, moving at 50 km/h. The bullet is a 100-g bullet, moving at 500 m/s. Which one has more kinetic energy?

3. How large a force is required to accelerate a 1,300-kg car from rest to a speed of 20 m/s in a distance of 80 m?

The Least You Need to Know

- The kinetic energy of an object is equal to half its mass times its velocity squared.
- Kinetic energy, like work, can be measured in joules or ergs.
- Bernoulli's principle deals with fluids that have a nice smooth flow for equal volumes.
- The more massive an object, and the faster it is moving, the more kinetic energy it possesses.

Chapter 11

Potential Energy

In This Chapter

- Defining potential energy
- Kinetic and potential energy combined
- How energy gets transferred
- The potential energy in springs

The word *potential* can be defined as something that exists in possibility, or is capable of becoming actual. This definition is pretty close to how we understand the use of the word in physics. We've already defined work as the result of a force acting over some distance. And energy we've also already defined as the ability to do work. If something has a large amount of energy, it has the ability to do a lot of work.

If Hercules lifts a large boulder over his head and holds it there threateningly, the boulder has the ability to do work. But it is in a state of potential work. The boulder has what is called potential energy. Let's examine this idea in more detail.

When Does an Object Have Potential Energy?

An object's *potential energy* (PE) depends on its position. For example, the boulder held over your head by Hercules has more potential energy than a boulder resting on the ground. The boulder held at a height 2 m over your head has an amount of PE that depends on its weight (mass times the acceleration of gravity) and its height. We'll talk about this idea in detail in just a little while.

There are different types of potential energy. Gravitational potential energy is potential energy that is related to moving objects in a gravitational field and is calculated as the product of the object's weight and its height above some defined zero level. (More about that in a moment.) We often write gravitational potential energy (GPE) as

def•i•ni•tion

Potential energy is the energy an object has because of its position in a force field, such as a gravitational field.

$$GPE = Weight \times Height$$

where weight is the mass times the acceleration of gravity, or

$$GPE = mgh$$

The unit of gravitational potential energy is the newton meter (N × m), the same unit to measure work. If we do work on an object (by lifting it above our head, for example) we give it an amount of potential energy equal to the work we did.

Chemical potential energy is the stored energy in chemical bonds. The gas tank in your car contains chemical potential energy, because when the gas is burned in the engine the engine can do work, exerting a force to move your car through some distance.

Toward the end of the preceding chapter, we discussed the kinetic energy of a projectile. During that discussion, we noticed that the kinetic energy seemed to disappear and then return. Let's examine this in light of potential energy and get a more complete picture of what was going on.

An object launched into the air has a certain velocity and therefore a certain kinetic energy as it leaves the thing that launched it (a hand, cannon, catapult, etc.). As it climbs upward, its velocity decreases because of the acceleration imposed upon it by the earth's gravitational field. Kinetic energy is therefore constantly disappearing and eventually, when the object reaches maximum height and comes to a halt, its kinetic energy is zero and has therefore entirely disappeared.

We could suppose that the kinetic energy has disappeared because work has been done on the atmosphere and that the kinetic energy has been converted into work. However, this explanation doesn't work because the same thing would happen in a vacuum.

We might also suppose that the kinetic energy had disappeared completely and is beyond redemption, without any appearance of work. But this isn't a working scenario either. So let's see what really happens.

After an object has reached maximum height and its kinetic energy has been reduced to zero, it begins to fall again, still under the acceleration of gravitational force. It falls faster and faster, gaining more and more kinetic energy, and when it hits the ground (neglecting air resistance), it possesses all the kinetic energy with which it started. Are we there yet? Almost!

Rather than lose our chance at a conservation law (something you'll be learning about in the next chapter), it seems reasonable to assume that energy is not truly lost as an object rises upward, but that it is merely stored in some form other than kinetic energy. Work must be done on an object to lift it to a particular height against the pull of gravity, even if after it has reached that height it is not moving. The work must be stored in the form of an energy that it contains by virtue of its position with respect to the gravitational field.

Kinetic energy is thus little by little converted into "energy of position" as the object rises. At maximum height, all of the kinetic energy has become energy of position. As the object falls once more, the energy of position is converted back into kinetic energy. Because the energy of position has the "potentiality" of kinetic energy, the Scottish engineer William J. M. Rankine (1820–1872) suggested in 1853 that it be termed potential energy, and this suggestion was eventually adopted.

Earlier we mentioned the concept of zero level. Let's talk about that in some more detail. An object has potential energy with respect to some level of reference. Usually the lowest level involved in a given situation is chosen as the zero level of potential energy. In this section, we choose ground level, or the level with zero potential energy, as our reference point or level. That means that anything having mass and located at a position above ground level has positive potential energy. So anything at ground level has zero potential energy, and anything below ground level, negative potential energy. We'll find the same idea when working with electrical circuits; the ground has zero potential. Circuits also have positive and negative potential energy. Our work here with potential energy will prepare us to deal with the same notion in other contexts.

Work and Potential Energy

We now have a basic idea of what potential energy is and we can relate that idea to something already familiar to us. Let's use the diagram in Figure 11.1 to establish the relationship among the quantities of work, kinetic energy, and potential energy. Notice that we've marked the zero level (0) for the level where potential energy is zero on the diagram.

Let's expand on some of the information that we've already presented. Suppose the object of mass *m* initially at the foot of the cliff is lifted straight up to the top of the cliff by a force equal in magnitude to the force of gravity on the object. The object is then placed on top of the cliff by moving it along a straight line horizontally. If there is no friction, the stone is raised at a constant speed in order to place it on the cliff. The work done to raise the mass is as follows:

$W_k = Fh$, where F is applied upward and is equal in magnitude to the force of gravity.

$W_k = mgh$ By Newton's second law.

Figure 11.1

The stone and cliff are used in relating potential energy to kinetic energy and work.

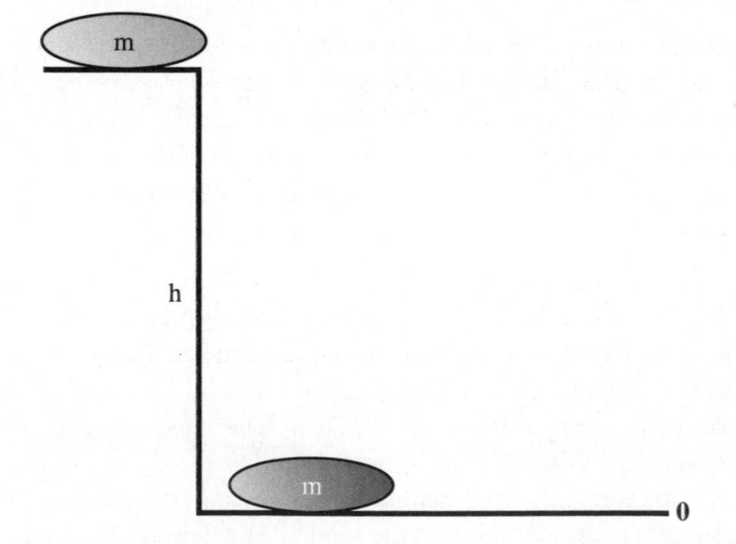

At the top of the cliff, $W_k = PE_{top}$, so $PE_{top} = mgh$; remember, this quantity is positive because work was done on the object to get it to its new position and its height is now above the zero level. We can see that the object on top of the cliff has zero kinetic energy because it is not in motion. If we shove the object over the side, it will gain kinetic energy and just before it touches the ground the potential energy will be zero. At that point, on the other hand, the kinetic energy is at a maximum. Of course, when it hits the ground it has neither.

Johnnie's Alert

The expression *mgh* is the amount of work done on an object by an applied force upward, equal in magnitude to the force of gravity on the object as it is lifted to a height *h* above an arbitrary zero position in a gravitational force field. That means that the expression *mgh* is correct only for objects near the surface of the earth, where the acceleration due to gravity is constant.

It is reasonable to believe that, under ideal conditions, the sum PE + KE = some constant. In this case, the constant will be *mgh*. We calculate that sum at two places other than the top and bottom of the cliff to see if that is true.

Let's suppose that we're able to freeze the object momentarily in space and time the instant it has fallen $\frac{h}{2}$ units, which means that it is $\frac{h}{2}$ units above the ground:

$-\frac{h}{2} = \frac{v'^2 - v^2}{-2g}$ Because it is uniformly accelerated motion.

$gh = v'^2$ Because $v = 0$.

So $KE = \frac{1}{2}m(gh)$ by substitution and $KE = \frac{mgh}{2}$ and $PE = mg\left(\frac{h}{2}\right)$ because the object is $\frac{h}{2}$ units above ground. Now $KE + PE = \frac{mgh}{2} + \frac{mgh}{2} = mgh$, the constant that we predicted!

Let's try one more position just to verify that that is true at all heights. Suppose that we can freeze the object again when it has fallen $\frac{2}{3}h$ units, which means it is $\frac{h}{3}$ units above the ground:

$-\frac{2}{3}h = \frac{v'^2 - v^2}{-2g}$ Because the object has fallen while accelerating at the acceleration of gravity.

$\frac{4}{3}gh = v'^2$ Because $v = 0$.

So $KE = \frac{1}{2}m\left(\frac{4}{3}gh\right) = \frac{2}{3}mgh$ and $PE = mg\frac{h}{3}$, therefore PE + KE = *mgh* again! Any time that friction is negligible for an object in vertical motion near the surface of the earth, the sum of the kinetic energy and potential energy is constant.

Physics Phun

1. Water at the top of Niagara Falls has potential energy with respect to the pool 167 ft below. How much potential energy does 5.0 ft^3 of water have at the top of the falls? What is the speed of the water just before it touches the pool at the foot of the falls?

   The weight density of water is $62.4\ \frac{lb}{ft^3}$, which means 1.00 ft^3 of water weighs 62.4 lb.

   Hint: the potential transfers to kinetic energy as the water falls.
2. A 20.0-kg crate slides down an inclined plane that is 3.00 m high and 20.0 m long. If friction is negligible, what is the speed of the crate when it reaches the bottom of the plane? Hint: use energy considerations in your solution.

Potential Energy of Objects Above the Earth

At the beginning of this chapter, we explained what happens to a projectile that is launched upward and falls back to earth and how kinetic energy and potential energy are converted back and forth through its motion. Let's look at that example again and use the projectile found in Chapter 7, Figure 7.3. Because we've already done the math for that problem, let's stick to the original values to save us extra work. Using an initial velocity of 30.0 m/s at an angle of 40.0°, we did a geometric construction whose horizontal and vertical components of velocity were $v_{0H} = 23.0 m/s$ and $v_{0V} = 19.3 m/s$. We also found that

$$KE = \frac{1}{2}(0.114kg)(30.0m/s)^2 = 51.3\ joules$$

for the kinetic energy of the projectile when it was launched. In addition, you found that the kinetic energy had become less by the time the object reached the top of its arc.

Newton's Figs

You're encouraged to check any and all ideas presented to see if they agree with current methods of science. As a methodology, science is always changing and occasionally some theories and concepts are changed or revised. Skepticism is a good attitude to cultivate, even in science!

Beginning with that information, we can now fill in some of the blanks we had before and see what happened from the point of view of kinetic and potential energy. Let's start by noting that the potential energy at the instant of launch is zero by using the level of the launch as zero potential energy.

If the projectile behaves like the stone, we would expect the sum of the kinetic energy and potential energy to be constant. After all, the projectile is still near the surface of the earth where the acceleration due to gravity is constant. Let's check to see whether the sum of the kinetic energy and potential energy at the maximum height is equal to the kinetic energy at launch.

$$KE = \frac{1}{2}(0.114kg)(23.0m/s)^2 = 30.2\,joules$$

$$PE = (0.114kg)(9.8m/s^2)(maximum\ height)^2$$

Hmmm! What is the maximum height of a projectile? We can derive that from a formula that you're already acquainted with. The maximum height is $Y = \frac{v_{0v}^{\ 2}}{2g}$. For this problem $Y = \frac{(19.3m/s)^2}{2(9.80m/s^2)} = 19.0m$.

Now let's check the sum of kinetic energy and potential energy at Y.

$$KE + PE = 30.2\,joules + (0.114kg)(9.80m/s^2)(19.0\text{m}) = 30.2\,joules + 21.2\,joules = 51.4\,joules$$

That is within about 0.2 percent of our initial kinetic energy plus potential energy, so that's good enough for this particular check. We encourage you to look at KE + PE when the projectile is at $\frac{Y}{2}$ for some extra practice and to see how close that final answer is, too.

Instantaneous Velocity of a Projectile

Having looked at the math for the motion of a projectile and how that's related to its kinetic and potential energy, let's turn our attention to another concept that you've already learned, instantaneous velocity, and see how that pertains to potential energy. We can develop an expression for $\frac{Y}{2}$ in terms of v'_V, the vertical component of the instantaneous velocity at $\frac{Y}{2}$, and use that along with the horizontal component of the initial velocity, which remains the same throughout the flight, to determine the instantaneous velocity at that point.

$\frac{Y}{2} = \frac{v'^2_V - v^2_{0V}}{-2g}$ This is the expression we wanted.

$-Yg = v'^2_V - v^2_{0V}$ By multiplying both members by $-2g$.

$\frac{-v^2_{0V}}{2} = v'^2_V - v^2_{0V}$ By substituting $\frac{v^2_{0V}}{2g}$ for Y and then simplifying.

$\frac{v^2_{0V}}{2} = v'^2_V - v^2_{0V}$ By adding v^2_{0V} to each member.

$$v'_V = \pm\sqrt{\frac{v^2_{0V}}{2}} = \pm\frac{v_{0V}}{\sqrt{2}} = \pm\frac{19.3m/s}{\sqrt{2}} = \pm 13.6m/s$$

Did you notice that this solution gave us two correct answers? That's because we have one value for the trip upward and one for the trip back.

The instantaneous velocity at $\frac{Y}{2}$ is found by

$$v_{\frac{y}{2}} = \pm\sqrt{(13.6m/s)^2 + (23.0m/s)^2} = \pm 26.7m/s$$

Why? Well, we know that the vertical velocity decreases as the projectile rises until it becomes zero at the maximum height. The instantaneous velocity is tangent to the path at every instant and has components v'_V and v_{0H} that are mutually perpendicular. (It may help to refer to Figure 7.3 to follow along in this explanation.) That means we can use the Pythagorean theorem to calculate its magnitude. Remember that the length of the hypotenuse of a right triangle is equal to the square root of the sum of the squares of the legs of the triangle.

Newton's Figs

A handheld calculator can save you hours of laborious arithmetic. Of course, who doesn't use one these days? But a graphing calculator is even better to check some of the suggestions we've made about the path of a projectile or other objects in motion.

The magnitude of the instantaneous velocity is equal to the square root of the sum of the squares of its components. That is why we found $v_{\frac{y}{2}} = \pm 26.7m/s$, the magnitude of v'_V at height $\frac{Y}{2}$, which is half the maximum height the projectile reaches. We can check these ideas at any height we wish by remembering that the height at any time it is in the air is given by $y = v_{0V}\tau - \frac{1}{2}g\tau^2$.

Let's do another check to see whether we obtain the same answers for the two new heights of 19.0 m and 26.7 m.

$$(KE+PE)_{\frac{y}{2}} = \frac{1}{2}mv_{\frac{y}{2}}^{\ 2} + mg\left(\frac{Y}{2}\right) = \frac{1}{2}(0.114kg)(\pm 26.7m/s)^2 + (0.114kg)(9.8m/s^2)\left(\frac{19.0m}{2}\right) = 51.2\,joules$$

This is essentially the same as all other checks so far, and this check also covers both places in the trajectory where the height is $\frac{Y}{2}$. When we calculate the kinetic energy, squaring the velocity, whether it is positive or negative, will give us a positive value. Because we're so into doing checks right now, we might as well look at the situation just before the projectile touches the ground. There is only one slight difference—we explain verbally what you need to do, but this time you try the math on your own.

The velocity is negative because it is going down. The potential energy is zero because the height is zero. So far, so good. The magnitude of the instantaneous velocity is the same as when the projectile was launched. Therefore, the kinetic energy is the same as the initial kinetic energy the instant of launch.

This means that for baseballs, footballs, and all other projectiles, the sum of the kinetic energy and potential energy is constant so long as friction is negligible. We wanted to include that disclaimer because there are some projectiles that may be launched fast enough that they slow down or even begin to melt because of friction! Some projectiles, if launched high enough and fast enough, will fall around the earth in a circular path! Let's take a closer look at those.

Newton's Figs

In case you haven't done so already, now would be a good time to review the process of resolving a vector into mutually perpendicular components. By reversing that process either by a graphical solution or by using the Pythagorean theorem, you can calculate the magnitude of the original vector. If you need the angle associated with the original vector, just construct the components to scale.

Orbital Velocity

Do you remember when we compared the weight of an object near the surface of the earth with its weight at any altitude in Chapter 8? Newton's universal law of gravitation was used to make that comparison, and we can use that law to clarify an idea we mention here, the acceleration due to gravity at any altitude. Newton's universal law of gravity enables us to calculate the acceleration due to gravity at any altitude, g_A, by calculating the weight of an object at any altitude.

$W_A = G\frac{m_e m}{R^2}$, where m_e is the mass of the earth and R is the distance from the center of the earth to the object.

$mg_A = G\frac{m_e m}{R^2}$ By Newton's second law.

$g_A = G\frac{m_e}{R^2}$ By dividing both members by m.

LandSat is a satellite with a polar orbit and is at an altitude of 570 miles. Let's calculate the acceleration due to gravity at that altitude.

The radius of the earth is about 3960 miles, so $R = 3960mi + 570mi = 4530mi$. The acceleration due to gravity is this:

$$g_A = \frac{6.67\times 10^{-11} N\cdot m^2 / kg^2 (5.96\times 10^{24} kg)}{(7.29\times 10^6 m)^2} = 7.48 m/s^2$$

Newton's Figs

The acceleration due to gravity at the altitude of an orbiting satellite is the centripetal acceleration of the satellite as it moves in its nearly circular path at a constant speed.

We can see that this value is considerably different than 9.8m/s², the constant value we've used for the acceleration due to gravity near the surface of the earth. If we had a tower high enough—570 mi is pretty high—then we could launch a satellite horizontally at a speed that would cause the satellite to fall around the earth in a circular orbit.

Many of us have seen rockets launch satellites as well as manned spacecraft into space on television. The rocket lifts the payload vertically, and when it reaches the peak of its path and is horizontal, other rockets are fired to give the payload the speed it needs to orbit. This is an oversimplified version of what takes place, but that's the basic idea. We can calculate the orbiting velocity of a satellite so that we'll know what speed the final stage must provide to the payload. The centripetal force that keeps the satellite in circular orbit is equal to its weight. We can state that as follows:

$f_g = f_c$

$mg_A = \frac{mv^2}{R_A}$ Substituting and using Newton's second law.

$g_A = \frac{v^2}{R_A}$ Because the mass remains constant, it is the same in both places.

$v^2 = g_A R_A$ Multiplying both members by R_A.

$v = \sqrt{g_A R_A}$

That means the orbiting speed is equal to the square root of the product of the acceleration due to gravity at that altitude and the distance the satellite is from the center of the earth at that altitude. The orbiting speed of the LandSat is calculated to be $v = \sqrt{(7.48m/s^2)(7.29\times10^6 m)} = 7.38\times10^3 m/s$.

That may not be a familiar figure to you, so you can change it to mi/hr:

$$7.38\times10^3 \frac{m}{s}\times\frac{1mi}{1610m}\times\frac{60s}{\min}\times\frac{60\min}{hr} = 16{,}500mi/hr$$

That figure may be a little more familiar. Usually a commentator will give you a figure of about 17,000 mi/hr for the orbital velocity of a spacecraft.

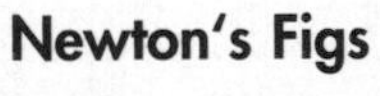

Newton's Figs

When we know the velocity required to keep a satellite in orbit, we can calculate its period. Using that information, we can calculate the mass of the earth. We can also calculate the mass of the earth using the orbiting velocity, $m_e = \frac{R_A v_V^2}{G}$, which we can derive from the statement that the centripetal force on the satellite at any altitude is equal to the gravitational force between the earth and the satellite at the same altitude above the earth.

Here, as we can see, the acceleration due to gravity gets less and less the greater the distance from the earth. We will need more mathematics to calculate the potential energy of objects as far away as satellites. We can calculate the kinetic energy of the satellites because we know how to calculate their velocities. We can also use that same information to calculate the period of motion of the satellite. Just to help you to recall how that is done, here is the algebra:

$$g_A = \frac{4\pi^2 R_A}{T^2} \text{ and } T^2 = \frac{4\pi^2 R_A}{g_A} \text{ so } T = 2\pi\sqrt{\frac{R_A}{g_A}}$$

Transfer of Potential Energy

Let's review what we've learned so far and then we'll explore how kinetic and potential energy are transferred back and forth in the swinging motion of a pendulum. We begin by analyzing the energy of the projectile that we've used as our example in the preceding sections.

The projectile had a maximum amount of kinetic energy initially and no potential energy. Our analysis also showed that as the projectile traveled up to its maximum height, the kinetic energy got less with the distance it traveled upward (height) and its potential energy increased. We can think of this as a decrease of kinetic energy and a corresponding increase of potential energy. During this process, even though no energy disappeared, it seemed that it did when we first observed its kinetic energy.

To continue, the potential energy reached a maximum value when the projectile was at its maximum height, at the time when the kinetic energy was at its minimum. On the way down, the potential energy decreased and there was a corresponding increase in kinetic energy because the speed of the projectile increased.

When the projectile was about to touch the ground, the potential energy was zero and the kinetic energy was at a maximum value. As we discovered in several examples, the sum of the kinetic energy and potential energy was constant so long as friction was negligible. We'll find later that when friction is involved, part of the kinetic energy is used to counter friction and part is transferred to potential energy.

Let's take a look at Figure 11.2 to revisit the simple pendulum, which is another good example of the transfer of kinetic energy to potential energy and back again. By examining the figure we can see that when the pendulum is raised to a height h on the right side of the diagram, that position has potential energy but no kinetic energy. In fact, PE = mgh, which is the work done in lifting the pendulum to height h.

For small angles, the pendulum swings smoothly when released on the right, and as it swings toward the equilibrium position, part of the potential energy is transferred to kinetic energy. When it reaches the equilibrium position, notice that the value of the KE is the same as the value of mgh, the amount of potential energy it had at height h, because the potential energy is zero at that level.

As it swings through the equilibrium position, it has enough kinetic energy to keep it moving up to a height h at the left of the diagram. The pendulum is at rest momentarily at that position, with potential energy but no kinetic energy. In fact, all of the kinetic energy it had at the equilibrium position was transferred to the pendulum as potential energy to get it to height h at the left of the diagram. We know that as the pendulum swings, the sum of the kinetic energy and potential energy is constant. The constant in this case is the value of the original mgh, the amount of potential energy given to the pendulum when it was placed h units above the zero potential energy level in the diagram. That is how the pendulum swings. Just like Galileo's chandelier at church.

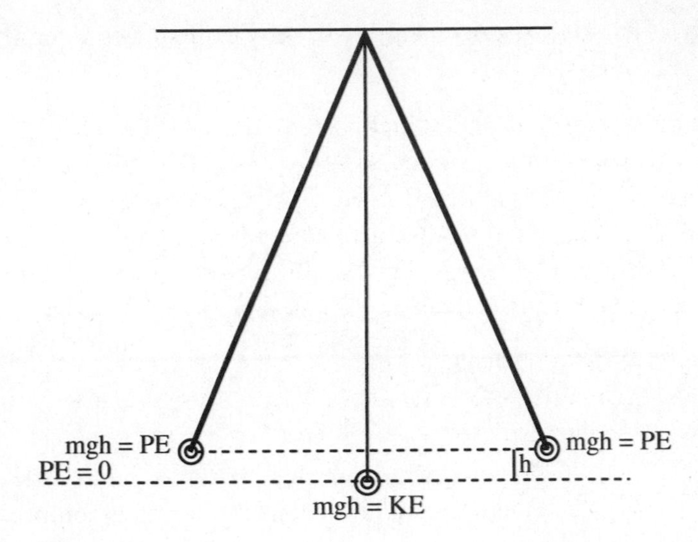

Figure 11.2
An example of energy transfer.

Hooke's Law

> **Johnnie's Alert**
>
> Remember that the force that was applied to stretch a spring in Chapter 6 is considered positive because there is work done upon the spring; that is, potential energy is being added to the spring.

The kinetic energy and potential energy of an object moving in the earth's gravitational field is fairly straightforward. However, there are other force fields that provide for potential energy and kinetic energy. One other force field familiar to you is that encountered in the stretching of a spring. Hooke's law states that the force required to stretch or compress a spring is directly proportional to the spring's elongation or compression. Each spring has a characteristic constant of proportionality. (Remember what that was? If not, reread the section in Chapter 8 that deals with the simple harmonic motion of a spring.) Hooke's law can be stated symbolically as $F = -kx$, where k is the constant of proportionality, x is the elongation or compression, and F is the force of the spring. Take a look at Figure 11.3 as we develop the energy considerations for a spring behaving according to Hooke's law.

An object with a mass m is on a horizontal frictionless surface and has an initial velocity $\vec{v}$ when the mass started stretching a spring. After the velocity of the object is reduced to zero, caused by an elongation of the spring by an amount $\vec{x}$, it reverses its direction. As the spring stretches, elastic potential energy is stored in the spring until all of the kinetic energy of mass m has been transferred to the spring. At that instant, the spring starts transferring the elastic potential energy to mass m and continues to do so until the mass has its original kinetic energy back. That happens when the elongation of the spring is again 0.

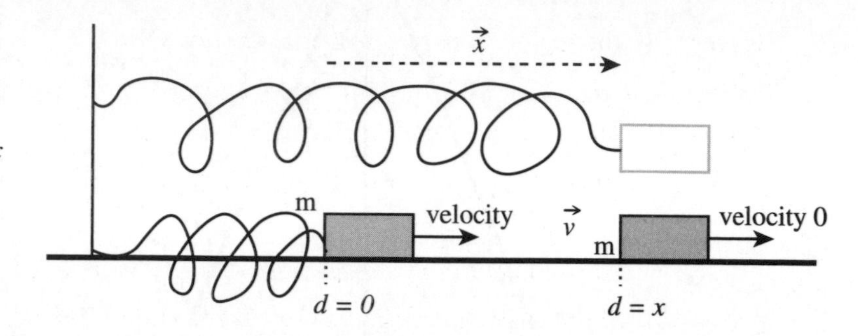

Figure 11.3

A spring stretched by a mass according to Hooke's law shows the interplay of kinetic energy and potential energy.

Newton's Figs

Hooke's law enables you to calculate the force exerted by the spring as simply $F = -kx$. Because of that relationship, the average force is calculated as the initial force plus the final force needed to elongate the spring, divided by two. Average force is not always that easy to calculate.

To perhaps explain this process in just a little more detail, we could say that the spring is compressed, storing energy as elastic potential energy until all of the kinetic energy is transferred to the spring as elastic potential energy. The elastic potential energy will return to the mass, and it will regain its original kinetic energy as it returns to its equilibrium position. The mass will oscillate back and forth along the horizontal surface. And as we've mentioned before, friction and the mass of the spring are negligible for the purposes of discussion here.

When mass m first begins to stretch the spring, it has traveled a distance 0, so no energy is stored in the spring. The total energy of spring and mass is the kinetic energy of mass m, $KE = \frac{1}{2}mv^2$. The force applied to the spring by the mass is 0 because the spring has not been elongated.

As the spring stretches, the mass applies a force equal to k times the amount of elongation until all of the kinetic energy has been stored as elastic potential energy. That means the force exerted on the spring is the maximum force, $F = kx$. The force varies from 0 to kx as it is stretched. In order to calculate the work done stretching the spring, we must calculate the average force that causes the elongation x. The average force is given by $F_{avg} = \frac{F_{initial} + F_{final}}{2} = \frac{0 + kx}{2} = \frac{kx}{2}$. The work done to stretch the spring is calculated as $W_k = F_{avg}x = \left(\frac{kx}{2}\right)x = \frac{kx^2}{2}$. The work done to stretch the spring is stored in the spring as elastic potential energy, $PE_{spring} = \frac{1}{2}kx^2$. We now have enough information to account for the transfer of energy from KE of mass to PE of spring and back again:

- When the elongation of the spring is zero, $KE + PE = \frac{1}{2}mv^2 + 0 = \frac{1}{2}mv^2$.
- When the elongation is x, $KE + PE = 0 + \frac{1}{2}kx^2 + 0 = \frac{1}{2}mv^2$. That means that the value of the elastic potential energy at the end of the stretching is equal to the value of the kinetic energy the mass had before it started stretching the spring.
- When the spring returns to the equilibrium position, the elastic potential energy of the mass is 0, $KE + PE = 0 + \frac{1}{2}mv^2 + 0 = \frac{1}{2}kx^2$. This means that all of the elastic potential energy stored in the spring is now returned to the mass as kinetic energy.
- Between 0 elongation and an elongation of x, PE + KE = constant. In this case, the constant is the kinetic energy the mass had initially.

The process may seem a little complicated at first; but if you run through it a few times, it will become clear. You observe the same process of transfer of energy for a spring as we observed for the projectile and the pendulum. Up next, the conservation of energy.

Problems for the Budding Rocket Scientist

1. Which has more potential energy, a 500-kg anvil suspended 10 m in the air, or a 3-kg bowling ball suspended at 200 m? How much work was required to get each object from the ground to their higher potential energy positions?
2. A 50-kg mass hangs from a Hookean spring. When 20 g more are added to the end of the spring, it stretches 7.0 cm more. (a) Find the spring constant. (b) If the 20 g are now removed, what will be the period of the motion?
3. A 25-kg child is on a swing. If her height is 2 m when she is at the maximum height of her swing, how fast will she be going at the lowest point of her swing? Describe in words any of the assumptions you make in this calculation.

The Least You Need to Know

- An object's potential energy (PE) depends on its position.
- Anything having mass and located at a position above ground level has positive potential energy. Anything at ground level has zero potential energy, and anything below ground level, negative potential energy.
- Any time that friction is negligible for an object in vertical motion near the surface of the earth, the sum of the kinetic energy and potential energy is constant.
- The magnitude of the instantaneous velocity is equal to the square root of the sum of the squares of its components.
- Kinetic energy and potential energy are the types of energy made use of by machines built from levers, inclined planes, and wheels, and the two forms of energy are therefore considered to be mechanical energy.

Chapter 12

Conservation of Energy and Momentum

In This Chapter

- Conservation laws
- If there is impulse, there is momentum
- Comparing energetic collisions
- Elastic and inelastic interactions
- Transferring energy

The law of conservation of energy is a fundamental law of physics. No matter what you do, energy is always conserved. The total amount of energy in the universe is, has been, and always will be the same as it is right now. And although energy may not be handled or even seen, it can be quantified. The law of the conservation of energy gives us the power to predict the workings of nature, and such power can provide something of certainty and control in the world.

Laws of nature are different from the laws of society. Although physicists argue with themselves over whether humankind will ever fully know nature's "truth," they agree that a fundamental logic seems to govern the physical world. And although that logic may please us, from a philosophical point of view, it can be argued as to whether or not this logical structure exists independent of the human mind and our desires. So accordingly, the laws of nature can't be arrived at by consensus or decree, as can statutory laws. They must be discovered by experiment and analysis. Because it is impossible to know whether we have complete knowledge of nature, we must regard the conservation of energy and all other "laws of nature" as simply our best current understanding of the world, subject to revision as warranted by new physical evidence.

How Conservation Laws Work

Sometimes the best way to explain how a physical law operates is to show how it works by means of an analogous example. To illustrate exactly what we mean by that, we'll tell you a story of how a conservation law operates.

A mother gives her child 28 blocks to play with. They are sturdy blocks and can't be split or chipped. One day, the mother comes home and sees only 26 blocks. She anxiously looks around the house and eventually finds the two missing blocks under the bed. The next day, only 24 blocks can be found. She looks under the bed and everywhere else, but she can't find any more blocks. Then she shakes her child's little locked box and hears a rattle. She knows that the box, when empty, weighs 16 ounces, and she also knows that each block weighs 2 ounces. She puts the box on the scale, and it registers 24 ounces. Aha! That accounts for the four missing blocks. One morning a week later, the mother sees only 8 blocks. She finds 3 under the bed and 2 in the box, making a total of 13, but no more anywhere in the house. Then she notices that the water in the fishbowl is dirty, and it seems a little high. The water level is normally 12 inches. She measures it now and finds it to be 13.5 inches. She carefully drops a block in the bowl and discovers that the water level goes up 0.1 inch. Another aha! If one block pushes the water level up 0.1 inch, and the level is up 1.5 inches higher than usual, there must be 15 blocks in the bowl. That accounts for all the missing blocks.

After several weeks, the mother decides that she has found all of the hiding places. To keep track of the blocks seen and unseen, she works out the following rule:

$$(\#\ \textit{blocks seen}) + (\#\ \textit{blocks under bed}) + \frac{(\textit{weight of box}) - 16oz}{2oz}$$

$$+\frac{(\textit{height of water}) - 12in}{0.1in} = 28 = \textit{constant}$$

This equation is an abbreviated but precise expression of what the mother has learned. She has discovered a conservation law. The total number of blocks remains constant, although it may take a little ingenuity to find them all. From one day to the next, the blocks may be parceled out and hidden in a variety of ways, but a logical system can always account for all 28.

All conservation laws are like this one. Some countable quantity of a system never changes, even though the system may undergo many changes. A conservation law does not reveal how a system operates or what causes it to change. Still, such a law can be extremely helpful for making predictions. For example, the mother could use her rule to figure out how many blocks would be in the box if she knew how many were under the bed and how many were in the fishbowl. From the known, she can reach to the unknown.

In most conservation laws of nature, the "blocks" are never visible. Counting them is therefore an indirect process, analogous to weighing the box and measuring the height of the water. In addition, the conserved quantity is usually not a tangible object, like a block, but something more abstract. However, all conservation laws express the result that some quantifiable property of system never changes. Got it?

Part of this chapter is devoted to the study of the conservation of energy and other parts will introduce you to the conservation of momentum. (We get to defining momentum in just a little while.) As we already have learned, energy is the capacity to do work. A moving hammer, a coiled spring, a paperweight on a desk, and boiling water all represent different forms of energy. A moving hammer can push a nail into a wall; a coiled spring can turn the hands on a clock; a falling weight can lift another weight or move paddles of an electric generator and light up a house; boiling water can cook an egg. Despite the many forms that energy can take, scientists have found that the total energy of an *isolated system* never changes. The conservation, or constancy of energy, is a strange and beautiful fact about nature.

Nature possesses other conservation laws, such as the conservation of momentum and the conservation of electrical charge, and we get to all of these at one point or another in this book. Why it is that any conservation law exists is a mystery, but if there were no conservation laws, the predictive power of science would be much reduced. The world might seem like a much more irrational place, at least from the point of view of physics.

def•i•ni•tion

An **isolated system** is an object or group of objects that has no contact with the surrounding world.

The Law of Conservation of Energy

In the preceding two chapters we've covered both kinetic and potential energy. And even though we didn't really define the law of the conservation of energy, we've been working with it all along. So without any additional discussion, let's get down to it.

If we ignore effects such as friction and air resistance, we can say that the total energy of a system, represented by the sum of its kinetic and potential energy, is conserved. Mathematically we can express the law of conservation of energy as

Total Energy + Kinetic Energy + Potential Energy

or

$$TE = KE + PE$$

The example of the projectile from Chapter 11 is a good model for demonstrating how the total energy of a system remains intact. Although the constant ebb and flow between kinetic and potential energy was taking place, the total energy of that system remained intact and was conserved. We should point out, however, that by not taking into account both friction and air resistance, we are talking about an ideal system. But before we go on, it would be a good idea to address those effects.

Let's imagine that we're holding two cups, one full of water and one empty. Label the full cup KE and the empty cup PE. Now, pour half of the water from the cup labeled KE into the cup labeled PE. The total amount of water that we have in both cups remains constant, although the amount in each cup may change as you pour water back and forth. You can see that you may have more potential energy or more kinetic energy at any instant, but the total (in an ideal world) does not change.

What happens, though, as you pour the water back and forth? Does a little water spill? If you were to pour the water back and forth many times, the total amount would decline slightly. That is more like how a real system works. Over time, energy is "lost" to processes such as friction and air resistance.

Newton's Figs

Conservation laws reflect a deep-seated human desire for order and rationality. That is not to say that human beings invented conservation laws. Nature invented conservation laws and lives by them. But the *idea* of a conservation law, of some permanent and indestructible quantity in the world, was conceived by humankind long ago and appeals to our minds and emotions.

Impulse and Momentum

We've all probably had some practice tossing a ball into the air and hitting it with a bat. So let's suppose that you use a baseball bat and toss a baseball into the air. Also suppose that you are good enough to hit it. Your result would be a quick way to launch a projectile.

What do you suppose would happen if you tossed a 4-kg shot up and hit it with a bat? The bat would most likely break, but the shot would probably move a bit, too. The thing that you would notice in both cases is that the bat would apply a force for a short time to the bat and the shot. Instead of using a bat, launch the shot by "putting" it like an athlete does with one heave (hence the event name *shot put*)—then do the same thing with the baseball.

You probably pushed each ball into the air with the same force and held each for about the same amount of time. The baseball had a much larger change in velocity than did the shot because of their different masses. In order to give the shot the same change in velocity as the baseball, you would either have to apply a much larger force for the same amount of time or the same force for a longer period of time. This experiment is somewhat unrealistic because the shot weighs about 10 pounds. We used these two objects because of the obvious difference in the way they behave when you push them both with the same force for the same amount of time.

In physics, the idea that is involved here is called *impulse*. The bat gave both balls the same impulse and each behaved in a much different way. Impulse is a vector quantity and is calculated by multiplying the applied force by the time interval during which it is applied, or simply, impulse = force times time.

In symbols, $\vec{I} = \vec{F}\,\Delta t$, where the direction of the impulse is the same direction as the force applied. The impulse of a force is measured in Newton seconds or $N \cdot s$ in the MKS system, $dyne \cdot s$ in the CGS system, and $lb \cdot s$ in the FPS system.

We know that when an object receives an impulse, it undergoes a change in velocity. The magnitude of the change in velocity depends upon the mass of the object. The more massive the body, the smaller the change in velocity it will realize from a given impulse. The physical property that is involved here is called *momentum*.

def•i•ni•tion

Impulse is a physical quantity that results from a force being applied to a body for a certain amount of time.

Momentum is a physical vector quantity that has its magnitude determined by the product of the mass and velocity of the object being described.

Newton's Figs

A change in momentum usually implies a change in velocity, but keep in mind that the mass can change also as in the case of a rocket launched from the earth. Not only is the fuel being used up, making the mass of the rocket less, but also the fuel tanks themselves are jettisoned after a short period of time.

Johnnie's Alert

Any time an object experiences an impulse, the object gives an impulse to the source at the same time. That is, the bat gives the ball an impulse and the ball gives the bat an equal and opposite impulse.

Momentum is a vector quantity, and its magnitude depends on the mass and change in velocity of the object being described. Its direction is in the direction of the velocity vector. Momentum is important enough to have a special symbol, $\vec{P}$, assigned to it.

Momentum is calculated as $\vec{P} = m\vec{v}$. The units of momentum are $\frac{kg \cdot m}{s}$ in the MKS system, $\frac{g \cdot cm}{s}$ in the CGS system, and $\frac{slug \cdot ft}{s}$ in the FPS system. You have probably already noticed that these units are a disguise of $N \cdot s$, $dyne \cdot s$, and $ft \cdot s$. The reason is the relationship between impulse and momentum.

Remember how the same impulse gives a different change in velocity to two objects of different mass? When an object is given an impulse, it undergoes a change in momentum. We can see that momentum involves both mass and change in velocity. The mathematical statement of this relationship is $\vec{F}\Delta t = \Delta(m\vec{v})$.

Understanding the Relationship Between Impulse and Momentum

Our first look at the relationship between impulse and momentum will involve expressions modeled after $\vec{F}\Delta t = m\Delta\vec{v} = m(\vec{v}' - \vec{v}) = m\vec{v}' - m\vec{v}$, and you should focus on the magnitude of these quantities unless it is explicit that they are involved in motion that is not along a straight line. Now we can see why the units of impulse are the same as the units of momentum even though the units for impulse are $N \cdot s$ and for momentum $\frac{kg \cdot m}{s}$ in the MKS system, that is, $N \cdot s = \frac{kg \cdot m}{s^2} \times s = \frac{kg \cdot m}{s}$. You can do the same analysis of the other systems of measurement to find that the units for impulse are the same as the units for momentum.

If we take a closer look at this new relationship, it will reveal to us why it's been thought that this is the relationship Newton originally used as his second law. That is,

$\vec{F}\,\Delta t = m\vec{v}' - m\vec{v}$ can be stated as $\vec{F} = \frac{m\vec{v}' - m\vec{v}}{\Delta t} m\left(\frac{\vec{v}' - \vec{v}}{\Delta t}\right) = m\vec{a}$. In the following discussion, we see how his third law could be derived directly from the second.

If the bat gives the ball an impulse, the ball gives the bat an impulse—you can feel it. Therefore, when the bat applies a force on the ball, the ball applies a force on the bat, equal in magnitude, but in the opposite direction, for the same period of time.

And if we suppose that we have two objects in which the first gives the second an impulse, then the second will give the first an impulse in return. That means that $\vec{I}_1 = -\vec{I}_2$ and $\vec{F}_1\Delta t = -\vec{F}_2\Delta t$, because the time of the impulse is the same for $\vec{F}_1 = -\vec{F}_2$. The forces are equal in magnitude but opposite in direction, so for every action there is an equal and opposite reaction. When Newton used the words *action* and *reaction* in his third law, he meant force and reactive force.

Physics Phun

1. A projectile weighs 151 lb. What net force is required to accelerate the projectile at a rate of 925 ft/s²?
2. (a) If a force of 5.00 N acts on an object for 3.00 s, calculate the change in momentum of the object. (b) If the object has a mass of 5.00 kg, calculate its change in velocity. (c) If its original velocity were 1.0 m/s, what would be its final velocity?
3. A force of 35.6 N acts on a body for 0.250 seconds. What is the impulse that acts on the body?

Conservation of Momentum

Suppose that you have two steel balls on a table, and you roll one of the balls, m_1, toward the other ball, m_2, that is initially at rest. Eventually they'll collide. You've probably done that many times before, with either marbles or billiard balls. And depending on the angle at which the two balls collided, a number of different things can occur. Those various collisions are almost too numerous to describe here, but you get the general idea. Let's take a look at what's going on during these collisions. We know that any two objects that collide, or interact, apply impulses one to the other. We can summarize what happens as follows:

$\vec{F}_1\Delta t = -\vec{F}_2\Delta t$ Their impulses are equal in magnitude and opposite in direction.

$\Delta\vec{P}_1 = -\Delta\vec{P}_2$ Substituting the change in momentum for impulse.

$(m_1v'_1 - m_1v_1) = -(m_2v'_2 - m_2v_2)$ Substituting for change in momentum.

$m_1v'_1 - m_1v_1 = m_2v_2 - m_2v'_2$ Simplifying.

$m_1v'_1 + m_2v'_2 = m_2v_2 + m_1v_1$ Adding $m_1v_1 + m_2v'_2$ to both terms.

Newton's Figs

When you read a problem in physics, it's always a good idea to record the given information first and make sure all data are in the same system of measurement. If different systems are employed, use the techniques of unit analysis to express everything in the same system before you begin your solution.

Did you notice that the last equation has in the left term the momentum of ball one and ball two after the interaction and the right member has the momentum of ball one and ball two before the interaction? In other words, the sum of the momenta (plural of momentum) after the interaction is equivalent to the sum of the momenta before the interaction. That is, the total momentum before the interaction is the same as the total momentum after the interaction; the total momentum is constant and therefore conserved. There is no change in the total momentum.

Newton's third law (for every action there is an equal and opposite reaction) is sometimes referred to as the law of conservation of momentum. Let's go through a simple example together so that you can see why his third law and the conservation of momentum are often called the same thing.

Consider the following situation. It is a hot day. You're standing on a motionless boat in your swimsuit. You are part of a system with zero momentum. You decide to jump out of the boat into the water to cool off. You go in one direction at a given velocity, with a certain momentum. You and the boat represent a "closed system," so in order for momentum to be conserved (in order for the total momentum in the system to be zero), the boat has to go in the opposite direction. If the boat is large, like an ocean liner, its velocity in the opposite direction will be small, probably imperceptible. If the boat is small, like a canoe, its velocity away from you will be about the same as your velocity away from it (if your masses are about equal). To restate this situation as a conservation of momentum problem, we write that the total momentum is

$$P_{total} = 0 = m_1v_1 + m_2v_2$$

or

$$m_1v_1 = -m_2v_2$$

indicating that the two objects (the boat: m_1, and the person: m_2) will move off in opposite directions. If the mass of the boat (m_1) is 100 kg, and the mass of the person (m_2) is 50 kg, then the velocity of the boat (v_1) will be

$$v_1 = (m_2/m_1)v_2$$

or

$$v_1 = -0.5v_2$$

indicating that the boat will move away from the jumper at half the velocity of the jumper moving away from the boat. Does this help to explain why Newton's third law is often called the conservation of momentum? Let's run through some more examples so that you get what conserving momentum is all about.

Example 1: Conserving Momentum

A 50.0-kg girl rides on a 15.0-kg skateboard that is moving in a straight line at a constant velocity of 3.0 m/s. The girl jumps off the skateboard. What is the change in velocity of the skateboard for each of the following cases?

Newton's Figs

An algebraic solution to a physics problem serves many purposes, the main one being that it provides you with a template you can use to find as many numeric solutions as you like.

(a) She jumps off at three times the original velocity of the skateboard.

(b) She jumps off with the same velocity as the skateboard had originally.

(c) She jumps off not moving with respect to the ground.

The outline of the solution is one that is familiar to us by now, but we'll review it to clarify the nature of the type of thinking involved in the solution of a physics problem: (1) List all information given and implied, and the information you are to solve for. (2) Assign appropriate algebraic symbols to all information in (1). (3) Solve the problem algebraically first, then substitute numerical values to identify a numerical solution. Sometimes a diagram that is properly labeled will help you to devise a strategy for a solution. Vector diagrams can provide a numerical solution by either a graphical construction or a calculation involving geometrical techniques.

Solution:

$m_g = 50.0kg$

$m_s = 15.0kg$ $\quad$ $(a)\ v'_g = 3v_s$

$v_g = v_s = 3.0m/s$ $\quad$ $(b)\ v'_g = v_s$

$\Delta v_s = v'_s - v_s = ?$ $\quad$ $(c)\ v'_g = 0$

$v_g = 3.0m/s$

The sum of the momenta after the interaction is equal to the sum of the momenta before the interaction.

$m_g v_g + m_s v_s = m_g v'_g + m_s v'_s$ Conservation of momentum.

$m_g v_g - m_g v'_g = m_s v'_s - m_s v_s$ Adding $(-m_g v'_g - m_s v_s)$ to each term.

$m_g(v_g - v'_g) = m_s(v'_s - v_s)$ Simplifying by factoring common factors in each member.

$v'_s - v_s = \frac{m_g}{m_s}(v_g - v'_g)$ Solving for $v'_s - v_s$ by dividing both members by m_s.

$\Delta v_s = \frac{m_g}{m_s}(v_g - v'_g) = \frac{m_g}{m_s}(v_s - v'_g)$ Substituting Δv_s for $(v'_s - v_s)$ and v_s for v_g because the velocity of the girl is the same as the velocity of the skateboard before the interaction.

This completes the algebraic solution. And now for numeric solutions:

(a) $\Delta v_s = \frac{50.0kg}{15.0kg}(3.0m/s - 3(3.0m/s)) = -2.0 \times 10m/s$

(b) $\Delta v_s = \frac{50.0kg}{15.0kg}(3.0m/s - 3.0m/s) = 0$

(c) $\Delta v_s = \frac{50.0kg}{15.0kg}(3.0m/s - 0) = 1.0 \times 10m/s$

Johnnie's Alert

Vector quantities are represented by symbols with an arrow over the top. The magnitude of the vector is represented by the same symbol as the vector but no arrow is over the top of the symbol.

Do these results make sense? This is a question that should occur to you anytime you solve a problem. In this case, in (a), the girl jumped off the front of the skateboard, sending it in the opposite direction. In (b), the girl had to have jumped off the side in order to have no change in velocity. In (c), the girl jumped off the back of the skateboard, sending it ahead faster than before she jumped. You can also find the numeric sum of momenta before the interaction and compare that to the numeric sum of momenta after the interaction. If the problem is solved correctly, the sums should be the same. Are they?

Example 2: A Total Momentum of Zero

An object at rest is torn into three pieces by an explosion. Two pieces fly off along paths that are perpendicular to each other. One of the two pieces has a momentum of $28.0 kg \cdot m/s$, and the other has a momentum of $15.0 kg \cdot m/s$. What is the mass of the third piece if its speed is 20.0 m/s?

Solution:

Let's use the same outline as before. In this case, a scale drawing is required because the vector nature of momentum is implied in the statement of the problem when you are given that two pieces fly off along paths that are perpendicular to each other. Refer to Figure 12.1 for the vector solution.

$$\vec{P}_1 = 28.0\frac{kg \cdot m}{s}E$$

$$\vec{P}_2 = 15.0\frac{kg \cdot m}{s}N$$

$$m_3 = ?$$

$$v'_3 = 20.0m/s$$

$$v_1 = v_2 = v_3 = 0$$

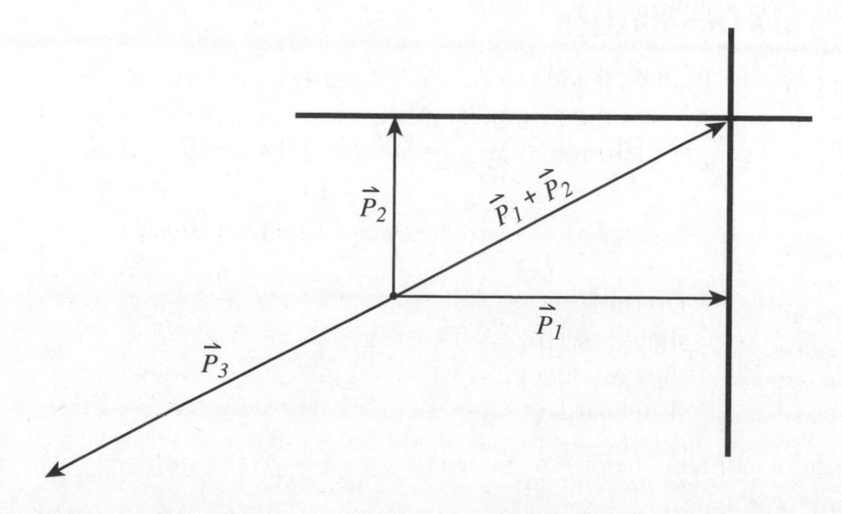

Figure 12.1

The vector diagram of momenta for an explosion shows that momentum is conserved in an interaction.

The sum of the momenta before the interaction, the explosion, must be equal to the sum of the momenta after the interaction. The sum of the momenta before the interaction is zero because the object is at rest. That means the sum of the momenta after the interaction must be zero. The sum of the momenta of two pieces after the interaction is $\vec{P}_1 + \vec{P}_2 = 32.0\frac{kg \cdot m}{s} E28°N$.

In order to have a sum of zero, the third piece must have a momentum equal in magnitude to the sum of the first two but opposite in direction, $\vec{P}_3 = 32.0\frac{kg \cdot m}{s} W28°S$.

That means that $\vec{P}_3 = m_3 v_3$ and $m_3 \frac{P_3}{v_3}$, so $m_3 = \frac{32.0\frac{kg \cdot m}{s}}{20.0m/s} = 1.6kg$. Did you notice that our final answer has two significant figures even though the data is expressed in three significant figures? Because we constructed a vector diagram, our measurements with the ruler had only two significant figures.

We can't calculate more accurately than we can measure, so our measurement limited the number of significant figures in our answer. We calculated the answer also, and found the mass to be 1.59 kg. The two answers are the same, considering methods used in each case.

Newton's Figs

If you use a scale drawing as part of your solution to a vector problem, the number of significant figures used in your calculations will be limited by the precision of the measuring instruments used in constructing the solution.

Example 3: Momentum of a Combination

A freight car having a mass of 14,200 kg and traveling at 15.0 m/s bumps into an identical car that has the same mass as the first one, initially at rest on a straight and level railroad track. If they couple and travel away together, what is the speed of the combination?

Solution:

$m_1 = 14{,}200kg$ $\quad m_2 = 14{,}200kg$

$v_1 = 15.0m/s$ $\quad v_2 = 0$

$v'_1 = v'_2 = ?$

We know that the sum of momenta before the interaction is equal to the sum of momenta after the interaction.

$m_1v_1 + m_2v_2 = (m_1 + m_2)\, v'_1$ Conservation of momentum reflecting that there is one mass after the interaction.

$v'_1 = \dfrac{m_1v_1 + m_2v_2}{m_1 + m_2}$ Dividing each member by the total mass.

$v'_1 = \dfrac{m_1}{m_1 + m_2}(v_1 + v_2)$ Because $m_1 = m_2$. The algebraic solution is complete.

We can substitute numeric values to find the numeric solution.

$$v'_1 = \frac{1}{2}(15.0m/s + 0) = 7.5m/s$$

We now have several examples that we can use as models for solving momentum problems. The important thing is for you to develop your own strategy and follow your own plan to arrive at a solution that helps you feel confident. A lot of practice will help. So here's a few more problems to give you just that.

Johnnie's Alert

Models for solving problems in this book are just that. They should not be a pattern for constructing solutions. You should use the general outline for solving problems given in this book to create your own solutions.

Physics Phun

1. A person threw a 2.50-kg object horizontally from a canoe with a speed of 15.0 m/s. The mass of the person is 50.0 kg, and the mass of the canoe is 102 kg. What is the velocity of the canoe after throwing out the object if the canoe were originally at rest? The 2.50-kg object was not part of the mass of the canoe.
2. A freight car with a mass of 12,200 kg traveling at 12.0 m/s strikes another freight car that was originally at rest. The cars stick together and move away at a constant speed of 5.00 m/s. Calculate the mass of the second car.
3. A bullet of mass 10.0 g leaves the muzzle of a stationary gun of mass 5.0 kg with a velocity of 355 m/s. What is the velocity of the gun as a result of the bullet leaving the muzzle? The shooter experiences the velocity of the gun as a "kick," which is sometimes referred to as the recoil velocity of the gun. Hint: watch out for units.

The solution to the third problem in the preceding Physics Phun sidebar gives you another reason why there's such a close correlation between Newton's third law and the conservation of momentum. We know that a tiny gas particle leaving the rocket gives the rocket an impulse, causing a change in momentum in the opposite direction. When all of the millions of tiny particles are fired out the end of the rocket, a tremendous change in momentum is given to the rocket in the opposite direction. The same thing happens when the rocket changes direction in space, but usually a burst of gas in one direction will give the rocket a change in momentum in the opposite direction because of the impulse of the gas.

Head-On Collisions

We've looked at several incidences of objects interacting with each other. We're now going to take a look at head-on collisions. It is often thought that the objects must touch in order to collide. However, in physics, objects can collide without touching; two objects with the same magnetic poles or the same electrical charge never touch, but they can certainly interact, as you have probably witnessed. In this book, we generally use the term *interaction* to make it clear that collisions with and without physical touching are included in the discussion.

The interaction of two billiard balls is the vehicle we use to discuss head-on collisions. There are some ideas needed in this discussion that are treated first. The collision of two billiard balls or two steel balls is much different from the collision of two mud balls or two balls of putty.

We've all probably made mud balls at some time in our lives; you may have even thrown them at something like a wall. When a mud ball hits an object, it tends to splatter. It becomes deformed and never returns to the shape it had before the collision. A ball of putty will do the same thing. A car tends to behave in the same way when it runs into a tree. It becomes distorted and remains that way until you pay a lot of money to get it smoothed out.

The mud ball or ball of putty had to be close enough to the wall to experience the impulse of the wall. The fact that the mud ball splatters means that there are forces within the ball causing parts of the ball to move over other parts and remain displaced. The force of the interaction depends on something other than the distance between the ball and the wall.

Newton's Figs

When two objects interact by touching, the collision is often referred to as a hard sphere collision. The important thing to keep in mind is that objects can collide without actually touching.

The steel balls hanging from strings bounce off each other and do not remain distorted. They are distorted so little that you would have to be an atom-sized person to notice. The same thing is true of billiard balls. The force of the interaction is experienced as soon as the balls touch. The force of the interaction depends only on the distance between the balls. There is no interaction when they are not touching; but as soon as they are close enough together, they experience the impulse of each other.

Charged objects are the same way. Charged objects do not affect each other until they are close enough to be affected by another charged object. The force of the interaction depends on the distance between the charged bodies. If you think about why the balls in the first example do not go through each other, you realize that each ball's outermost electrons are repelling the other ball's outermost electrons. So when the balls get too close, the forces of repulsion push them apart. They have hit and bounced off each other.

Magnetic poles behave the same way. You have probably brought one north pole of a magnet close to the north pole of another and felt the repulsive force pushing on the magnet you hold. The force of the interaction depends only on the separation of the magnets, charged bodies, billiard balls, or steel balls. There are no other forces involved in the interaction like the attraction of the wall for the mud that sticks to it after a mud ball–wall interaction. The idea that is involved here is elasticity.

Johnnie's Alert

Surprisingly, a rubber band or a rubber ball is not a very elastic object. However, such objects are more elastic than a mud ball. The closest thing we can come to as a model for a perfectly elastic object is a billiard ball or an ivory ball.

Understanding Elasticity

You may have thought that a rubber band was highly elastic. Guess what? It's just the opposite, at least in the way physics defines elasticity. If an object interacts with another object in such a way that the force of interaction depends only on the separation of the objects, the objects are elastic. Of course, the billiard balls and steel balls are not perfectly elastic; that kind of an object is ideal. A billiard ball is highly elastic, whereas a mud ball is highly inelastic because …

- A highly elastic object may be slightly deformed momentarily but is immediately restored to its original shape.
- An inelastic object may be permanently deformed and never returns to its original shape.

A rubber band is more on the end of the scale with the mud ball; the billiard ball is on the other end of the scale. You can stretch a rubber band and lay it on a table and see it relatively quickly move toward its original shape. However, that's really very slow compared to the speed at which the surface of a billiard ball returns to its original form. After a bit of stretching, the rubber band will loose its shape and remain stretched out considerably from its original form. If two billiard balls interact with each other, they can behave in a lot of ways depending on the force of interaction. They can fly off at different angles, or in some cases bounce straight back or stop. A little later, we take a closer look at the behavior of two billiard balls interacting along a straight line. But for now, the reason we need to understand the concept of elasticity is because of its pertinence to our discussion of conservation laws. Kinetic energy is conserved in elastic interactions. It is not conserved in inelastic interactions. Momentum is conserved in all interactions, but not kinetic energy.

Johnnie's Alert

Even though a small amount of kinetic energy may be transferred as sound or thermal energies in an interaction, the interaction may still be treated as elastic because these transfers are negligible, in most cases too small to measure.

As with many examples that we have analyzed throughout all of the preceding chapters, we are usually describing conditions in our problems that are ideal. This is true of elastic interaction, too. There is no such thing as a perfect elastic interaction, so in our discussions we are talking about ideal situations. Billiard ball to billiard ball interaction is about as close to an elastic interaction as we can get. If the billiard ball is distorted at all, even though it flies back to its original shape, some kinetic energy is changed to thermal energy through the action of friction. There is a click when they hit, so some kinetic energy is changed to sound. The amount of kinetic energy lost due to heat or sound is negligible compared to the energy of the system of interacting bodies under consideration. So when we say that the interaction is elastic, that simply means that it is highly elastic and not perfectly elastic.

Let's go through a simple problem involving the head-on collision of two balls and see what happens. Suppose that ball 1 has an initial speed of $v_{1i} = 7 ft/s$ and ball 2 has an initial speed of $v_{2i} = 5 ft/s$. Ball 1 has a mass half as big as ball 2, so $m_1 = \frac{m_2}{2}$. If ball 1 has a final speed of $v_{1f} = \sqrt{65}\, ft/s$, what is the final speed of ball 2?

Using our equations for kinetic energy, we have

$$\frac{1}{2}m_1v_{1i}^2 + \frac{1}{2}m_2v_{2i}^2 = \frac{1}{2}m_1v_{1f}^2 + \frac{1}{2}m_2v_{2f}^2$$

then, multiplying everything by 2 gives us $m_1v_{1i}^2 + m_2v_{2i}^2 = m_1v_{1f}^2 + m_2v_{2f}^2$.

Dividing by m_2 and rearranging terms, we get $v_{2f}^2 = v_{2i}^2 + (m_1/m_2)(v_{1i}^2 - v_{1f}^2)$.

Substituting in the given values yields $v_{2f}^2 = 25ft/s + \frac{(49-65)\,ft/s^2}{2} = 17ft/s^2$, and taking the square root of this gives $v_{2f}^2 = \sqrt{17} = 4.12ft/s$.

From this example, what can we say about the conservation of kinetic energy and momentum? First of all we can see that the final velocity of ball 2 is almost exactly half of the final velocity of ball 1 ($\sqrt{65} = 8.03ft/s$). Does that sound like an appropriate answer given that the mass of ball 1 is half the mass of ball 2? We can always plug all of our known values into our initial equations to see what the difference is in the initial kinetic energy of the system and the final kinetic energy of the system to see if all of the energy is conserved and if all of the momentum is, too. We'll leave those calculations for you to do and see how much is lost or not in an elastic interaction.

Understanding Inelasticity

Because we have had an example of what a possible elastic interaction is like, we might as well take a look at what happens to kinetic energy in an inelastic problem too. Let's use the information and solution from problem 2 in the last Physics Phun of the two freight cars and see what we can learn about the conservation of energy from that problem. We found the mass of the second freight car to be 17,100 kg. And as we asked you to do at the end of the last problem, let's compare the kinetic energy of both cars before and after the interaction.

$$m_{c1} = 12{,}000kg \qquad v_{c1} = 12.0m/s$$

$$m_{c2} = 17{,}100kg \qquad v_{c2} = 5.00m/s$$

$$KE_{before} = \frac{1}{2}(12{,}200kg)(12.0m/s)^2 = 878{,}400\ joules$$

$$KE_{after} = \frac{1}{2}(12{,}200kg + 17{,}000kg)(5.00m/s)^2 = 366{,}250\ joules$$

We can see that kinetic energy is not conserved in an inelastic interaction; in fact, about 58 percent of the kinetic energy is not accounted for. The trains stuck together in much the same way that two mud balls stick together when they interact. If you plugged the values into the formulas for kinetic energy in our discussion of the two balls interacting, how much energy was lost? Would you consider it to be an elastic or inelastic interaction? (If you plug arbitrary values in for mass, as long as ball 1 is half the mass of ball 2, you should find that they are pretty close and could be defined as elastic.)

Newton's Figs

To see how much energy in an interaction is lost, calculate the kinetic energy before and after the interaction has taken place to see if the interaction is elastic. If there is a large difference in the two figures, the interaction is inelastic; momentum is conserved, but kinetic energy is not.

A better example of an inelastic collision is one between a ball of putty and a billiard ball. They stick together because the force of the interaction depends on something other than the separation of the objects.

We can think of the collision of two billiard balls as being much different from the collision of two cars. The calculations we make about those collisions shows us that even though momentum is conserved in a car crash, kinetic energy is not conserved. We've also gained some knowledge of elastic and inelastic collisions that will guide our thinking in general about objects interacting with each other. The loss of kinetic energy in inelastic interactions can be accounted for by the way other types of energy manifest during that interaction, such as sound and heat. And a final insight that we learned is that you can account for the transfer of kinetic energy from one elastic object to another when they collide.

Johnnie's Alert

The nature of science dictates that there is no such thing as "the right solution." There are many correct solutions, and it is important that you construct your own solution using the relevant physical principles and an answer that makes sense to you. The solution is the path that you take to your answer. Test your answer, as it's been modeled for you in example solutions, to see whether it makes sense, and whether the units of measurement are correct for your answer.

Most of us have a great deal of information from our everyday life about elastic collisions. You've seen billiard balls collide and have some understanding of their behavior.

Two elastic objects with the same mass in a head-on collision will have the object originally in motion stop after the collision and the object originally at rest will take off with the original velocity of the first object. The transfer of energy is remarkable; the first object transfers all of its original kinetic energy to the second object. The more elastic an object is, the more complete the transfer of energy between them. That's very cool!

Problems for the Budding Rocket Scientist

1. While waiting in his car at a stoplight, an 80-kg man and his car are suddenly accelerated to a speed of 5 m/s as the result of a rear-end collision. Assuming the time taken to be 0.3 s, find (a) the impulse on the man and (b) the average force exerted on him by the backseat of his car.
2. A ball of 0.4 kg mass and speed of 3 m/s has a head-on, completely elastic collision with a 0.6-kg mass initially at rest. Find the speeds of both bodies after the collision.
3. An 8-g bullet is fired horizontally into a 9-kg block of wood and sticks in it. The block, which is free to move, has a velocity of 40 cm/s after the impact. Find the initial velocity of the bullet.

The Least You Need to Know

- The law of conservation of energy is a fundamental law of physics. No matter what you do, energy is always conserved.
- Impulse is a vector quantity and is calculated by multiplying the applied force by the time interval during which it is applied, or simply, impulse = force times time.
- Momentum is a physical vector quantity that has its magnitude determined by the product of the mass and velocity of the object being described.
- The total momentum before an interaction is the same as the total momentum after an interaction.
- To see how much energy in an interaction is lost, calculate the kinetic energy before and after the interaction has taken place to see if the interaction is elastic.

Part 4

Heat, Sound, and States of Matter

In Part 4, our exploration of physics takes us into direct contact with the most familiar elements of our world. Everything we come in contact with is either a solid, a liquid, or a gas. We delve into the physical laws that explain why matter exists in these different forms, and we examine how they move under pressure. We also look at two other forms of energy, heat and sound. Many attempts had been made since the time of Aristotle to understand heat, but without much success until the nineteenth century. To understand the physics of sound, we explore the anatomy of waves to see just how the fundamental properties of waves are determined.

Chapter 13

Solids, Liquids, and Gases

In This Chapter

- Cohesive and adhesive forces
- Density, diffusion, and specific gravity
- Pressure versus force
- Gas laws

Our bodies are composed mostly of oxygen (about 65 percent by weight), carbon (about 18 percent), and hydrogen (about 10 percent), three elements that are abundant on the surface of the earth. We are made of tiny building blocks that have been reused an enormous number of times. The hundreds of billions of atoms that constitute all of us have been a part of the earth for billions of years. These atoms were undoubtedly a part of other animals and plants that walked the earth eons ago. The saying is, "Ashes to ashes and dust to dust," but for all practical purposes, atoms are eternal. The atoms that make us up will be part of the earth long after we're gone.

The hydrogen atoms in our bodies (there are two in every water molecule, and our bodies are about 75 percent water) have been around since the Big Bang. The carbon, oxygen, and sodium in our cells were made in the core of a star that had as fuel only hydrogen and a little helium; and the atoms of nickel, copper, zinc, and selenium that are essential to the functioning of our cells were generated in the cataclysmic explosion of a star much larger than our sun. We are physical beings, and the particles that make up our bodies connect us to the universe in very profound ways.

The atoms and molecules that define our physical structure are also the fundamental building blocks that make up all solids, liquids, and gases. They are just arranged in ways that are unique to particular states of matter. Let's begin our look into these states of matter by discussing what it means to be a solid.

Solids

Objects such as desks, chairs, billiard balls, and most of what we see around us are solid. The materials in these objects have atoms that are fixed (meaning that they don't move around very much relative to other atoms) owing to bonds and certain forces. In general, solids can be characterized as either amorphous or crystalline. The arrangement of atoms in an amorphous solid is random, whereas the arrangement in a crystalline solid is highly ordered. The atoms in glass (a compound containing silicon) are randomly arranged and can even shift very slowly, and the atoms in a diamond are highly ordered. That ice is slightly less dense than liquid water is the result of the randomness of the molecular arrangement in a liquid, and the more ordered arrangement of water molecules in ice.

The particles in solids are thought to vibrate randomly about a point of equilibrium much like a simple pendulum or the vibrations of a string. The amplitude of the movement of the particles is quite small, about 0.40 nanometers apart with periods of about $10^{-11}s$.

The *cohesive forces* between the particles are the causes of the vibrating motion. However, because the distances moved by the particles are small, that force is strong enough to cause the particles to stay together and hold a definite shape for the solid. There are also *adhesive forces* involved when solids of different kinds come into contact. For example, when you accidentally scrub your shoe against the floor, some of the finish on your shoe adheres to the floor and requires a little effort to scrub it off.

Even though the motion of the particles is somewhat restricted, some particles can escape the surface of one solid and enter the surface of an adjacent solid.

For example, if you lay a bar of gold on a bar of lead, you find after a long period of time some gold particles in the lead and some lead particles in the gold. That process is called *diffusion.* Diffusion in solids occurs slowly; a flood of particles will not diffuse as can occur in other states of matter. Because the particles cause the solid to have a definite shape, the particles also have a definite volume. That means that the solid has mass and inertia. If it is within a gravitational field, it will also have a defined weight. Because a solid has definite weight and mass in a definite volume, it has a definite *density.* The spacing between atoms in a substance also determines its density.

def•i•ni•tion

Cohesive forces are the attractive forces between particles of the same kind. **Adhesive forces** are forces of attraction between particles of different kinds of matter. **Diffusion** is the movement of particles of one kind of matter into the empty space of a different kind of matter because of the random motion of the particles. **Density** is the amount of matter in a unit volume. Because matter is measured in two different ways, there are two types of density. Mass density is the amount of mass in a unit volume of matter and weight density is the amount of weight in a unit volume of matter. The unit measure of mass density might be kg/m^3, and the unit measure of weight density might be N/m^3. **Specific gravity** of a substance is the ratio of the density of the substance to the density of water. There is no unit measure for specific gravity because the two measurements cancel.

Because density is a general property of all types of matter, now would be a good time to discuss it in a little more detail. Density is a derived quantity and it is a scalar quantity. We use the symbol ρ_w to represent weight density and ρ_m to represent mass density. The definition of each, then, is $\rho_w = \frac{weight}{volume} = \frac{w}{v}$ and $\rho_m = \frac{mass}{volume} = \frac{m}{v}$. The relationship between the two follows from Newton's second law: $\rho_w = \frac{w}{v} = \frac{mg}{v} = g\rho_m$. A quantity that is closely related to density is *specific gravity.* In symbols, specific gravity is expressed as $spgr = \frac{\rho_{ws}}{\rho_{ww}} = \frac{\rho_{ms}}{\rho_{mw}}$. It's important to note that weight density, mass density, and specific gravity are all scalar quantities and are derived quantities, too. Specific gravity has no units of measurement because it is a ratio of two quantities having the same units of measurement.

The units of weight density are: $\frac{N}{m^3}$ in the MKS system, $\frac{dynes}{cm^3}$ in the CGS system, and $\frac{lb}{ft^3}$ in the FPS system. Mass density is measured in $\frac{kg}{m^3}$, $\frac{g}{cm^3}$, $\frac{slugs}{ft^3}$ in the MKS, CGS, and FPS systems of measurement, respectively. Before going any further, we think that a practice problem here will help you to tie the idea of density to your world of daily life. You'll get more practice with these ideas later in this chapter.

Newton's Figs

Working with a model can sometimes be confusing. Using the particle (as a word we use in place of molecules and atoms) model, as is done here, means that if matter is portrayed as being made up of particles as described in the kinetic theory of matter, then certain things can be explained in terms of that model. If it is a good model, certain things can be predicted with a lot of success.

def•i•ni•tion

A **proportion** is an equation each of whose members is a ratio.

Some solids such as steel springs display the property of elasticity. Elasticity in a steel spring is the ability to stretch or distort within reason and then return to the original length. Elasticity depends on the cohesive forces of the particles of the solid. If the spring is stretched beyond its elastic limit, it will remain distorted and never return to its original shape. This is the same property of elasticity that we worked with on larger objects such as the billiard balls.

We also observed the stretching of a spring when we developed the concept of spring potential energy. Hooke's law (see Chapter 11) is the basic idea for the elasticity discussed here, too. The force required to stretch the spring is directly proportional to the elongation of the spring as long as you remain within the elastic limit of the spring. You can hang a spring vertically and calibrate it to measure weight. Let's suppose that we have a spring supported vertically and hang 4.9 N of weight on it and observe an elongation (or stretch from the original length) of 0.14 m.

Using Hooke's law, what would you expect the elongation to be for 9.8 N? We can make this statement using Hooke's law: $\frac{4.9N}{0.14m} = \frac{9.8N}{y}$, where y is the unknown elongation. Did you recognize that statement as a *proportion?* Now let's solve the proportion for y.

$$(4.9N)y = 9.8N(0.14m), \text{ then } y = \frac{9.8N}{4.9N}(0.14m) = 0.28m.$$

Let's try another one. What is the elongation for 14.7 N? Did you get an elongation of 0.42 m? And what about solving for the spring constant in the same problem? In other words, what is the spring constant in the equation $F = ky$? Did you find that the spring constant is $35\frac{N}{m}$? We hope you did!

This is a little different from the application in which you were calculating spring potential energy, but both deal with elasticity and Hooke's law.

Liquids

Like a solid, the particles of a liquid vibrate rapidly, but in a random way. Unlike the vibration of the particles of a solid, which are about a point of equilibrium as if the solid particles are tied to that point by little elastic springs, the particles of the liquid vibrate in all directions haphazardly. The particles of a liquid are about as close to each other as those of a solid. However, they have more freedom of movement around and over each other with enough mobility to flow and take on the shape of their container. Liquids diffuse just as solids do, but whereas just a few particles of solids move from one solid to another, practically all the particles of both liquids will diffuse.

> **Physics Phun**
>
> A quantity of aluminum is 21 cm wide, 1.05 m high, and 4.03 m long. How much does the quantity of aluminum weigh if the mass density of aluminum is $270 \frac{kg}{m^3}$? Hint: remember that $\rho_w = \frac{w}{V} = \frac{mg}{V} = g\rho_m$.

We can demonstrate the diffusion of liquids by placing two liquids in a container with only the surfaces of the liquids in contact. That surface separates the liquids.

Placing a more dense liquid such as oil in a container first and then a less dense liquid such as water on top of the dense substance can accomplish that. It's helpful if one of the liquids is colored so that you can observe the diffusion that will be nearly complete in a few days. The more dense liquid diffuses upward and the less dense downward even in the presence of the force of gravity.

The particles of liquids cause cohesive forces as well as adhesive forces much like particles of a solid, except you will find that the forces are not as strong in liquids. As we observed with the stretch of a spring, a considerable force was required to stretch the spring and even more to exceed the elastic limit of the spring or to break it. We can observe the strength of the cohesive forces of a liquid by poking a finger in a thick liquid such as paint or motor oil. When you remove your finger, you experience a noticeable tug of the paint particles on your finger (the attraction of paint particles for paint particles) and the paint continues to adhere to your finger even after wiping (the attraction of paint particles for your finger).

You can see the relative strengths of the cohesive and adhesive force of water by pouring water into a graduated cylinder. The water particles tend to attract the glass particles more than they attract other water particles and cause a concave or crescent-shaped water surface. The water particles creep an observable distance up the sides of the cylinder, causing that part of the water surface to be higher than the middle of the column of water. If you observe a column of mercury in a glass tube, you find a convex surface on the top of the column because mercury particles attract each other more than they attract glass particles. The middle of the mercury surface is noticeably higher than the sides touching the glass.

The cohesive forces of particles in a liquid cause a liquid to have a free surface that is characterized by *surface tension.* If you've ever floated a needle on the surface of water, you've seen an example of surface tension. The water surface supports the needle so that it stays on the surface. We can explain surface tension by thinking of the cohesive forces of the particles in the liquid acting on one particle. The particles must be within a certain distance from each other to exert the cohesive force. If they are too close, the force is repulsive; if they are too far away, the force is negligible.

This single particle can be thought of as being affected by forces within a certain region surrounding the particle; call it the sphere of force. For a particle completely surrounded by particles of the liquid, the particles in its sphere of force will cancel each other out so the one particle has no net force. Also this particle near to the surface will have its sphere of force decreased on top so there will be a slight unbalanced force downward, because more particles attract it toward the liquid than those attracting it upward. A single particle at the surface, however, experiences only particles in its sphere of force pulling it toward the liquid. All particles at the surface experience this same pull, causing the surface to form a condition that resembles a membrane stretched taut over the free surface. Because of the tension, the surface tends to be as small as possible.

When the needle is placed on the inflexible filmlike surface, it makes a dimple in the surface, increasing the area. The cohesive forces tend to restore the minimum area, a taut horizontal surface, by exerting an equal but opposite force upward. The needle is denser than water, but it floats because of the surface tension. The water pushes up on the needle with as much force as the earth pulls down on it. Small bugs use this phenomenon to walk on lake water when it is still. Surface tension also causes small amounts of water such as dew or rain to form small spheres. The spherical surface is found to be the minimum surface for a given volume. The particles of the liquid attract the particles at the surface, causing the surface to have the minimum surface possible with the appearance of an inflexible film stretched over the droplet.

def•i•ni•tion

Surface tension is a quantity or condition of the surface of a liquid that causes it to tend to contract. **Pressure** is a quantity determined by the force on a unit of area.

Liquids share another property with solids: they exert *pressure*. Pressure is not force; it is force per area. It is a scalar quantity with units that you may have heard, such as pounds per square inch (psi).

Pressure is exerted by a liquid in all directions. You'll learn much more about pressure in Chapter 14, but we wanted to discuss it briefly now before making a very detailed study of it later. To begin with, force is a vector quantity and pressure is not.

Next, let's look at the definition of pressure in symbols, $P = \frac{Force}{Area} = \frac{F}{A}$, and determine the units of measurement of this new quantity. In the FPS system, pressure is measured in $\frac{lb}{in^2}$, sometimes referred to as psi. The units of measurement in the MKS system are $\frac{N}{m^2}$ (also called a pascal, or Pa) and in the CGS system $\frac{dynes}{cm^2}$. As you can see, the units of force and pressure are conceptually different.

An example of the type of problem you can encounter involving pressure might be helpful here, especially if it emphasizes the difference in force and pressure.

Example 1: High-Heel Pressure

Suppose that a 110-lb woman is wearing high-heeled shoes. The heels have an area of about $\frac{1}{16} in^2$ at the base. Her full weight is first on one heel and then the other as she walks across the floor. How much pressure does she exert on the floor?

Solution:

$$P = \frac{F}{A} = \frac{110 lb}{\frac{1}{16} in^2} = 1760 lb / in^2$$

That's quite a bit of pressure applied to such a small area. And notice that it's the 110-lb force that exerts the pressure in this small area. It emphasizes a difference in the magnitude of force and pressure. There is also a difference in units of measurement as well as the fact that force is a vector quantity and pressure is a scalar quantity. So as you can see, high-heeled shoes exert the force of a woman's weight over a small area, resulting in higher pressures. Snowshoes have the opposite effect, spreading the weight of a person out over a large area, resulting in a lower pressure.

In a liquid, pressure increases with depth. In fact, the pressure exerted by a liquid at a given depth depends on two quantities: the density of the liquid and the depth of the liquid. Submarines and other submersible ships have to withstand very high pressures, for the deeper they go, the more external pressure pushes on them. So let's work through another problem and see how pressure is created by the depth of a liquid.

Example 2: Swimming Pool Pressure

Suppose a swimming pool is in the shape of a rectangular solid 4.00 m wide, 8.00 m long, and 2.00 m deep. It is completely filled with water. What is the pressure of the water on the bottom of the pool? Hint: the mass density of water is $10^3 \frac{kg}{m^3}$.

Solution:

$$p = \frac{F}{A_b} = \frac{W_w}{A_b} = \frac{\rho_{mw} V}{A_b} = \rho_{mw} g \frac{V}{A_b}$$

$$= \frac{(10^3 kg/m^3)(9.8 m/s^2)(4.00 m)(2.00 m)(8.00 m)}{(8.00 m)(4.00 m)} = 1.96 \times 10^4 N/m^2$$

Now that we've worked through a few problems involving pressure, why don't you try a few on your own and see how well you can do?

Physics Phun

1. A tank is 9.0 m long and 5.0 m wide. The tank is filled with milk to a depth of 4.5 m. The mass density of milk is 1.03 g/cm^3. How much pressure does the milk exert on the bottom of the tank?
2. A water tank is 10.0 ft in diameter and 20.0 ft tall. What is the pressure on the bottom of the tank when it is completely filled with water? Hint: the volume of a cylinder is $V = \pi r^2 h$, and the mass density of water is 1g/cm^3. Hint: 3.28 ft = 100 cm.
3. A man weighs 192 lb. When he stands still supporting his weight on both shoes, the total area of his shoes in contact with the floor is 484 cm^2. What pressure does he exert on the floor?

Gases

Let's turn our attention to the third common state of matter, gas. The particles of a gas are so far apart that they do not exert force on each other until they bump into each other. Obviously, the particles are very far apart compared with those of solids and liquids. In fact, particles of matter in the gaseous state occupy about 103 times the volume they occupy in the liquid state. That doesn't mean that the particles are bigger now that they are gaseous—just that they take up more space.

At room temperature, gas particles travel at about 500 m/s and travel about 100 nanometers before bumping into another particle or the sides of the container such as a sealed balloon or glass beaker. The sides of the container experience between 4 and 10 billion collisions each second. Gases have density, exert pressure, and diffuse. We've all probably sprayed air freshener in a room. The particles in the air freshener move randomly in all directions and mix with the gas particles of the air rather quickly. Before you know it, the room's former odor has been replaced by a fresher scent.

Johnnie's Alert

Because the words *force* and *pressure* are often interchanged incorrectly, it's good to know the difference between the two. Remind yourself, with notes if necessary, that force is a vector quantity and pressure is a scalar quantity. Be able to identify the units of both in all systems of measurement.

The density of a gas is expressed in the same way as the density of a solid or liquid. Air has mass and weight. The mass density of air near sea level is found to be $1.29 \times 10^{-3} g/cm^2$. The atmosphere that we're referring to begins at or near sea level, but the atmosphere of the earth extends upward for about 25 miles (although most of it is below an altitude of 20 miles). Above 25 miles, the atmosphere gets pretty thin. The pressure of the air we breathe near the surface of the earth is often reported as the pressure of a column of air 1 square inch in area and reaching to the top of the atmosphere.

The pressure at sea level of our atmosphere is 1 atmosphere, or 1 atm. That is the same as $14.7 lb/in^2 = 1.010 \times dynes/cm^2$. There are other measures of pressure, and some of them are listed here for your information:

$$1 \text{ Pa} = 1 \text{ Pascal} = \frac{N}{m^2}$$

$$1 bar = 1.000 \times 10^5 \; N/m^2 = 10^5 \; Pa$$

$$1 atm = 1.013 \times 10^6 \; dynes/cm^2 = 1.013 \times 10^5 \; N/m^2 = 1.013 \times 10^5 \; Pa = 101.3 kPa = 1.013 bar$$

Johnnie's Alert

Unless stated otherwise in a problem, the pressure near the earth's surface means the pressure at sea level at standard temperature 0°C.

The unit of measure of pressure used depends on the application. Physicists use the units we reviewed earlier. Chemists use the same units as the physicist along with millimeters of mercury, or Torr. The weatherman uses millimeters of mercury, inches of mercury, and bars or millibars when he discusses pressure in his weather reports. The pressure in millimeters of mercury or inches of mercury will be meaningful after a more complete discussion of pressure in Chapter 14.

Changing State

At this stage, we have examined three different states of matter. We're all familiar with three states of water, for instance, and know that they exist under certain conditions. To get ice to change state, some drastic changes must take place to alter the activity of the particles of the ice. That is, the particles of ice must be caused to vibrate so wildly that they start to move one over the other to become water. Another possibility is that, under the right conditions, the particles of ice become gas particles! The special name for that change is *sublimation.* Perhaps you've observed the results of that process when you see what appears to be steam rising from the frost on a neighbor's roof in the early sunshine of a wintry morning. In reality what you see is condensed water vapor forming from the cold water molecules rising from the frost.

A similar process is observed when the sun pops out after a summer rain shower. Steam appears to rise from the asphalt on rain-slick city streets. That occurs when the particles of a liquid suddenly become particles of a vapor. The process is called *evaporation.* And although there are terms that define each of these changes in states of matter, all of them are also considered to be *phase transitions.*

The ice skates of the figure skaters glide across the ice with such ease that it appears the skates are lubricated. A closer look at the tracks of the skates reveals that they are riding on a thin layer of water. You would notice a similar action if you observe a piece of bare wire with heavy weights on each end hanging across a block of ice. The action is not as fast as a gliding ice skate, but in time the wire cuts completely through the ice. The water immediately freezes above the wire, though, so that the block appears to remain intact even after the wire has passed through it. The name for this phenomenon is *regelation.* We explore some of the causes of these changes in greater detail in the next two chapters.

def•i•ni•tion

Sublimation is the direct change of a solid to a vapor without going through the liquid state. **Evaporation** is the production of a vapor or a gas from a liquid as some of the fastest-moving molecules escape from the liquid. Evaporation happens at any temperature. **Regelation** is the melting under pressure and then freezing again after the pressure is released. A **phase transition** is a change in a physical system from one state or phase to another without any change in its chemical composition, such as water freezing into ice or heating into steam.

Boyle's Law

Around the time that Newton was at Cambridge (1660), another Englishman, Robert Boyle, was at Oxford investigating the properties of compressed air and other gases. He made a U-shaped glass tube with one end open and the other sealed. He found that as he poured mercury into the open end, the volume of air contained in the tube decreased. When he doubled the amount of mercury, the volume of air was cut in half. When he doubled the amount of mercury again, the volume of the air became one quarter of what it was.

What Boyle discovered is now known as Boyle's law (go figure). This law states that the volume of a given amount of any gas at a set temperature is inversely proportional to the pressure applied. Do you remember which one of Newton's laws is described in terms of an inverse proportion? (Big hint: it has to do with gravity.)

About a century later, Jacques Charles discovered a second law about gases, which was called, surprisingly enough, Charles' law. He found a similar relationship between temperature and pressure. But instead of keeping a set temperature and increasing the pressure, Charles fixed the pressure and increased the temperature. Stated in proper science lingo, the law states that the pressure of any gas contained in a set volume increases by $1/_{273}$ of its original value for every degree the temperature is raised. Boyle's law and Charles' law were combined to form what's called the ideal gas law. An ideal gas is a gas in which the molecules or atoms can be thought of as individual, pointlike particles that don't exert intermolecular forces on one another. Of course, no gases are truly ideal, but air in our atmosphere and many gases under normal conditions behave like ideal gases. That is, they expand when heated and contract when cooled, much like solids and liquids do.

But now is a good time to take a look at some aspects of Boyle's law because it deals with two properties that we have discussed for all states or phases of matter: volume and pressure. The particles in a gas are so far apart that gases have no fixed shape or volume. The particles of a gas can be made to occupy a smaller volume with relative ease. In the larger scheme, gases differ from solids and liquids in that gases can be compressed much more easily than liquids and solids. As a matter of fact, for all intents and purposes, liquids and solids are virtually incompressible.

The particles of solids differ from liquids and gases in that the particles of solids vibrate randomly about a point of equilibrium. The particles of solids do not have the freedom of motion that the particles of liquids and gases do. Because of the freedom of motion of the particles of liquids and gases, liquids and gases are referred to collectively as fluids. The streamlined flow of fluids was mentioned when we looked at the difference in pressure on an airplane wing as an application of Bernoulli's principle (see Chapter 10). That picture should become clearer now that you have a better understanding of pressure, and it will become even more clear as we explore fluids in more detail in the next chapter.

A known mass of gas in a closed container and constant temperature will behave according to the mathematical model shown by the graph in Figure 13.1. The graph essentially shows us the inverse proportion that Boyle's gas law defines.

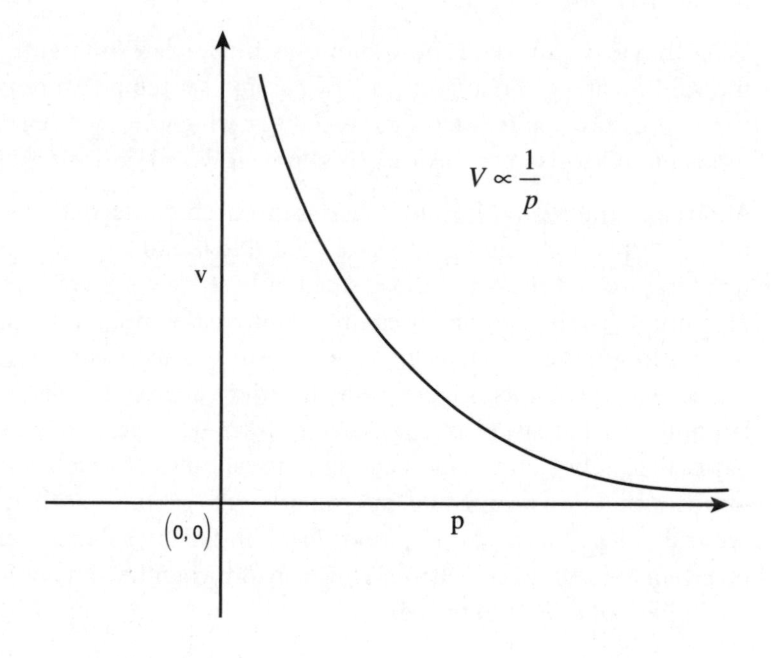

Figure 13.1

A graph of Boyle's law reveals the mathematical relationship between volume and pressure.

We can write the inverse relationship of volume to pressure as $V \alpha \frac{1}{p}$. A direct correlation that can be derived from this formula is that at a constant temperature for a given quantity of gas, the product of its volume and its pressure is a constant: $PV = k$.

This can lead us to one more handy formula that states that at a constant temperature for a given quantity of gas, $P_iV_i = P_fV_f$, where P_i is the initial pressure, V_i is its initial volume, P_f is its final pressure, and V_f is its final volume.

Using this formula, let's work through a gas problem so that you can see how Boyle's law can be applied.

Suppose a certain mass of gas occupies a volume of 2.5 L at 90 kPa pressure. What pressure would the gas exert if it were placed in a 10.0 L container at the same temperature?

Solution:

Using our previous formula, $P_iV_i = P_fV_f$, we can plug in the known values and solve for our unknown pressure.

$P_i = 90kPa \qquad V_i = 2.5L$

$P_f = ? \qquad V_f = 10.0L$

$$P_iV_i = P_fV_f$$

$$90 \times 2.5 = P_f \times 10.0$$

$$225 = P_f \times 10.0$$

$$\frac{225}{10.0} = P_f = 22.5kPa$$

That was a pretty straightforward example. And that also brings us to the end of this chapter. In Chapter 14, we deepen our exploration of pressure and see how it affects things such as pistons, more high heels, and baseballs.

Problems for the Budding Rocket Scientist

1. A person has a mass of 75 kg and a volume of 0.064 m^3. Will she sink or float in water? (Hint: the density of water is 1,000 kg/m^3.)
2. Find the density and specific gravity of ethyl alcohol if 63.3 g occupies 80.0 mL.
3. 4.5 L of gas at 125 kPa is expanded at constant temperature until the pressure is 75 kPa. What is the final volume of the gas?

The Least You Need to Know

- The atoms and molecules that define our physical structure are also the fundamental building blocks that make up all solids, liquids, and gases.
- Pressure is not force; it is force per area. It is a scalar quantity with units such as pounds per square inch (psi).

- The pressure exerted by a liquid at a given depth depends on two quantities: the density of the liquid and the depth of the liquid.
- Force is a vector quantity and pressure is a scalar quantity.
- A phase transition is a change in a physical system from one state or phase to another without any change in its chemical composition.

Chapter 14

Pressure

In This Chapter

- A brief history of hydraulics
- Pressure and force revisited
- Hydrometers and barometers
- Bernoulli's principle and baseballs
- The art of compressing air

Although the title of this chapter is "Pressure," the word itself is intimately connected to the science of hydraulics. The first half of this chapter is devoted to examining some aspects of that science. Then in the second half, we look at atmosphere pressure and the concept of buoyancy and see how Bernoulli's principle explains a curveball in baseball. But before we get into all of that, we'd like to provide you with some background information about the science of hydraulics.

A Brief History of Hydraulics

The word *hydraulic* is sometimes understood as the use of water for the benefit of humanity, and in practice it can be considered to be older than recorded history itself. Traces of irrigation canals from prehistoric times still exist in Egypt and Mesopotamia (current-day Iran and Iraq); the Nile is known to have been dammed at Memphis some six thousand years ago to provide the necessary water supply, and the Euphrates River was diverted into the Tigris even earlier for the same purpose. Ancient wells still in existence reach to surprisingly great depths; and underground aqueducts were bored considerable distance, even through bedrock. In what is now Pakistan, houses were provided with ceramic conduits for water supply and drainage some five thousand years ago; and legend tells of vast flood-control projects in China barely a millennium later. All of this clearly demonstrates that people must have begun to deal with the flow of water countless times before these.

The science of hydraulics had its origins more than two thousand years ago during the course of Greek civilization. The Greek who made the most lasting contribution to hydraulics was the Sicilian mathematician Archimedes (287–212 B.C.E.), who reasoned that a floating or immersed body must be acted upon by an upward force equal to the weight of the liquid that it displaces. This is the basis for *hydrostatics* and also of the apocryphal story that Archimedes made this discovery in his bath and upon doing so ran naked through the streets crying "Eureka!" Nevertheless, even though Archimedes' writings, like those of his fellow Greeks, were faithfully transmitted to the West by Arabian scientists, further progress in hydrostatics was not to be made for another 18 centuries.

def•i•ni•tion

A simple definition of the word **hydraulic** is when something is operated, moved, or affected by the movement of water. Hydraulics is the science that deals with the practical application of liquids in motion. And if hydraulics is the study of liquids in motion, then **hydrostatics** by definition is the study of liquids at rest and the pressures they exert or transmit.

In the course of the millennium following the time of Archimedes, the science of hydraulics regressed rather than advanced. Even though the Romans developed extensive water supply and drainage systems, and windmills and water wheels appeared on the scene in increasing numbers, these represented the art of hydraulics rather than the science.

The Greeks tended to reason without recourse to observation. It was the Italian genius Leonardo da Vinci (1452–1519) who first emphasized the direct study of nature in its many aspects. Leonardo's hydraulic observations extended to the detailed characteristics of jets, waves, and eddies, not to mention the flight of birds and comparable facets of essentially every other field of knowledge. In particular, it was Leonardo who first correctly formulated the basic principle of hydraulics known as continuity: the velocity of flow varies inversely with the cross-sectional area of a stream. Unfortunately, not only were his copious notes written in mirror image (for reasons of secrecy), but in addition, most of them were lost for several centuries after his death. Thus his discoveries had little effect on the growth of science.

The second essential contribution to hydrostatics was made by the Dutch hydraulic engineer Simon Stevin (1548–1620) in 1586, nearly two millennium after the time of Archimedes. Stevin showed that the force exerted by a liquid on the base of a vessel is equal to the weight of a liquid column extending from the base to the free surface. That this force doesn't depend on the shape of the vessel became known as the hydrostatic paradox. After this time, the science of hydraulics began to increase in relation to the direct outgrowth of science in general.

A Simple Hydraulic Machine

One of the most fundamental principles learned in the study of hydraulics is that pressure is exerted in all directions in a fluid. Now let's suppose that we have a confined fluid in a container with a cylinder at each end both completely filled with the fluid like that found in Figure 14.1.

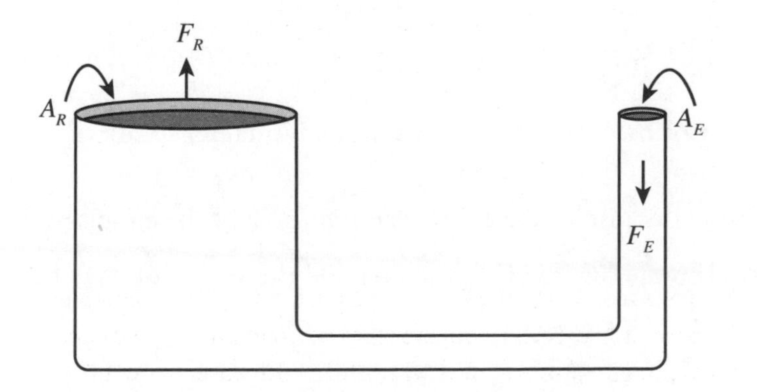

Figure 14.1

The hydraulic press operates as a simple machine.

On top of each cylinder is a piston that is free to move up and down. This arrangement constitutes a simple machine that will multiply force by using pressure on the confined motionless fluid. The machine is based on Pascal's principle, which states that any pressure applied to a confined fluid at rest will be transmitted undiminished to every point in the liquid. That means that the pressure will cause a force to act perpendicular to every unit of surface area exposed to the confined liquid.

So if you push down on A_E, the pressure transmitted undiminished throughout the fluid will cause a force upward on A_R. This arrangement acts like a simple machine and has applications like the lift that raises your car when the mechanic changes the oil or switches the tires. How does this machine compare to other simple machines?

Johnnie's Alert

Because pressure is transmitted undiminished throughout a liquid, liquid exposed to the atmosphere will have the pressure of the atmosphere transmitted undiminished throughout the liquid. That's an important fact to take into account when calculating total pressure in a liquid. Many problems involving pressure in liquids specify only the pressure of the liquid.

Notice that when the effort piston is pushed down a distance S_E, a volume of the fluid is displaced. Because the pressure is transmitted undiminished throughout the liquid, an equal volume of liquid is displaced in the resistance cylinder. That means that the resistance piston moves a distance S_R to displace an equivalent volume. Because the volume in each case is the volume of a tiny cylinder of fluid, multiplying the area times the distance moved defines the volume of the fluid displaced in both the effort cylinder and the resistance cylinder. How about if we display the above discussion in mathematical terms so you can see what we're talking about? And then we'll work through a problem so that you get some hands-on practice with these new formulas.

$$V_E = V_R$$

$$A_E S_E = A_R S_R$$

$$\frac{S_E}{S_R} = \frac{A_R}{A_E}$$

$IMA = \dfrac{A_R}{A_E}$ Definition of IMA, the ideal mechanical advantage.

$\dfrac{4.9N}{0.14m} = \dfrac{9.8N}{y}$ Because the pistons are circular and the area of the circular piston is proportional to the square of its diameter and the square of its radius.

$$p_E = p_R$$

$\dfrac{F_E}{A_E} = \dfrac{F_R}{A_R}$ Definition of pressure.

$$\frac{A_R}{A_E} = \frac{F_R}{F_E}$$

Johnnie's Alert

Your mechanic probably applies the effort force by releasing compressed air into a chamber that causes the effort piston to move. You never see the details discussed here in play at the service station, but you can see the giant cylinder move upward, raising your car on the service ramp.

Hopefully, you can see by these formulas that pressure can play an important role in the operation of a simple machine to multiply force even though pressure is not a force. Oh, you don't see that it multiplies force? Then let's consider the following example problem.

Pressure and a Simple Machine

Suppose your mechanic lifts your 2-ton SUV (a ton is the name of 2,000 pounds of weight) with his hydraulic lift. The area of the large piston is 1,130 in^2 and the area of the small piston is 13 in^2. How much effort force must be applied to the small piston?

Solution:

$$F_R = 4 \times 10^3 lb$$

$$F_E = ?$$

$$A_R = 1130 in^2$$

$$A_E = 13 in^2$$

$$\frac{A_R}{A_E} = \frac{F_R}{F_E}$$

$$F_E = \frac{A_E}{A_R} F_R$$

$$F_E = \frac{13}{1130}\ (4 \times 10^3 lb)$$

$$F_E = 46 lb$$

So to lift a 2-ton SUV requires a force of only 46 lbs. Remember, to multiply force with a simple machine, distance must be sacrificed. That means the liquid in the small diameter pipe moves a much greater distance than in the large diameter pipe. Also, the areas of the pipes are directly proportional to the square of their diameters or radii. Pressure is not a force, but it is closely related to force, as you see here and in the rest of this chapter.

Liquid Pressure and Depth

When you dive into a swimming pool and swim to the bottom, your ears can experience a great deal of pressure. (Of course, it all depends on how deep the pool is. The deeper you go, the more the pressure.) It doesn't matter how you turn your head, you experience the same pressure.

Newton's Figs

A fluid exerts pressure in all directions at any given point. The amount of pressure depends on the height of liquid above that particular point.

Pressure is exerted equally in all directions at a given depth. If you refer to Figure 14.2, we'll use that diagram to develop a way of calculating the pressure at any depth of a liquid.

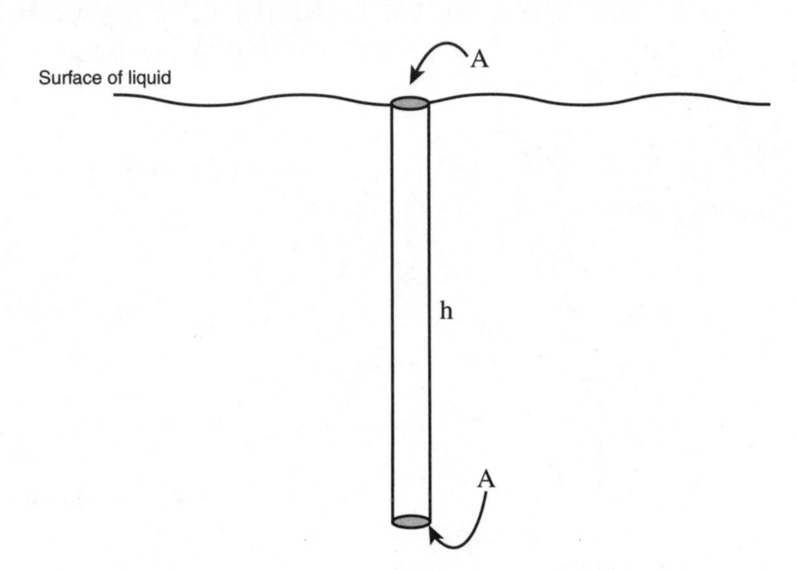

Figure 14.2

Calculating the pressure on the base of a cylinder of liquid.

> **Newton's Figs**
>
> Checking the units of measurement implied by an algebraic solution or resulting from an arithmetic solution can often reveal errors in reasoning and/or omissions of factors or terms of sums. For example, if your solution should be dynes and you find that grams are the actual units, you possibly omitted a factor of the acceleration due to gravity, 980 cm/s^2.

The top of the cylinder is at the surface of the liquid. The bottom of the cylinder is at a depth h in the liquid. The cylinder is just a cylinder of the liquid itself. Do you remember from the last chapter that the pressure on the bottom of the cylinder is the weight of the liquid divided by the area of the base of the cylinder? The cylinder is uniform, with the area of the base the same as the area of the top of the cylinder. We can calculate the pressure on the bottom in the following way:

$$p = \frac{F}{A}$$

$$p = \frac{W_L}{A}$$

$$p = \frac{\rho_L V_L}{A}$$

$$p = \frac{\rho_L A h}{A}$$

$$p = \rho_L h$$

The final result indicates that you can calculate the pressure of any liquid at any depth for which the weight density is constant. The pressure at any depth can be calculated by multiplying the weight density of the liquid by the depth.

Pressure and Total Force

Earlier we calculated the pressure on the bottom of the pool as the product of the density times the volume divided by the area. Now you can calculate the total force on the bottom of the pool as follows:

$$p = \frac{F}{A}$$

$$F = pA$$

$$F = \rho_W hA$$

$F = \rho_W V$ Where the height (h) times the area (a) is substituted with (V) volume.

What do you suppose the total force would be on the sides? As long as the swimming pool is in the shape of a rectangular solid, we can calculate it with algebra. However, if the shape is an ellipse or if the pool varies in depth, we would need calculus to find the total force on the sides.

Newton's Figs

The average value of a quantity that is directly proportional to another quantity can be calculated by adding its smallest value to its largest value, then dividing the sum by two.

But regardless of the shape, we would still use the same procedure on a swimming pool with a simple shape as we would for a weird shape. Because the pressure varies with the depth, you must use the average pressure to find the total force on a side.

$$p_{avg} = \frac{F_{total}}{A_{side}}$$

$$F_{total} = p_{avg} A_{side}$$

$$F_{total} = \rho_w h_{avg} A_{side}$$

Let's work through a problem that incorporates this equation as the solution.

> **Johnnie's Alert**
>
> The pressure of a liquid varies directly as the depth, and the constant of proportionality is the weight density of the liquid. The weight density of a liquid varies in some liquids, such as water in the ocean.

Total Force

Suppose you have a swimming pool full of water. The swimming pool is 20.0 ft long, 10.0 ft wide, and 6.00 ft deep. Calculate the total force on the bottom of the pool and the total force on the ends and the sides due to the water in the pool.

Solution:

$\rho_{ww} = 62.4 lb/ft^3$ The weight density of water.

$l = 20.0 ft$

$w = 10.0 ft$

$h = 6.00 ft$

$F_{total\ end} = ?$

$F_{total\ sides} = ?$

$F_{total\ bottom} = ?$

$F_{total\ bottom} = \rho_{ww} h A_{bottom}$ Our previously defined formula.

$F_{total\ bottom} = \rho_{ww} hwl$

$F_{total\ bottom} = (62.4 lb/ft^3)(6.00 ft)(10.0 ft \times 20.0 ft) = 7.49 \times 10^4 lb$

$F_{total\ end} = p_{avg} A_{end}$

$F_{total\ end} = \rho_{ww} h_{avg} A_{end}$

$F_{total\ end} = \rho_{ww} \frac{h}{2} wh$

$F_{total\ end} = (62.4 lb/ft^3)(10.0 ft)\frac{(6.00\ ft)^2}{2} = 1.12 \times 10^4 lb$

$F_{total\ side} = \rho_{ww} h_{avg} A_{end}$

$F_{total\ side} = \rho_{ww} \frac{h}{2} lh$

$F_{total\ side} = (62.4 lb/ft^3)\frac{(6.00\ ft)^2}{2}(20.0 ft) = 2.25 \times 10^4 lb$

And that's our solution for all of the forces acting on the bottom, sides, and end of the pool.

Did you understand how we arrived at the average depth? The average depth is calculated by adding the shortest depth to the largest depth and dividing by two; that is, $h_{avg} = \frac{0+h}{2} = \frac{h}{2}$. The force on the ends and sides starts out as zero at the minimum depth and increases as the depth increases to the greatest value at maximum depth. Because the pressure of the liquid is exerted in all directions, you can think of the force due to the pressure as being perpendicular to the ends and sides all the way from top to bottom. Now let's turn our attention to another aspect of hydraulics, buoyancy.

Pressure and Buoyant Force

Take a look at Figure 14.3 and think of the small uniform cylinder as completely submerged in a liquid.

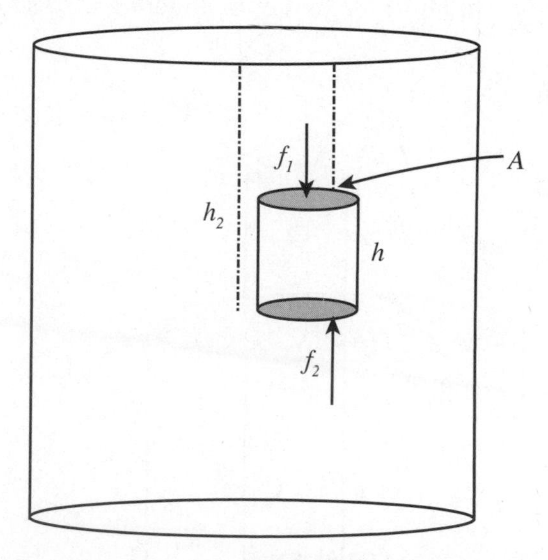

Figure 14.3

The cylinder completely submerged in water has a buoyant force.

We know that, due to the pressure of the liquid, there is a force, f_2, acting upward on the bottom of the submerged cylinder that is at a depth h_2. For the same reason, there is a force, f_1, with direction downward on the top of the submerged cylinder. Remember the story of Archimedes running naked down the street yelling "Eureka!" when he discovered that there is a net force on a submerged object because of the pressure differences in the liquid? You are about to find whether Archimedes was justified in declaring that naked truth. Because pressure varies directly as the depth of a liquid, we know that $f_2 > f_1$, so there must be a net force due to the liquid.

$f_2 - f_1 = f_{net}$

$\rho_{ww} h_2 - \rho_{ww} h_1 = f_{net}$

$\rho_{ww}(h_2 - h_1) = f_{net}$ Because $h_2 - h_1 = h$, the height of the submerged cylinder.

$w_{liquid\ displaced} = f_{buoyant}$

Eureka! The weight of the liquid displaced by the submerged cylinder, or any object for that matter, is equal to the buoyant force on the submerged object due to the pressure differences in the liquid.

Now let's suppose the object is not completely submerged. We know that the object must float in the liquid and is not moving vertically. That means that there is no net force on the floating object. Our diagram in Figure 14.4 shows us an example of this situation. Here you see a cylindrical object that has length, h, and uniform cross section, A_{cyl}, floating in a liquid with h_a out of the liquid and h_L submerged in the liquid.

Figure 14.4

The floating cylinder in a liquid that provides a buoyant force.

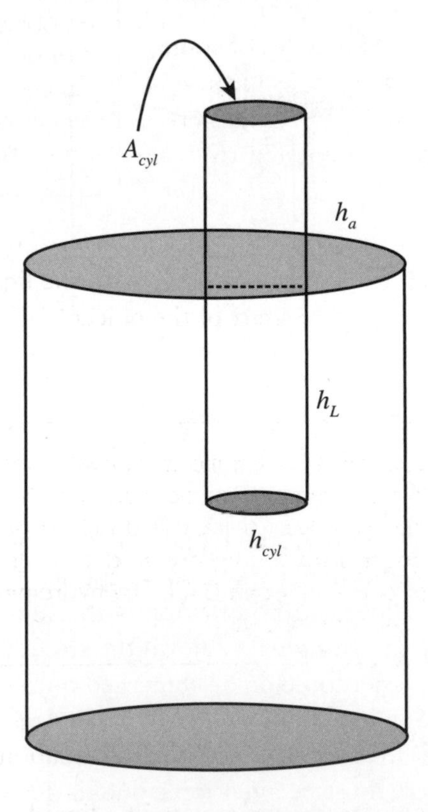

The buoyant force on the object must have the same magnitude as the weight of the object. That is, if the liquid is water,

$$f_b = W_{cyl}$$

$$\rho_{ww} V_{w\ displaced} = \rho_{w\ cyl} V_{cyl}$$

$$\rho_{ww} A_{cyl} h_L = \rho_{w\ cyl} A_{cyl} h$$

$$\rho_{ww} h_L = \rho_{w\ cyl}\ h$$

$$\frac{h_L}{h_{cyl}} = \frac{\rho_{w\,cyl}}{\rho_{ww}}$$

That means that as long as the ratio of the amount of the object submerged to the total amount of the object is less than one, the object will float. That will happen as long as the weight density of the object is less than the weight density of the liquid, water in this case. Examples of some floating objects are the oil-laden ships in the Persian Gulf and the workhorse LSTs that ferried soldiers along with their equipment to shore during World War II.

Newton's Figs

The LST is made of steel and so are submarines and other ships, and yet they float very well. It does not matter what an object is made of as long as it displaces enough water so that the buoyant force of the water is greater than the weight of the object floating in the water.

The really cool thing about the last equation is that if the liquid is water, $\frac{h_L}{h_{cyl}} = \frac{\rho_{w\,cyl}}{\rho_{ww}}$ enables us to determine the *specific gravity* of the object.

def•i•ni•tion

The **specific gravity** of a substance is the ratio of the weight density of the liquid to the weight density of water. It is also the ratio of the mass density of the liquid to the mass density of water. The specific gravity of a liquid is a ratio using water as a standard. The specific gravity of a gas is a ratio using dry air at S.T.P. (standard temperature and pressure—that is, 760 mm of mercury and 0°C). The **hydrometer** is an instrument used to measure the specific gravity of a liquid.

That is, $\frac{h_L}{h_{cyl}} = \frac{\rho_{w\,cyl}}{\rho_{ww}} = spgr_{cyl}$. We now have all the information we need to construct an instrument that will measure the specific gravity of a liquid with a uniform object like the cylinder in Figure 14.4, given that it floats in water. You can use symbols that enable you to generalize the discussion a bit such as h_{sL} to represent the amount submerged in a liquid and h_w to represent the amount the same object is submerged in water. The expression then becomes $\frac{h_{sL}}{h_{cyl}} = \frac{\rho_{w\,cyl}}{\rho_{wL}}$. Then:

$\rho_{w\,cyl} = \frac{h_{sL}}{h_{cyl}} \rho_{wL}$ Multiplying both members by ρ_{wL}.

$\frac{\rho_{w\,cyl}}{\rho_{ww}} = \left(\frac{h_{sL}}{h_{cyl}}\right)\left(\frac{\rho_{wL}}{\rho_{ww}}\right)$ Dividing both members by ρ_{ww}.

$spgr_{cyl} = \frac{h_{sL}}{h_{cyl}} spgr_L$ Definition of specific gravity.

$spgr_{cyl} = \frac{h_{sL}}{h_{cyl}}$ If the liquid is water.

$$\frac{spgr_{cyl} = \frac{h_{sL}}{h_{cyl}} spgr_L}{spgr_{cyl} = \frac{h_{s\,water}}{h_{cyl}}}$$ Dividing both members by $spgr_{cyl}$.

$1 = \frac{h_{sL}}{h_{s\,water}} spgr_L$ Simplifying.

$spgr_L = \frac{h_{s\,water}}{h_{sL}}$ Multiplying both members by $\frac{h_{s\,water}}{h_{sL}}$.

Therefore, we may use a uniform rod, glass tube, floating cylinder, etc. to determine the specific gravity of any liquid by dividing the length submerged when placed in water by the length submerged when placed in any other liquid. The device you have just designed is called a *hydrometer*. The hydrometer has several useful applications, from the determination of the condition of your car battery to the determination of the correct content of alcohol in the brewing of wine or beer. And now that all of this is fresh in your mind, try working a few problems.

Physics Phun

1. Suppose that the surface of the lake behind Grand Coulee dam is 445 ft above the base of the dam. What is the pressure of the water at the base of the dam? Would a lake 10 miles long behind the dam have twice the pressure as one 5 miles long? What is the approximate total force on 1 square inch of the dam at the base? Now you know why that dam is 500 ft wide at the base and 30 ft wide at the top.
2. A bundle of crushed tin cans weighs 10.5 lb in air and 9.10 lb in water. Calculate the specific gravity of tin.
3. A rectangular flat river barge is floating empty in a river. The barge is 25 ft wide and 105 ft long. How much deeper will it sink into the water when 255 tons of wheat are loaded on it?

Gas Pressure and Altitude

We know that a gas is a fluid but not the same as a liquid. They are both fluids in that they are free to flow and take on the shapes of their containers. A liquid has a definite volume but a gas expands and completely fills its container and has neither a definite shape nor volume. A liquid exerts pressure, and a gas can exert pressure.

We also already know that a gas exerts pressure but in a different way than a liquid. The particles of the gas collide with the sides of the container many times each second. In each collision, the particles and sides exchange impulses. The pressure of a gas is the result of the billions of impulses of the bouncing particles. Another aspect of our study of gases has revealed to us that you can increase the pressure of a gas by placing more particles into the container. And one final bit of knowledge that we've gained is that we know how to measure the pressure of a liquid, and because a gas is a fluid, maybe we can use that knowledge to find a way to measure the pressure of a gas.

Johnnie's Alert

Gases, like liquids, exert pressure in all directions at any given point. The pressure of a gas is the same at all points on a horizontal plane at a given level. That means that everything, even the human body, near the surface of the earth is experiencing 14.7 lb/in^2 of pressure on every square inch of the body.

Sipping Liquids Through a Straw

The atmosphere (atm) is a good source of gas, or a mixture of gases, with which to begin. We found earlier that at sea level the pressure of the atmosphere is 14.7 lb/in^2. Do you know how that was measured? We will find a way to do that, but let's begin with an activity with which you are familiar.

We've all sipped some delicious drink through a straw at some time in our lives. We did that by removing the air from the straw. If you think about it, the liquid doesn't rise up through the straw but is pushed up by the atmospheric pressure on the surface of the liquid outside the straw. When you remove the air inside the straw with your mouth and cheeks, there is little or no opposition to the pressure of the atmosphere, so the liquid is pushed up the straw to fill the region of reduced pressure. That is an idea for measuring the pressure of the atmosphere—measure the pressure of the atmosphere by relating it somehow to the pressure of a liquid.

We live at the bottom of an ocean of air many miles high. Its pressure is considerable and is equivalent to the pressure produced by a column of water ten meters high. This pressure was first measured by the Italian physicist Evangelista Torricelli (1606–1647) in 1644.

Torricelli took a long tube, closed at one end, and filled it with mercury. He then upended it in a dish of mercury. The mercury in the tube poured out of the tube, of course, in response to the downward pull of the gravitational force. There was a counter force, however, in the form of the pressure of the atmosphere against the surface of the mercury in the dish. This pressure was transmitted in all directions within the body of mercury, including a pressure upward into the tube of mercury.

As the mercury poured out of the tube, the mass of the column, and therefore the gravitational pull upon it, decreased until it merely equaled the force of the upward pressure due to the atmosphere. At that point of balancing forces, the mercury no longer moved. The mercury column that remained exerted a pressure (due to its weight) that was equal to the pressure of the atmosphere (due to its weight). The total weight of the atmosphere is, of course, many millions of times as large as the total weight of the mercury, but we are concerned with pressure, which (and we know you know this) is weight per unit area.

It turns out that the pressure of the atmosphere at sea level is equal to that of a column of mercury 30 inches (or 760 millimeters) high; Torricelli had, in effect, invented the barometer. Air pressure is frequently measured, particularly by meteorologists, as so many inches of mercury or millimeters of mercury, usually abbreviated HG or cm Hg, respectively. (Hg is the chemical symbol for mercury.) So 30 (it's really exactly

29.92) inches of Hg or 760 mm Hg is equal to 1 atmosphere. A millimeter of mercury (mmHg) has been defined as 1 torricelli, in honor of you know who. We can then say that one atmosphere is equal to 760 torricellis.

The mercurial barometer is diagrammed in Figure 14.5, showing that 1 atm of pressure on the reservoir of mercury causes the mercury to rise to a height of 760 mm inside the glass tube.

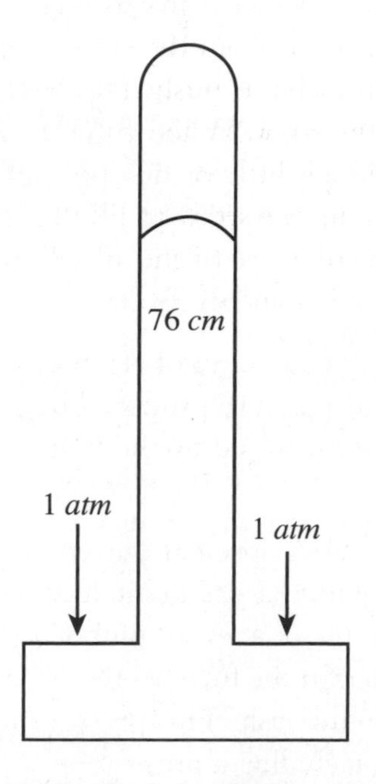

Figure 14.5

The mercurial barometer measures the atmospheric pressure directly.

As we already know, air pressure may also be measured as weight per area. In that case, normal air pressure at sea level is 14.7 pounds of weight per square inch, or 1,033 grams of weight per centimeter ($gm[w]/cm^2$). Expressed in more formal units of force per area, 1 atmosphere is equal to 1,013,300 $dynes/cm^2$. One million dynes per square centimeter has been set to equal 1 bar (from a Greek word for "heavy"), so one atmosphere is equal to 1.10133 bars.

Johnnie's Alert

The greater the distance from sea level, or at increased altitude, the smaller the reading on a barometer. The lower pressures at increased altitudes result from a lower density of air particles and the fact that the weight of the column of air above you is now less, so the pressure is less. The nearly zero density beyond our atmosphere in space is very nearly a vacuum.

Naturally, if it is pressure of the atmosphere that balances the pressure of the mercury column, then when anyone carrying a barometer ascends a mountain, the height of the column of mercury should decrease. As one ascends, at least part of the atmosphere is below, and what remains above is less and less. The weight of what remains above, and therefore its pressure, is lower and so is the pressure of the mercury it will balance.

This was checked by the French mathematician Blaise Pascal (1623–1662) in 1658. He sent his brother-in-law up a neighboring elevation, barometer in hand. At a height of a kilometer, the height of the mercury column had dropped by 10 percent, from 76 to 68 centimeters.

In addition to that, the atmosphere is not evenly distributed about the earth. There is an unevenness in temperature that sets up air movements that result in the piling up of atmosphere in one place at the expense of another. The barometer reading at sea level can easily be as high as 31 inches Hg or as low as 29 inches Hg. (In the center of hurricanes, it may be as low as 27 inches Hg.) These "highs" and "lows" generally travel from west to east, and their movements can be used to foretell weather. The coming of a high (a rising barometer) usually means fair weather, whereas the coming of a low (a falling barometer) promises storms.

After all of this discussion, it's probably worthwhile for us to make a few calculations to clarify your understanding of the quantity called atmospheric pressure.

$$p = \rho_{ww} h_w = (62.4 lb/ft^3)(34 ft) = (2122 lb/ft^2)\left(\frac{1 ft^2}{144 in^2}\right) = 14.7 lb/in^2$$

$$p = \rho_{wm} h_m = \rho_{ww} spgr_m h_m = (980 dynes/cm^3)(13.6)(76.0 cm) = 1.013 \times 10^6 dynes/cm^3$$

The above calculations show us exactly what the discussion in the above paragraph reveals, simply, the way in which you can calculate the pressure of one atmosphere. Remember that 1 atm = 14.7 lb/in^2 = $1.013 \times 10^6 dynes/cm^3$ = 1.013 bars.

Air Pressure and Buoyant Force

The atmosphere exerts pressure, and it also has density. The density of a gas is measured at S.T.P. (standard temperature and pressure = 760 mm of mercury and 0°C), and at S.T.P. the mass density of dry air is $1.29 \times 10^{-3} \frac{g}{cm^3}$. There are times when the density of a gas is $1.29 \frac{g}{l}$ (read 1.29 grams per liter). That means that you need to be familiar with the liter as a measure of volume, too.

We list some defining equations here for you to use in comparing familiar measures of volume with this new measure.

$1ml = 1cm^3$ Read 1 milliliter equals 1 cubic centimeter

$10^3 ml = 10^3 cm^3$ Multiplying each member by 10^3

$1l = 10^3 cm^3$ One liter is equal to 1,000 milliliters

You can use these defining equations to interpret the new unit in terms of a familiar unit of measurement as follows:

$$\left(1.29\frac{g}{l}\right)\left(\frac{1l}{10^3 cm^3}\right) = 1.29\frac{g}{10^3 cm^3} = 1.29 \times 10^{-3}\frac{g}{cm^3}$$

Notice that a defining equation was used to define unity in order to change the looks of $\frac{g}{l}$ to $\frac{g}{cm^3}$.

Because a gas has density, it behaves like other fluids in that objects submerged in a gas experience a buoyant force. Let's consider a safe gas to handle such as helium, and use the buoyant force of air to calculate how much a liter of He, the symbol for helium, will raise or lift. We need the specific gravity of helium or the weight density of helium. Usually it is easy to look up the specific gravity of a solid, a liquid, or a gas in tables in the back of engineering or physics books. But it may be difficult to find the weight density of those substances.

Johnnie's Alert

Properties of gases are often given in terms of S.T.P. 760 mm of mercury and 0°C, and volume is often given in liters. We've given you information in this part of the text that enables you to interpret these quantities in terms of more familiar quantities discussed earlier.

The specific gravity of a gas is defined as the ratio of the weight density of a gas to the weight density of air at S.T.P.

The weight density of helium can be expressed in terms of the specific gravity of helium, which is given as 0.138. Specifically, $\rho_{wHe} = spgr_{He}\rho_{wA}$, or the weight density of helium is equal to the product of the specific gravity of helium and the weight density of air.

Newton's Figs

Buoyant force is calculated for gases the same way it is in liquids. Because they are both fluids, you can say the buoyant force is the weight of the fluid displaced by the object immersed in the fluid.

$$F_B - W_{He} = F_L$$

In words, the equation states that the buoyant force minus the weight of the helium (actually helium and container) is equal to the lifting force. Below are our calculations to show you what the lifting force is.

$$\rho_{wA}V_{He} - \rho_{He}V_{He} = F_L$$
$$\rho_{wA}V_{He} - \rho_{wA}spgr_{He}V_{He} = F_L$$
$$\rho_{wA}V_{He}(1 - spgr_{He}) = F_L$$
$$(1l)\left(1.29\frac{g}{l}\times 980\frac{cm}{\sec^2}\right)(1-0.138) = F_L$$
$$F_L = 1.09 \times 10^3 \; dynes$$

That means that 1 liter of helium will lift 1,090 dynes of weight at S.T.P. How many newtons is that? How many pounds would that be? To help clarify those questions for you, we've listed a couple of mass and weight relations and a defining equation here and made some conversions:

454 grams weigh 1 pound on Earth.

1 kilogram weighs 2.2 pounds on Earth.

Therefore, 9.80 N = 2.2 lb, or 1 pound equals 4.45 N.

$$1dyne = 1\frac{g\cdot cm}{s^2} = \left(\frac{1g\cdot cm}{s^2}\right)\left(10^{-3}\frac{kg}{g}\right)\left(10^{-2}\frac{m}{cm}\right) = 10^{-5}\frac{kg\cdot m}{s^2} = 10^{-5}N$$

$$1dyne = 10^{-5}N = \left(10^{-5}N\right)\left(\frac{2.2lb}{9.80N}\right) = 2.24\times 10^{-6}lb$$

So, to answer the questions we posed above, we have this:

$$1.09 \times 10^3 \; dynes = 1.09 \times 10^3 \times 10^{-5}N = 1.09 \times 10^{-2}N = 2.45 \times 10^{-3}lb$$

How Does a Baseball Curve?

Remember when we talked about the lift of an airplane wing at the end of Chapter 10? The idea is that particles of fluids move at different speeds around or through objects suspended in the fluid, immersed in the fluid, or containing a fluid. Bernoulli's principle helps you to understand the behavior of such objects subjected to a streamline flow of fluid. Remember, the principle states that $\frac{KE}{V} + p$ = constant. That means that particles traveling at a higher speed will have a larger *KE*, so the first term of the equation $\frac{KE}{V}$ is larger then the second term, *p*, which must be smaller. As an example, let's consider the flight of the baseball in Figure 14.6, which shows the ball traveling toward the left of the diagram, indicated with the large arrow.

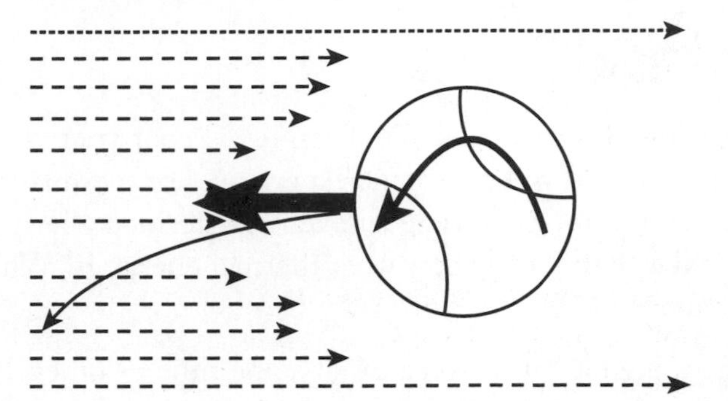

Figure 14.6

The baseball in flight tends to move toward the bottom of the diagram.

The ball experiences a current of air moving toward the right, shown as dotted lines with arrowheads pointing to the right of the diagram. If you swing your arm through the air, you can feel the breeze the ball is experiencing. The ball is spinning counterclockwise as indicated in the diagram. Particles of air near the baseball are rotating counterclockwise with the baseball as the seams drag the air particles along with the spinning ball. The particles near the top of the baseball, that part of the ball near the top of the page, are traveling in a direction opposite the breeze and are slowed down by the opposing motion of the breeze. The particles near the bottom of the baseball are traveling in the same direction as the breeze and are sped up by those particles. That means that the pressure at the bottom is less than the pressure at the top of the baseball, and so the baseball is pushed toward the region of lower pressure. The baseball travels in a curved path as indicated down and to the left in the diagram. The diagram is oversimplified but shows how a baseball curves. The direction of the curve depends on your frame of reference.

> **Newton's Figs**
>
> Think of Bernoulli's principle as being composed of two terms: the first term increases with the speed of particles in a streamline flow, and the second term, pressure, is added to the first term to yield a constant value. If the first term increases because of higher speeds of particles, the pressure must decrease for the sum to be constant.

If the flight of the ball is viewed from above while the ball spins in a horizontal plane, the baseball would curve away from a right-handed batter. If viewed from the side while the ball rotates in a vertical plane, the baseball would curve downward as it approaches home plate. The curved path of the ball is exaggerated in the diagram to make the point but the ball does curve enough so that a curveball with this flight path would be pretty tough to hit.

Pressure and Thermal Energy

Thermal energy is discussed in more detail in Chapter 15, but your general knowledge of the world has provided you with some experiences of thermal energy. If you have ever removed a steamy apple pie from an oven and accidentally spilled some of its savory juice on a naked hand, you know what thermal energy is! What does that have to do with pressure?

You've probably experienced a flat tire on a car at some time or other. If you have a flat tire in some lonely spot and must inflate the tire with a handheld air pump, you not only get a fantastic workout, but you also cause the pump to increase in temperature in the region where the air is forced into the tire. After removing the nail that caused the flat, and inserting enough air into the tire to get you on your way, you probably will have to do the same thing over several times until you reach a service station. The air pump gets hotter each time you use it.

Other examples of using an air pump could be an air mattress, football, basketball, or just about anything that needs to be inflated. The point we're trying to make is that an air pump increases in temperature for a good reason. Because a gas is compressible, the pump takes in air at atmospheric pressure and compresses it, decreasing its volume significantly.

Newton's Figs

Thermal energy is the result of doing work upon a gas when it is compressed. The more work done compressing the gas, the more thermal energy is developed.

To decrease the volume, the particles of the gas must be forced to be closer together. You force the particles to be closer together by doing work on them with the pump. Much of the work done appears in the form of thermal energy. Sure, there is some thermal energy as a result of the friction caused by the operation of the air pump, but most of the thermal energy is generated by the work done to compress the air that is forced into the leaking flat tire. That being said, try out some more problems and then you'll be ready to move on to the next chapter, on heat energy.

Problems for the Budding Rocket Scientist

1. The weight density of water is $62.5 lb/ft^3$. A piece of cork has a specific gravity of 0.25 and weighs 4 lb in air. Find the volume in cubic feet of this piece of cork.

2. Atmospheric pressure is about 100 kPa. How large a force does the air in a room exert on the inside of a window pane that is 40 cm × 80 cm?

3. A gas bubble, whose volume is 33 cm^3 when it is at the bottom of a lake 45.7 m deep, rises to the surface. What is its volume at the surface of the lake if a mercury barometer stands at 762 mm?

4. In a hydraulic press the large piston has cross-sectional area $A_1 = 200cm^2$ and the small piston has cross-sectional area $A_2 = 5cm^2$. If a force of 250 N is applied to the small piston, what is the force F_1 on the large piston?

The Least You Need to Know

- One of the most fundamental principles in the study of hydraulics is that pressure is exerted in all directions in a fluid.
- The pressure of a liquid varies directly as the depth, and the constant of proportionality is the weight density of the liquid.
- As long as the ratio of the amount of the object submerged to the total amount of the object is less than one, the object will float. That will happen as long as the weight density of the object is less than the weight density of the liquid.
- Gases, like liquids, exert pressure in all directions at any given point.
- The specific gravity of a gas is defined as the ratio of the weight density of a gas to the weight density of air at S.T.P.

Chapter 15

Heat Energy

In This Chapter

- What exactly is heat?
- Common scales of temperature
- The laws of thermodynamics
- Phase changes and latent heat
- Heat transference

In Chapter 14, you were introduced to some of the basic principles found in hydraulics, the study of the movement of liquids. Now we explore the fundamentals of thermodynamics, the study of the movement of heat. More specifically, thermodynamics is the field of physics that studies the properties of systems that have temperature and involve the flow of thermal energy from one place to another. Within this context, we examine the laws that govern the conversion of energy from one form to another, the direction it will flow, and the availability of that energy to do work. In other words, you'll become familiar with some of the laws of thermodynamics.

Temperature and Thermometers

One of the central principles of *thermodynamics* (and one of the most noticed aspects of the weather) is *temperature*. *Heat* and cold are two names given to how we perceive temperature. They are imprinted on our consciousness at a very early age, and we can usually tell the difference between the two very easily. In some cases, however, the difference isn't clear. If you were blindfolded and touched with a hot iron and with a piece of dry ice, you'd have a very hard time distinguishing them. This example demonstrates that physiological response isn't a dependable method for measuring temperature.

Galileo invented the first scientific instrument for measuring temperature in 1592. He used a glass flask with a very narrow neck that was half filled with colored water. He placed the flask upside down into a bowl of colored water. When the temperature changed, the air in the flask would expand or contract. The column of water in the neck would then move up or down. This invention was called a thermoscope. (*Thermos* is the Greek word for heat, which is why you use a thermos to keep your coffee hot.) It didn't have a temperature scale to give a measurement of how warm or cold it was, so it wasn't called a thermometer. That invention would come almost half a century later, around 1640.

A group of scientists in Italy built the first thermometer. This group's prototype of the modern thermometer used mercury in a partially sealed tube. This thermometer utilized the expansion and contraction of gases. Thus, it was natural that some of the first studies of heat dealt with the nature of gases. That's one of the reasons we studied them in previous chapters.

def•i•ni•tion

Thermodynamics involves the study of the reversible transformation of heat into other forms of energy, such as mechanical energy, and also covers the laws governing those transformations. **Heat** is a transfer of energy from one body to another as a result of a difference in temperature or phase change. **Temperature** is the property of a body or region of space that determines whether or not there will be a net flow of heat into or out of it from a neighboring body or region—and in which direction (if any) the heat will flow. If there is no heat flow, the bodies or regions are said to be in thermal equilibrium and at the same temperature. If there is a heat flow, the direction of the flow is from the body or region of higher temperature.

The Mystery Behind Heat

Heat had never been thought of as a measurable quantity until a Scottish physician named James Black (1728–1799) came along. Black thought that heat was a colorless, invisible fluid that was able to permeate or penetrate whatever it came in contact with. He called this liquid *calor*, which is Latin for "heat."

If you mix a gallon of boiling water with a gallon of ice water, the temperature of the mixture will be halfway between the two temperatures you started with. Black thought that when the two waters mixed, the calor in the hot water was equally distributed between the two portions. After a number of experiments in which Black mixed different liquids at different temperatures, he defined a unit of heat as the amount of heat it takes to raise the temperature of one pound of water by 1 degree Fahrenheit. Today, that unit is called a calorie and is equal to the amount of heat it takes to raise 1 gram of water 1 degree Celsius. (The kind of calorie that you would eat would raise 1,000 grams of water 1 degree Celsius and is called a kilocalorie; but they are still units of energy and the energy content of food could just as easily be measured in joules.) We talk more about calories and different scales of temperature later in this chapter.

To illustrate how heat acts like fluid, a Frenchman by the name of Sadi Carnot compared a water wheel to a steam engine. For a water wheel, as water falls over a wheel, the wheel turns. In a steam engine, the hot steam flows through the engine, turns a shaft, and comes out cooler. So the water fell from a high point to a low point and in the process, turned the water wheel. The calor, or heat fluid, fell from a high temperature to a low temperature to turn the steam engine shaft.

Although Carnot was getting close to understanding what heat was, it just plain wasn't a fluid. A steam engine transforms some of the heat that flows through it into mechanical energy, and the amount of heat that comes into the condenser is less by the amount of heat that is thus transformed. In other words, the amount of heat that is no longer found in the condenser is the amount that was used to do the work of the steam engine. Heat can do work!

A man by the name of Benjamin Thompson grew up in Massachusetts during the Revolutionary War. He was the first to realize that heat is a form of internal motion, not a fluid as James Black and Sadi Carnot thought it was. Thompson became the minister of war of Bavaria and was given the title of Count Rumford for training and reorganizing the German army. As the director of the arsenal, he supervised the boring of cannons for the army. He noted that the water that was used to cool the cannon during the boring process continually needed to be replaced, because the boring tool boiled it away. What was causing the cannon to get so hot? Heat was being introduced into the boring tool by the friction caused by the movement of the boring tool against the metal cannon; heat wasn't being transferred as a fluid from some other source.

Someone you've already met in an earlier chapter, James Prescott Joule, developed the ideas of Count Rumford in the 1840s. Joule performed a number of experiments with an ingenious setup. He attached some weights to a string that ran over a pulley to a paddle wheel inside an insulated container of water. Figure 15.1 gives you an idea of what his setup looked like. As the weights fell, the paddle wheel turned, and the friction caused by this turning heated up the water. The temperature change corresponded to the amount of heat put out into the water. Joule discovered that the same amount of heat was put into the water whenever the same weights fell the same distance.

Figure 15.1

Joule's experiment on the transformation of mechanical energy into heat.

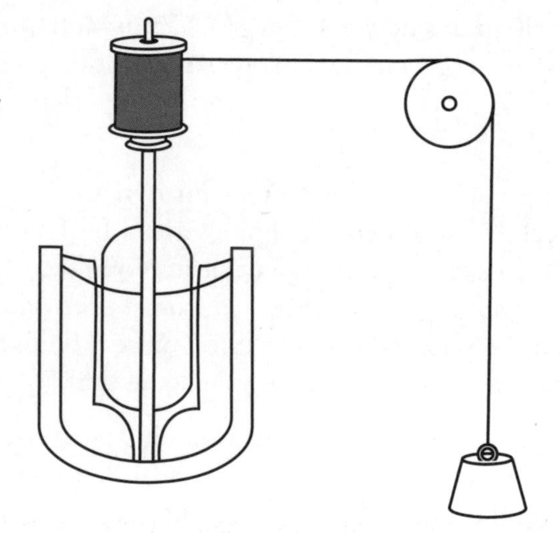

Before Joule released the weights that turned the paddle wheel, the weights had potential energy. They were ready to fall. As the weights fell and turned the paddle wheel, the potential energy was converted into kinetic energy and from there into heat. During the entire transformation process, guess what happened? Does the conservation of energy ring a bell? Yes, no energy was lost. The energy was conserved. And although you've already been introduced to the law of conservation of energy, you'll now also know it as a version of the *first law of thermodynamics*. And if you remember your laws, this simply states that the total energy content of a closed system remains constant. The second law of thermodynamics will explain what goes on inside open and closed systems, but that's coming up later in the chapter.

def•i•ni•tion

The **first law of thermodynamics** states that adding heat to a system or doing work on it results in an increase in the internal energy of the system; conversely, if the system does work or if heat is removed from it, its internal energy decreases. The basic idea here is that energy must be conserved when heat (energy) is added to or removed from a system. In a simple sense, the first law of thermodynamics is a version of the law of conservation of energy.

Temperature Scales

There are three scales in which temperature is generally measured, the *Celsius* (c) scale, the *Fahrenheit* (F) scale, and the *Kelvin* (K) scale. In the United States, temperatures are generally reported in degrees Fahrenheit (°F), whereas in the rest of the world they are reported in degrees Celsius (°C). As with units of length and weight, the United States has largely ignored the metric system used by the rest of the world.

def•i•ni•tion

The **Fahrenheit** scale, created by Daniel Fahrenheit, is based on the melting point of ice and the temperature of the human body and is the most popular temperature scale in America. Anders Celsius created the **Celsius** or Centigrade scale in 1742. It's based on the division of 100 equal degrees, with 0 being the freezing point of water and 100 being the boiling point. Most scientists and most of the world use this scale. The **Kelvin** or absolute scale is based on the existence of absolute zero—the temperature at which, in theory, atomic motion stops.

On the Celsius scale, used in scientific work, at a pressure of 1 atmosphere, the temperature of a mixture of water and ice (the freezing point) is designated as 0 degrees (°C), and the temperature of water boiling to produce steam (the boiling point) is 100 degrees (100°C). The temperature range between these two marks is divided into 100 equal parts, each being equal to 1 degree Celsius (1°).

The system more familiar to U.S. readers is the Fahrenheit scale. In this system, the temperature of the freezing point of water is 32 degrees (32°F), and the temperature of the boiling point of water is 212 degrees (212°F). One can easily convert between these two temperatures using the following formulas. Use this equation if you want to convert from Celsius to Fahrenheit:

$$\text{(degrees) } F = \frac{9}{5} \text{ (degrees) } C + 32$$

And to convert from Fahrenheit to Celsius use this formula:

$$\text{(degrees) } C = \frac{5}{9} \text{ (degrees) } F - 32$$

And an important point to remember is that if the temperature on either scale is below zero, place a minus sign in front of that temperature in the equation. Let's use an example so you can see just how the formula works. Suppose we want to know the temperature at which carbon dioxide freezes (also known as dry ice) in degrees Fahrenheit. If it freezes at –80°C, what is the temperature equivalent in degrees Fahrenheit? Using the conversion formula given, and inserting the temperature as –80°C,

$$°F = \frac{9}{5} (-80)\ C + 32 = -122°F$$

So the freezing temperature of carbon dioxide is –122°F. And the fact that frozen carbon dioxide has been found at the north and south poles of Mars should tell you something about the surface temperatures of that planet.

The other temperature scale most commonly used in science is called the Kelvin or absolute scale. This scale is based on the existence of absolute zero, the temperature at which, in theory, atomic motion would stop. The zero point on the Kelvin scale, abbreviated 0 K, is set at absolute zero. The graduations on the Kelvin scale are the same size as a degree on the Celsius scale; only the zero point is different.

Fundamentally, temperature is a measure of how much atoms or molecules are jiggling around, how much random motion they have. Atoms and molecules that have a lot of random motion are hotter than those that have less, and when random motion of a given element gets sufficiently small (that is, that the material is sufficiently cold), then it may be able to solidify or "freeze" into an amorphous or crystalline solid. Now if this random jiggling stopped entirely, the material would be at 0 K. At this temperature, a gas would also exert no pressure.

Absolute zero is the lowest temperature possible in the universe—in principle there is no upper limit to temperature. As a result, on the Kelvin scale there are no negative temperatures. The freezing point of water on the Kelvin scale is 273 K, and the boiling point is 373 K. By international agreement, the increments on the Kelvin temperature scale are simply called kelvin; a reading in Kelvin can be converted to degrees Celsius simply by adding 273. For example, a summer day's temperature of 27°C would be 300 kelvin (27 + 273), or 300 K. Similarly, if you wish to convert from Kelvin to Celsius, simply subtract 273 from the Kelvin reading to arrive at a Celsius reading. A temperature of 373 K is what in Celsius? The answer is (373 – 273) = 100°C. That also tells us the boiling point of water on the Kelvin scale is 373 K, which is the same as the boiling point on the Celsius scale of 100°C.

Johnnie's Alert

Water has many properties that make it a unique substance. When liquid water releases heat, it becomes denser until it reaches about 4°C, and then gets less dense as it continues to release heat. As water changes to ice, it continues to release thermal energy. The ice expands with such a tremendous force that it can break the blocks of cars that are not properly protected by antifreeze. The expansion is so drastic that ice floats in water even though most of the ice is below the surface. This is not typical of most solids in contact with their liquids. Most materials are denser in their solid state than in their liquid state.

Specific Heat and the Second Law of Thermodynamics

Heat is thermal energy that can be transferred between two bodies that are at different temperatures. Heat naturally flows from a body at a higher temperature to a body at a lower temperature. The amount of thermal energy that a body contains depends on two quantities, its temperature and the amount of material it contains, that is, its mass. A tiny speck of molten metal may have less thermal energy than a bucket of hot water. And the molten speck, as a result, will cool off much more quickly. What we're talking about here is something called *heat capacity*.

def•i•ni•tion

The **calorie** is the quantity of heat needed to raise the temperature of 1 gram of water 1 Celsius degree. The **kilocalorie** is the quantity of heat required to raise the temperature of one kilogram of water one Celsius degree. The **British thermal unit** is the quantity of heat required to raise the temperature of 1 pound of water 1 Fahrenheit degree. The **heat capacity** of a body is the quantity of heat required to raise its temperature 1 degree. The **specific heat** of a substance is the ratio of its heat capacity to its mass or weight.

Objects that take longer to increase their temperature also take longer to cool off than others do. This property of a substance is called its heat capacity. The heat capacity of a substance is the heat necessary to raise the temperature of a material one degree. Heat capacity does not depend on the type of material or the mass of the substance.

The amount of heat required to raise the temperature of a substance can be calculated if you know the specific heat of the substance. In general, the amount of heat required to change the temperature of a substance is proportional to the mass of the substance and the change in temperature, according to the following relation:

$$Q = mc\Delta T$$

where Q is the heat required (in joules or calories), m is the mass of the substance, c is the specific heat of the substance, and ΔT is the change in temperature. Specific heat is usually represented with units $\frac{kcal}{kgC^\circ}$, $\frac{cal}{gC^\circ}$, $\frac{Btu}{lbF^\circ}$.

We now have a way to calculate a quantity of thermal energy. Let's suppose you have a 5.00-g piece of aluminum, $c = 0.214 \frac{cal}{gC^\circ}$, that is warmed up from 5.00°C to 55.0°C. How much heat is required to cause the change in temperature?

$$Q = c_{al}m_{al}(t_f - t_0), \text{ and } Q = (0.214 \frac{cal}{gC^\circ})(5.00g)(55.0^\circ - 5.00^\circ) = 53.5\ cal.$$

The problem you just completed had the change in temperature as part of the information given. You know that thermal energy leaves objects of higher temperature and enters objects of lower temperature until their temperatures are the same. In fact, that process is formalized in the *second law of thermodynamics.*

def•i•ni•tion

The **second law of thermodynamics** states that heat flows spontaneously from a hot body to a cold body. It is a statement about the natural direction in which heat (energy) flows. This law is also called the **law of entropy,** because in general, natural systems proceed to states of greater disorder. **Entropy** is a measure of the amount of disorder in a system—flowers whither, dishes break, and mountains erode—and the entropy of a system increases with time.

The Law of Entropy

In an open or unbounded system, the total amount of energy is constantly changing. It's just the opposite of a closed system. And within the boundaries of a closed system the *law of entropy* rules. *Entropy* simply means disorder or chaos. Within a closed or isolated system, entropy can only increase or occasionally remain constant. An egg in your hand can be an example of an ordered system; an egg smashed on the floor is an example of a disordered system. The entropy of the broken egg is higher than the entropy of the whole egg. The nature of entropy lies at the core of the second law of thermodynamics, which states that the entropy in a closed system always increases. Entropy increases as a function of time. In other words, the future is the direction in which entropy increases. It can't go the other way. This brings us to one of the central paradoxes in physics.

Newton's second law shows us that if we know the initial position of an object, as well as its mass and velocity, we can predict its future position. Using this law, we can predict the future behavior of an entire closed system. The interesting thing is that you can do this in reverse. If you know where an object is now, given the same information as before, you can also find out where it was in the past. For example, astronomers use this law to show the position of heavens at a certain time in the past. You can show the exact position of everything in the sky, say 3,000 years ago. Newtonian mechanics allows you to know how a system was in the past or what it's most likely to be like in the future.

If the molecules and atoms of an egg or piece of wood obeyed Newton's laws, they would distinguish between the future and the past. Yet did you ever see a broken egg become whole or a piece of burnt wood unburn? Of course not. Nature is not reversible, and that brings us to the paradox. You can't derive the second law of thermodynamics from Newton's laws. They are irreconcilable. Newton's mechanics are time reversible, but the second law of thermodynamics isn't. The difference might not seem like any big deal, but Newtonian mechanics and the laws of thermodynamics are the two cornerstones of classical physics, and the fact that you can't derive one law from the other makes for an incomplete picture in what's basically a very complete system.

In any system in which heat is converted into another form of energy, such as mechanical or electrical energy, there is always a loss of heat. This lowers the efficiency of the system. This lower efficiency is entropy in action. And as you know, this running down can't be reversed, so entropy continues to increase. In all natural processes, the organized motion of molecules has a tendency to become disorganized or random.

Okay, now that you have an understanding of the second law of thermodynamics and entropy, let's continue on and look at some more aspects of heat and how we can calculate a quantity of thermal energy other than with specific heat.

Latent Heat of Vaporization and Fusion

As we have mentioned, the phases of matter are solid, liquid, and gas. The phase of a material depends on two factors: its internal energy and the external pressure. Perhaps surprisingly, when substances change phase (for example, from a solid to a liquid), they absorb or give off energy in the form of heat, yet there is no temperature change. Thermal energy can be added to a solid and the temperature of the solid rises until it reaches a certain characteristic temperature. At that temperature, the solid changes to a liquid. The temperature at which this happens is called the *melting point.* The thermal energy continues to be absorbed, but the temperature does not increase. That must mean that the internal kinetic energy is not increasing. Where is the energy going, then?

Remember those little springlike forces that were holding the solid together? Those attractions are being stretched. Work is being done. The internal potential energy of the material is increasing just as it does if you pull on a spring and stretch it. The process of changing from a solid to a liquid is called *fusion.* Crystalline solids, such as ice, have a definite melting point, and the *freezing point* is the same temperature. A liquid becomes a solid during a process called *solidification.*

def•i•ni•tion

The **melting point** is the temperature at which a solid changes to a liquid. The process of changing from a solid to a liquid is called **fusion** or melting. Fusing in this case means the same as when it is used to mean the melting of metals together to make an alloy. The **freezing point** is the temperature at which a liquid changes to a solid. **Solidification,** or freezing, is the process of changing from a liquid to a solid.

Johnnie's Alert

The melting point and freezing point is the same temperature for most crystalline substances. Noncrystalline substances do not have a definite melting or freezing point. Such substances freeze slowly and melt slowly at no specific temperature.

The temperature at which solidification takes place is called the freezing point. The melting point of ice is 0°C and the freezing point of water is 0°C, but you know that even though the temperature is the same the processes are very different.

To change ice from your freezer to a liquid, in other words to melt ice, heat must be added to the ice until it is at 0°C; then you must continue to add thermal energy to the ice until it changes to water at 0°C. The amount of thermal energy you must add to ice at 0°C to change it to water at the same temperature is called the latent heat of fusion, or just the heat of fusion. Each crystalline solid has a characteristic heat of fusion. The heat of fusion for ice is 80 $\frac{cal}{g}$ or 144 $\frac{Btu}{lb}$.

That means that you must add 80 cal of thermal energy to every gram of ice at 0°C to change it to water at 0°C or 144 Btu to every pound of ice at 32°F to change it to water at 32°F.

In like manner, water must be cooled until it reaches 0°C, and then 80 cal of thermal energy must be removed from each gram of water to change it to ice at 0°C. Notice that you use the specific heat of ice to find the quantity of heat necessary to warm it up to the melting point, then calculate the quantity of thermal energy necessary to change it to water by using the heat of fusion. Similarly, you can use the specific heat of water to find the quantity of heat required to cool the water to the freezing point, and then determine the quantity of thermal energy necessary to change it to ice by using the heat of fusion. Let's see how we can apply this knowledge to a practical problem.

Johnnie's Alert

The heat of fusion is the quantity of thermal energy required to change a solid to a liquid or a liquid to a solid without changing the temperature of the final substance.

Suppose you have 50.0 g of ice, specific heat 0.500 cal/g C°, at –20.0°C and you change it to water at 20°C. How much thermal energy is required for the job?

Solution:

$m_i = 50.0g$

$c_i = 0.500 cal/gC°$

$t_i = -20.0°C$

$m_w = 50.0g$

$c_w = 1.00 cal/gC°$

$t_w = 20.0°C$

$L_f = 80 cal/g$ Latent heat of fusion

$Q = ?$

$Q = m_i c_i (0°C - t_i) + m_i L_f + m_w c_w (t_w - 0°C)$

In verbal summary: the quantity of thermal energy equals the amount of heat required to warm the ice to the melting point plus the amount of thermal energy required to change state plus the amount of heat required to warm the water from the melting point to the final temperature.

$$Q = (50.0g)(0.500cal/gC^{\circ})(20.0C^{\circ}) + (50.0g)(80cal/g) + (50.0g)(1.00cal/gC^{\circ})(20.0C^{\circ})$$

$$Q = 500cal + 4000cal + 1000cal = 5500cal$$

The calculation of the required thermal energy in this type of problem is straightforward, and the algebraic statement of the problem provides a good bookkeeping plan to make sure you don't leave anything out. We included the temperature at the melting point in the algebraic statement as a reminder that the ice becomes water at the melting point, and the water is warmed up from the melting point to the final temperature given in the problem. We know you can't wait to work some of these problems on your own, so here ya go.

Physics Phun

1. Calculate the final temperature of a mixture of 8.00 g of ice at 0.00°C and 60.0 g of water at 90.0°C.
2. Calculate the quantity of thermal energy required to change 15.0 lb of ice at 20.0°F to water at 40.0°F.

You can calculate the quantity of heat needed to change the temperature of a material using specific heat if you want to warm a substance up or cool it off. You can also calculate the quantity of thermal energy necessary to change a solid to a liquid or a liquid to a solid. If you use specific heat, remember that the quantity of heat depends on the substance, the mass of the substance, and the change in temperature. Calculating the quantity of thermal energy for changing from a solid to a liquid depends upon the substance, its heat of fusion, and its mass—there is not a change in temperature.

One other quantity of thermal energy you can calculate involves the changing of state (called a phase change, too) from a liquid to a gas or from a gas to a liquid. The new quantity of thermal energy depends on the type of material and the amount of it you have. This particular change of state takes place at the boiling point.

A liquid must absorb heat until it reaches its boiling point. Then you must continue adding thermal energy to the liquid at the boiling point until the liquid changes to a gas or a vapor at the boiling point. The process is called *boiling*. Changing from a gas to a liquid is just the reverse process and is called *condensation*.

def•i•ni•tion

Evaporation is the process of a liquid changing to a gas or a vapor. **Boiling** is the process of a liquid changing to a gas or a vapor at its boiling point. **Condensation** is the process of a vapor or a gas changing to a liquid. The **heat of vaporization** is the quantity of thermal energy required to change a unit mass or weight of a liquid to a gas or vapor at the normal boiling point.

In this reverse process, the gas is cooled off until it reaches the boiling point. By continuing to remove thermal energy from the gas or vapor at the normal boiling point (the boiling point at standard pressure), you cause it to condense, and it becomes a liquid at the boiling point. Calculating the quantity of heat removed from the gas to cool it to the boiling point requires the use of the specific heat of the gas.

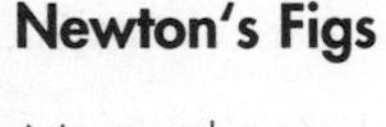

Newton's Figs

Notice that you can convert $\frac{cal}{g}$ to $\frac{Btu}{lb}$ by multiplying the magnitude of $\frac{cal}{g}$ by 1.8. If you make the conversion for yourself using unit analysis, you will discover this relationship. So $80 \times 1.8 = 144$, that is, $80\ \frac{cal}{g} = 144\ \frac{Btu}{lb}$ and $540 \times 1.8 = 970$; that means that $540\ \frac{cal}{g} = 970\ \frac{Btu}{lb}$.

The quantity of thermal energy required to change a liquid to a vapor or a gas is called the latent heat of vaporization, or *heat of vaporization*. The actual process during which a liquid changes to a vapor or gas is called *evaporation*.

The heat of vaporization is the quantity of thermal energy required to change a unit mass or weight of a liquid to a vapor or gas at the normal boiling point. The heat of vaporization for water is 540 cal/g or 970 Btu/lb. That means that when the temperature of water is increased to the normal boiling point of 100°C, you must continue adding thermal energy to change the water to steam at the same temperature. The normal boiling point for water is 100°C or 212°F at 1 atmosphere of pressure. Now let's work through another problem:

Suppose you have 50.0 g of water at 50.0°C, and you want to change it to steam, specific heat 0.500 cal/gC°, at 110°C. How much thermal energy is required to complete the task?

Solution:

$m_w = 50.0g$

$c_w = 1.00cal/gC^\circ$

$t_w = 50.0^\circ C$

$L_v = 540cal/g$ Latent heat of vaporization

$c_s = 0.500cal/gC^\circ$

$t_s = 110^\circ C$

$Q = ?$

$$Q = m_w c_w (100^\circ C - t_w) + m_w L_v + m_s c_s (t_s - 100^\circ C)$$

$$Q = (50.0g)(1.00cal/gC^\circ)(100^\circ C - 50.0^\circ C) + (540cal/g)(50.0g) + (50.0g)(0.500cal/gC^\circ)(110^\circ C - 100^\circ C)$$

$$Q = 2500cal + 27000cal + 250cal = 29800cal$$

We now have three ways in which we can calculate a quantity of thermal energy. Specific heat is used to calculate the heat required to warm or cool a substance. The heat of fusion, L_f, enables you to find the quantity of thermal energy required to change a solid to a liquid or a liquid to a solid at the melting point. Finally, the heat of vaporization, L_v, is used to determine the quantity of thermal energy required to change a liquid to a gas or to change a vapor or a gas to a liquid at the normal boiling point. Remember that the temperature does not change when a change of state occurs.

Johnnie's Alert

Many people lose terms in their calculation of a quantity of thermal energy by not relying on an algebraic statement of the problem. The algebraic statement is much like a recipe for your favorite dish—if you leave out an important ingredient, the whole thing is garbage.

It's a good idea to write an algebraic statement for the quantity of thermal energy to provide a plan to follow when you calculate the thermal energy involved. The algebraic statement can help you to see easily each quantity of thermal energy required. It also provides you with a method of accounting for each quantity of thermal energy so that every quantity of thermal energy required is included in your calculations. You know that the solids and liquids considered so far are in some kind of container. We haven't included a container before, but we'll do so in the next example so that you have a more realistic picture of handling some of the substances. Steam can be contained, too; the procedure for working with steam in a container is the same as that used for containing ice and liquids. So let's take a look at a problem that involves exactly that.

You are given ice in a container and both are at –15.0°C. The mass of the ice is 55.5 g and the aluminum container, specific heat 0.214 cal/gC°, has a mass of 88.4 g. The ice remains in the container as it is changed to steam. How much thermal energy is required to change the ice to steam?

Solution:

$t_i = -15°C$

$t_c = t_i$

$m_i = 55.5g$

$c_i = 0.500cal/gC°$

$c_w = 0.500cal/gC°$

$m_{al} = 88.4g$

$c_{al} = 0.214cal/gC°$

$L_f = 80cal/g$

$L_v = 540cal/g$

$Q = m_i c_i (0°C - t_i) + m_{al} c_{al} (0° - t_i) + m_i L_f + m_i c_w (100°C - 0°C) + m_{al} c_{al} (100°C - 0°C) + m_i L_v$

$Q = (55.5g)(0.500cal/gC°)(15°C) + (88.4g)(0.214cal/gC°)(15°C) + (55.5g)(80cal/g) + (55.5g)(1.00\ cal/gC°)(100°C) + (88.4g)(0.214cal/gC°)(100°C) + (55.5g)(540cal/g)$

$Q = 416.25cal + 283.764cal + 4440cal + 5550cal + 1891.76cal + 29970cal = 42600cal$

Physics Phun

1. Suppose you have 16.0 lb of steam at 212°F and it condenses, cools, and changes to ice at 32°F. How much thermal energy does this process release? Hint:
$$L_v = 970\ \frac{Btu}{lb},\ L_f = 144\ \frac{Btu}{lb},\ c_w = 1\ \frac{Btu}{lbF°}.$$
2. You are given 46.0 g of ice at –25.0°C, held by a 75.0-g copper container, specific heat 0.0921 cal/gC°, and it is to be changed to steam at 100°C. How much thermal energy is required? Hint:
$$c_{ice} = 0.500\ \frac{cal}{gC°},\ c_{water} = 1.00\ \frac{cal}{gC°},\ L_v = 540\ \frac{cal}{g}.$$
3. An aluminum container weighs 0.50 lb and holds 2.00 lb of water at 80.0°F. What is the final temperature of the mixture when .046 lb of steam at 212°F is added to the container of water?

Transfer of Heat

There are three ways in which thermal energy can be transferred: *conduction*, *convection*, and *radiation*. We sometimes hear the weatherman discuss the weather in terms of these different methods of thermal energy transfer.

def•i•ni•tion

Conduction is the transfer of thermal energy within a substance from one particle to the next while the particles are not moved from one place to another. Thermal energy causes the particles to vibrate more and the vibrating particles bump into neighboring particles, transferring the energy throughout the object. **Convection** is the transfer of thermal energy by the movement of matter. **Radiation** is the transfer of thermal energy by having only the energy transferred. No substance or convection currents are needed.

Transfer of Heat by Conduction

Have you ever held a needle in the flame of a match to purify it before removing a splinter from your finger? If you have, you probably have also experienced a painful finger that held the needle in the flame because it gets very hot. The thermal energy from the match is transferred to your finger through the needle. This type of transfer of thermal energy is called conduction. The collision between molecules in the substance transfers their energy to neighboring molecules, sending the thermal energy throughout the object.

All materials conduct heat at different rates, and metals are some of the best conductors. Stone is a moderately good conductor, and wood, paper, cloth, and air are relatively poor conductors. Materials that conduct heat poorly are called insulators. Air, for example, makes an excellent insulation material when it is trapped between spaces, as in double-pane windows that trap air between the two panes of glass. And vacuum bottles, like a thermos, contain a double-walled inner liner that is pumped clear of air then sealed. Thermal energy cannot be transferred through a vacuum barrier.

Transfer of Heat by Convection

Because of their low density and smaller number of molecular and atomic collisions, most liquids and gases are poor conductors, but they do transfer heat another way: through convection, which is the movement or circulation of parcels of heated liquid or gas. You know that hot air rises because it is less dense and cold air rushes down to take its place. Cold water entering your hot water heater moves to the bottom of the tank, where it is heated. The warm water in the tank rises and supplies warm or hot water to your home. The circulating motion of the water in the tank is an example of the transfer of heat by convection. The air that is warmed in your fireplace rises up the chimney, carrying away the smoke and much of the thermal energy in the warm air.

Transfer of Heat by Radiation

Some of the thermal energy from the fireplace is transferred to the room to warm you when you stand in front of it. The fire does not warm you by conduction because that usually means contact with the source. There is warming by convection because of the rising hot air and falling cold air, but that is not what causes you to get too hot if you stand in front of the fire very long. The fire in the fireplace warms you by a method of transfer of thermal energy called radiation. No substance moves for the transfer of thermal energy by this means to be effective.

Radiation is simply the transfer of energy from one point in space to another through the oscillation of electromagnetic fields. The motion (jiggling) of the electrons in the object that is emitting radiation generates electromagnetic waves. These waves are transmitted through empty space, and when they strike your body, the electrons in your body absorb the radiation; the atoms in your body start to move around at a greater velocity, and you warm up. The electrons act like a pebble tossed into a lake. We may detect the waves that the pebble makes far from where the pebble enters the water. In a similar fashion, we can see the electromagnetic waves caused by the jiggling electrons, even though the electrons doing the jiggling are far away.

In this way, electromagnetic waves can carry energy between two points that are not connected physically. So, unlike conduction and convection, radiation doesn't require direct contact between two substances to transfer heat.

How Will the Universe End?

In this last but brief section of our look at heat energy, we discuss some of the implications of thermodynamics and how they relate to our universe.

You know that the second law of thermodynamics states that heat is transferred from hotter to cooler substances, the basis for the law of entropy. The implication of this law has a direct connection to our universe as a whole. If the universe is a closed system and there is nothing outside of it, then there is a fixed amount of energy within it. The stars represent high concentrations of energy and reflect an ordered system. But you know what happens to order in a closed system: entropy increases. Eventually, after billions of years, all the stars will cool and burn out. There won't be enough heat energy left in the universe to sustain any life. This is known as the heat death of the universe, sometimes called the "Big Chill," as well. If, on the other hand, it is an open system, order can increase and entropy decrease, because there is input coming from somewhere else that will help bring this about. But like many things in physics, our understanding of the universe is always changing. So who knows?

Problems for the Budding Rocket Scientist

1. It is a summer day in Phoenix and the temperature is 100°F. What is the temperature in degrees Celsius? In Kelvin?
2. Which requires more energy—raising the temperature of a 1-kg block of aluminum (c = 920 J/kg°C) by 20°C, or raising the temperature of 500 g of water by 5°C? (c = 4,186 J/kg°C)
3. How many Btu are removed in cooling each of the following from 212 to 68°F? (a) 1 lb of water, (b) 2 lb of leather (c = 0.36 cal/gC°), (c) 3 lb of asbestos (c = 0.20 cal/gC°).

The Least You Need to Know

- Thermodynamics involves the study of the reversible transformation of heat into other forms of energy, such as mechanical energy, and also covers the laws governing those transformations.
- The first law of thermodynamics states that the total energy content of a closed system remains constant.
- There are three scales in which temperature is generally measured: the Celsius (c) scale, the Fahrenheit (F) scale, and the Kelvin (K) scale.
- In any system in which heat is converted into another form of energy, such as mechanical or electrical energy, there is always a loss of heat.
- The phase of a material depends on two factors: its internal energy and the external pressure.
- There are three ways in which thermal energy can be transferred: conduction, convection, and radiation.

Chapter 16

Sound Energy

In This Chapter

- Early concepts of sound
- A wave model for sound
- How fast does sound travel?
- The speed of sound and temperature
- Resonance of sound

At a comparatively early stage in the quest for knowledge, sound came to be thought of as resulting from a kind of wave motion. The ancient Greeks conducted the first experiments on sound, and these were rather remarkable in one way, for the study of sound was one branch of physics in which the Greeks seemed, by modern criteria, to start off in the right direction from the very beginning. Let's take a brief look at how sound was understood by some of these ancient thinkers before we move into a more modern analysis.

Vibrating Strings

As early as the sixth century B.C.E., Pythagoras of Samos (yes, the same Pythagoras who gave us his theorem) was studying the sound produced by plucked strings. It could be seen that a string vibrated when plucked. The plucked string's motion was only a blur, but even so, certain facts about that blur could be associated with sound. The width of the blurred motion seemed to correspond to the loudness of the sound. As the vibration died down and the blur narrowed, the sound grew softer. And when the vibration stopped, either by natural slowing or by an abrupt touch of the hand, so did the sound. Furthermore, it could be made out that shorter strings vibrated more rapidly than longer ones, and the more rapid vibration seemed to produce a shriller sound.

By 400 B.C.E., Archytas of Tarentum, a member of the Pythagorean school, was suggesting that sound was produced by the striking together of bodies—swift motion producing high pitch and slow motion producing low pitch. By about 350 B.C.E., Aristotle was pointing out that the vibrating string was striking the air; and that the portion of the air that was struck must in turn be moved to strike a neighboring portion, which in turn struck the next portion, and so on. To Aristotle, then, it seemed that air was necessary as a medium through which sound was conducted, and he reasoned that sound would not be conducted through a vacuum. Which he was absolutely correct about. (And Hollywood outer space explosions are totally wrong about.)

Because in rapid rhythm a vibrating string strikes the air not once but many times, not one blow, but a long series of blows must be conducted by the air. The Roman engineer Marcus Vitruvius Pollio, writing in the first century B.C.E., suggested that the air did not merely move, but vibrated, and that it did so in response to the vibrations of the string. It was these air vibrations, he held, that we heard as sound.

Finally, around 500 C.E., the Roman philosopher Anicius Manilius Severinus Boethius made the specific comparison of the conduction of sound through the air with the waves produced in calm water by a dropping pebble. Although this analogy has value, and although water waves can be used to this day (and often are in science classes) to serve as a preliminary to a consideration of sound waves, there are nevertheless important differences between water waves and sound waves.

Where Does Sound Come From?

Are you familiar with the question about a tree falling in the forest and whether there is sound resulting from the falling tree if no one is there to hear it? (That's a typical first-year philosophy question.) The physiologist might tell you that to have sound you must have a source, a medium to transmit the sound, and a receiver of sound. He might say that because there is no receiver, there is no sound. The physicist might say that sound is a special disturbance of matter to which the ear is sensitive. He might say that sound is there whether it is received by the ear or not. As an approach to explaining the disturbances of matter called sound, let's refer back to a topic you considered earlier in this book. Remember the simple pendulum.

We found that once set in motion the pendulum moves with simple harmonic motion. The motion is a vibratory motion that is repeated over and over. (You may want to refer to Chapter 8 and Figure 8.3 to help you to recall information discussed there.) One vibration can be traced from the equilibrium position to maximum displacement on one side to maximum displacement on the other side and back to the equilibrium position. The maximum displacement from the equilibrium position is the amplitude of simple harmonic motion. One vibration is also called one cycle. The time for the pendulum to complete one cycle is called the period of the motion. The reciprocal of the period is the frequency of the motion. The period is measured in seconds and the frequency in cycles/s or hertz (Hz). The disturbances in matter that are sensitive to our ears closely approximate the to-and-fro motion of the pendulum. Got it?

Newton's Figs

Any object undergoing simple harmonic motion has associated with it not only a period of motion but also a cycle. One cycle starts at equilibrium, goes to maximum displacement in one direction and back to equilibrium, then to maximum displacement in the other direction and back to equilibrium. Remember that the maximum displacement from equilibrium is called the amplitude.

The strip of metal in Figure 16.1 is secured to a desk so that one end is free to vibrate in a vertical plane.

You may have placed an object on a table in a similar fashion to that in the diagram and flipped it to set in motion and hear the hum it creates. A plastic ruler works fine if you want to try it. Like the simple pendulum, it vibrates with a motion that closely approximates simple harmonic motion.

The maximum disturbance is labeled *b* for bottom and *t* for top, and the equilibrium position is *e*. When the metal is set in motion, it vibrates from *e* to *t* back to *e* to *b* and back to *e*, then repeats the motion over and over. The path just completed is called one vibration or one cycle.

Newton's Figs

The frequency of simple harmonic motion is the reciprocal of the period of motion. That is, $f = \frac{1}{T}$ and $T = \frac{1}{f}$ so the period is measured in seconds per cycle or just seconds, and the frequency is measured in s^{-1}, $\frac{1}{s}$ or $\frac{s}{cycle}$. And the common unit for expressing the frequency of waves is the hertz, abbreviated Hz.

The time to complete one cycle is the period, and the reciprocal of the period is the frequency of the motion. This vibrating metal is much like the simple pendulum, but it is different in that you can hear the sound associated with the disturbance of the metal strip.

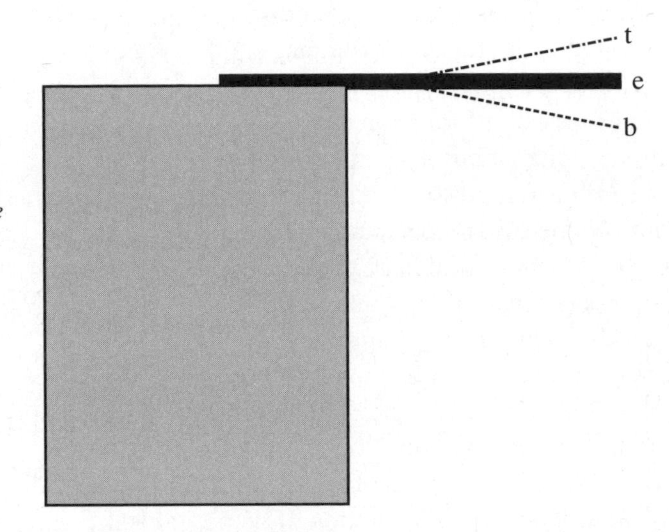

Figure 16.1

The vibrating metal strip compresses air particles on one side and then pulls away from them, leaving a void into which the particles move back.

The vibrating metal strip transfers mechanical energy to the air particles surrounding its motion. As it moves from *e* to *b*, the air particles below the strip are pushed downward, compressing them. That leaves a void in the space above the metal strip. The air particles above the strip are able to move into that space. That increases the volume that they have to occupy and thus decreases their pressure. We know that the strip has its maximum velocity when it passes through *e*, so it causes maximum compression of air particles below the strip at that time. While moving from *e* to *b*, the strip continues to compress the particles, but the compression becomes less and less as the speed slows to zero at *b*. As the metal moves from *b* to *e*, the particles beneath the strip get farther apart.

> **Newton's Figs**
>
> Sound waves require a medium to move through, and generally, sound waves are transmitted through the earth's atmosphere. The compressions and rarefactions that move through the atmosphere are compressing the molecules of nitrogen and oxygen all around us.

The particles continue to be separated as the strip moves from *e* to *t*, but they become less and less separated as the velocity slows to zero at *t*. From *t* to *e*, the particles beneath the metal strip are being compressed more and more until maximum compression is reached at *e* when the cycle begins again.

Types of Waves

There are many models used in science to explain complex ideas. One model that is useful to explain the concepts behind an understanding of sound is the wave. There are two basic types of physical waves: transverse and longitudinal.

Transverse Waves

The *transverse wave* is probably the most familiar type of wave to you. Have you ever held a jump rope in your hand and sent a loop down the rope by shaking your hand perpendicular to the rope? The loop traveled down the rope all by itself without having any particles of the rope making the trip with it.

The particles of the rope vibrate up and down, assuming that you created a vertical loop, perpendicular to the rope that determines the line of travel of the wave. A wave is a series of disturbances that travels through a medium because the particles of the medium vibrate. When the particles vibrate perpendicular to the direction the wave is traveling, the wave is called a transverse wave.

def•i•ni•tion

A **transverse wave** is a series of disturbances traveling through a medium in which particles of the medium vibrate in paths that are perpendicular to the direction of motion of the disturbances of the wave.

Have you ever been at a football or baseball game when the fans started doing a "wave"? This type of wave is a perfect example of a transverse wave. The wave moves around the stadium, while the "medium" (the fans) simply moves up and down. What do you do when the wave passes you? You stand (and scream) and then sit down.

Longitudinal Waves

Sound is not thought of as a transverse wave because of the behavior of the particles of the medium. If you think about the vibrating strip of metal, the particles of the medium, the air, are vibrating but vibrating in paths parallel to the direction of the motion of the disturbance. So sound does not fit the transverse wave model. Sound can be thought of as a *longitudinal wave* because of the vibrations of the particles of the medium.

def•i•ni•tion

A **longitudinal wave** is a series of disturbances traveling through a medium in which the particles vibrate in paths parallel to the direction the disturbances of the wave are traveling. A **pulse** is a wave of short duration.

A longitudinal wave is a series of disturbances moving through a medium in which the particles of the medium vibrate in paths that are parallel to the direction of travel of the wave. The loop for the transverse wave and the disturbance created by the part of the vibration of the metal strip moving downward are called pulses. A *pulse* is a single disturbance or a wave of short duration.

Another way of looking at a longitudinal wave is to understand that they are waves in which the motion of the medium is in the same direction as the motion of the wave. Sound waves arise when the atmosphere is alternately compressed and stretched—for example, by the backward and forward motion of a speaker, or the clapping of your hands.

Properties of Waves

Figure 16.2 shows you a diagram of a transverse wave and a longitudinal wave for comparison.

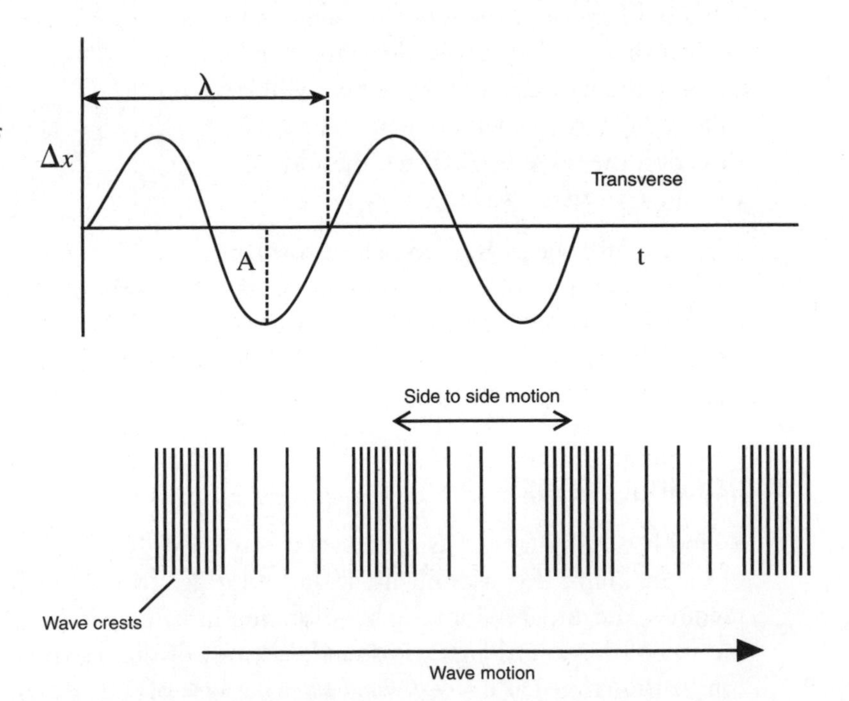

Figure 16.2

The diagram of the transverse and longitudinal waves displays their distinguishing characteristics.

If you look at the transverse wave you'll see that it is a graph of the displacement of the tip of the bottom side of the vibrating metal strip versus the time. The amplitude of the displacement, the maximum displacement of the tip, is labeled A in the diagram. The part that looks like a camel's hump is called a *crest*, and the bowl-shaped portion is called a *trough*.

The distance from the beginning of a crest to the end of an adjacent trough is the *wavelength*, labeled λ (the Greek letter lambda) in the diagram. Did you notice that the horizontal axis is time and not distance? The time to complete one full wave is what we've labeled λ on the graph.

The longitudinal wave is created by the motion of the air particles near the bottom of the metal strip. When the metal strip is passing through the equilibrium position in Figure 16.1 moving downward, a *compression* is being created.

As the strip continues down to *b* in that diagram, the particles get farther apart because the bottom of the metal is moving away from them. When the strip continues from *b* and passes through *e*, the particles are at a maximum distance apart and in the longitudinal wave this is called a *rarefaction*.

def•i•ni•tion

The **crest** is that portion of the graph of a transverse wave that lies above the horizontal time axis, and the **trough** is that portion that lies below the axis. The **wavelength** of a transverse wave is the distance from the beginning of a crest to the end of an adjacent trough. It can also be thought of as the distance from the point of maximum displacement in one crest to the point of maximum displacement in the next closest crest. One wave is made up of a crest and a trough. The distance from the beginning of a crest to the end of the crest is one half of a wavelength, and the trough measures the other one half of a wavelength. A **compression** is that part of a longitudinal wave where the particles of the medium are pushed closer together. A **rarefaction** is that part of a longitudinal wave where the particles of the medium are spread apart the most.

Continuing the motion of the strip toward *t*, the strip is slowing down so the particles near the bottom of the strip are not as far apart as they were. Traveling from *t* downward, the strip is speeding up, causing the particles beneath the metal to compress and reaching maximum compression at *e* as the strip continues its downward motion.

You have just traced the creation of one complete longitudinal wave that starts with a compression and ends with the next compression in a train of waves. That complete wave-pulse has a length that is called the wavelength of a longitudinal wave. It would be the distance that the beginning of the pulse moves out from the source during the time that it takes for the whole cycle to occur.

Stating that sound is a longitudinal wave means that the longitudinal wave is the best model we presently have for explaining the behavior of sound. Scientists often understand a model and borrow it to see how well-observed phenomena fit the model. There are several sites on the web where you can find dynamic models that illustrate this motion. You might enjoy finding some to get a moving picture of what is described in this chapter.

Newton and the Speed of Sound

At any given time in the development of physics, certain experiments or measurements are just barely possible. These aren't the most difficult experiments we can imagine, but they are the most difficult experiments that one can effect with existing equipment. These experiments become a challenge to the artistry and imagination of the most gifted experimenters. A typical example today might be the detection of gravity waves. Toward the end of the seventeenth century, one such state-of-the-art experiment was to measure the speed of sound.

Newton's Figs

The compressions and rarefactions of a longitudinal wave are regions where particles are closer together and farther apart, respectively. The action of pushing and spreading apart throughout the room from the source of sound is the wave nature of sound. It spreads like ripples on the surface of water (but in three dimensions) and remains circular because the speed of the wave's disturbances are constant.

Sound is some sort of disturbance (we know that it is a longitudinal wave, but back then no one was sure what it was) that most often travels to our ears through air. Sound also travels through liquids and solids, but in any case a medium is needed in order for the sound waves to travel. In a vacuum, there is no sound.

Although sound travels very fast, its speed is not infinite. We can observe this fact very simply. For example, you hear echoes because it takes a finite amount of time for sound to travel to a distant object, which reflects it back to you. In addition, we see lightning before we hear thunder because light travels much faster than sound. The speed of sound is a very useful quantity to know.

In the seventeenth century, experimenters tried to determine the speed of sound and found a wildly ranging set of values, from 600 ft/s to 1,474 ft/s. In view of this discrepancy, Isaac Newton chose to make a measurement of his own to substantiate better his mechanical derivation of the speed of sound. He didn't make the best measurement; experimental physics is a special art, and wasn't Newton's strongest point. Nonetheless, his method demonstrated his ingenuity.

At Trinity College in Cambridge, where Newton lived and worked, there was a long arcade, which was known to produce an echo. For a timing device, he used a pendulum, whose period he could adjust by changing its length (recall that $L = 2\pi\sqrt{L / g}$). He arranged to have a sharp noise go off just as the pendulum started its swing. If the echo returned before the pendulum did, the length was too long, the period being greater than the time of a trip down the corridor and back (416 ft). If the pendulum returned first, the length was too short. In this way he could make increasingly better estimates of the time sound takes to travel a known distance, and therefore its speed, which is equal to the distance divided by the time.

Newton determined that the echo was faster in returning than one oscillation of an 8-in.-long pendulum and slower than a pendulum 5 ½ in. long. These measurements narrowed the speed of sound to be between 920 ft/s and 1,085 ft/s. In the first edition of his *Principia*, Newton reported better measurements, placing the speed of sound between 984 and 1,109 ft/s. As it turns out, the speed of sound is about 331.5 m/s at 0°C; that is, about 1,087 ft/s at 32°F or about 740 mi/hr.

Wavelength and Frequency

Suppose we pursue the notion of a sound wave just a bit further by looking at the relationship between wavelength and frequency.

Take a look at the graph in Figure 16.3. Notice that the graph suggests an inverse variation, that is, $\lambda \propto \frac{1}{f}$. That's the same type of inverse variation we've found in two other relationships. Do you remember what they were? The first was the inverse variation found in Newton's universal law of gravitation; the second was Boyle's law, which described the inverse variation of volume and pressure. In this instance, we can express it as: the longer the wavelength, the slower the frequency, or the shorter the wavelength, the faster the frequency. That's graphed for you in Figure 16.4, where you can see that

Newton's Figs

Outer space, like the region above the surface of the moon, is about as close to a vacuum as you can get. Remember that astronauts dropped a feather and a hammer and both hit the surface of the moon at the same time. The hammer and feather experienced no opposition to their motion from air as they fell to the moon's surface.

the graph confirms the notion because the straight line means that $\lambda = \frac{k}{f}$ or $f\lambda = k$, where k is the constant of proportionality. Can you guess what that constant might represent? What is it that is staying constant?

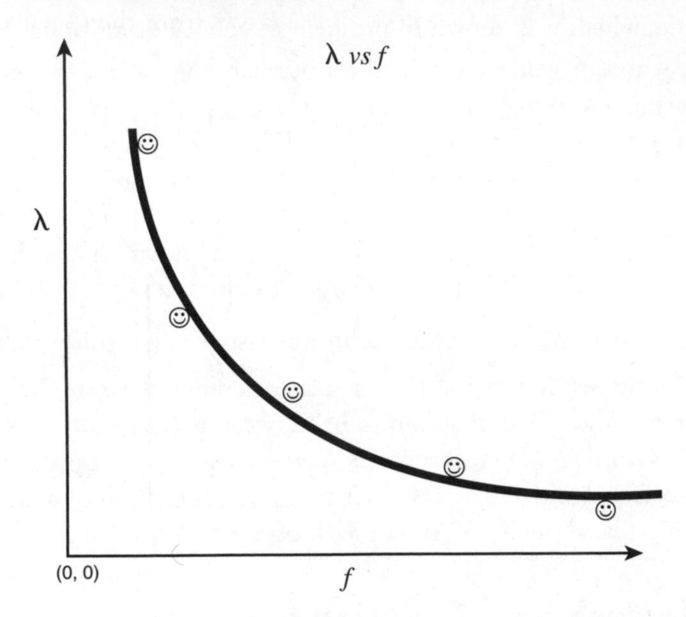

Figure 16.3

The graph of wavelength versus frequency suggests an inverse relationship.

Looking at the graph, we can see that k is the slope of the straight line for that graph. What do you suppose that constant k is? You can get a hint by checking the units of the slope if you have not already guessed what it is. The units of k turn out to be $\frac{meters}{\frac{1}{\frac{1}{s}}}$ and that simplifies to $\frac{m}{s}$, the units of speed! The graphical analysis of the hypothetical data suggests that the speed of the wave in the medium is equal to the product of the frequency of the wave and the wavelength of the wave, $v = f\lambda$. That relationship is one that you can expect any time you are discussing waves or the wave nature of physical phenomena. And now's probably a good time to apply what we've learned so far, so let's take a look at a simple problem and then you can try your hand at a few.

You're sitting in a rowboat in the middle of a lake, fishing, at 9 A.M. There haven't been any nibbles on your line all morning, and the lake is completely calm. In the distance, you hear the high-pitched whine of a speedboat. From its motion, you can see that the boat will come close to you. Sitting there, you see the ripples of the speedboat's wake approaching you; the crests are about 2 m apart. As the waves pass you,

you bob up and down in the water, a lot at first, and then less and less. You notice that you are bobbing up and down (one full cycle) once each second. The height of your bob is the amplitude of the wave passing you through the medium of the water, and the time it takes for you to bob up and down is related to the wavelength of the water wave. What is the speed of the waves from the speedboat's wake?

Solution: $v = \lambda f$

Using $\lambda = 2m$ and $f = \frac{1}{s}$, we have $v = (2m)(1/s) = 2m/s$.

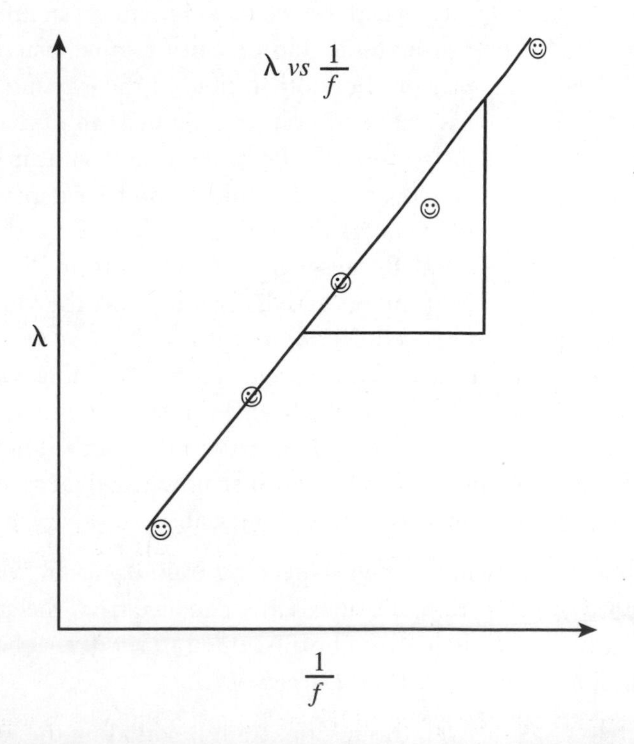

Figure 16.4

The graph of wavelength versus the reciprocal of frequency.

Physics Phun

1. A source attached to a rope vibrates at 45 cycles/s, 45 Hz, and creates waves in the rope traveling at 15 m/s. What is the wavelength and period of the waves in the rope?
2. A train of waves moves along a wire with a speed of 25 ft/s. What is the frequency and period of the waves if they are 5.0 inches in length?
3. Audible sound ranges from about 20 cycles/s, or 20 Hz, to 20,000 Hz. What are the wavelengths for these frequencies? Note: vibrations below 20 Hz and above 20,000 Hz are not detectable by the human ear.

Characteristics of Sound

Did you anticipate that temperature plays a role in the behavior of sound because the speed of sound was given at 0°C? At a given temperature, the speed of sound is constant in a given medium. Even though you may have available a solid, a liquid, or air at the same temperature, the speed of sound is constant in each medium, but the speed has a different value in each medium.

Newton's Figs

Like other phenomena that can be described as waves, sound waves diffract (bend around corners), refract (change direction when traveling from one medium to another), and reflect (bounce off surfaces).

At some point in our lives, we've all had an opportunity to experience sound manifest in different ways at a given temperature. For example, listening to a band play on the football field in the evening you may have noticed that you can hear all of the different instruments at the same time. You hear high notes and low ones. The pitch you hear depends on the frequency—the higher the frequency, the higher the pitch. Because all of the notes from the various instruments arrive at your ear at the same time, all frequencies must travel at the same speed. Recall that because the speed is constant, each frequency has associated with it a definite wavelength. All of this occurs the way it does because the speed of sound is constant at a given temperature. If there were sections of drastic temperature change going on down on the football field where the band was playing, the sound from all the different instruments could reach your ears at slightly different speeds and would sound very unusual.

Have you ever stood at a train station when a speeding train passes by with horn blaring? You probably noticed the warning horn at some distance from the station. When the train passed the station, the pitch of the horn seemed to suddenly change to a lower tone. That phenomenon is called the *Doppler effect.*

def•i•ni•tion

The **Doppler effect** is the apparent shift in the pitch of a source of sound because of the relative motion between the source and the observer.

Even though the train is sounding the same tone at all times, when it is coming toward you the frequency appears higher than it does when the train is standing still at the station. The source of the sound sends out a series of waves, one after the other. When the locomotive is coming toward you, you are receiving more waves each second than you do when the train is standing still. Even though the speed of sound is constant, you are observing the speed of sound plus the speed of the train for the waves you observe with your ears.

As the train passes by, you experience the difference in the speed of sound and the speed of the train. Therefore your ears perceive fewer waves each second, resulting in a lower frequency and lower pitch. The apparent shift in pitch as a source of sound passes you is the Doppler effect. And like the sound coming from the band on the football field in the previous example, the experience of the Doppler effect would be different with significant changes in temperature.

Interference and Reflection of Sound

Waves do all sorts of interesting things when they meet one another. When waves encounter a barrier, there are several possible results. A wave may be diverted in the opposite direction. For example, a sound wave may strike a wall and bounce back so that you hear it as an echo. This is called a reflection. Reflections of sound can be troublesome if you are sitting in the wrong place in an auditorium that does not have good acoustics. If the acoustics of a room are bad, reflections of sound from walls or other surfaces can meet in such a way as to cancel sound altogether. We often have the tendency to think of a sound wave as linear but it can also be compared to a ripple of a water wave. Actually a sound wave is three-dimensional spreading away from a source of sound in the shape of a sphere. If two sound waves meet at a point so that a rarefaction from one source meets a compression from another source of the same frequency, they will completely cancel each other. If your ear is at that point, it detects no sound!

When two waves occupy the same physical space, the result is called interference. How two waves interfere depends on where each wave is in its cycle. There is a general rule that says that when two waves interfere, the combined wave will be the sum of the amplitudes of each of the interfering waves.

Johnie's Alert

Vibrating sources of sound have images in reflecting surfaces just like you have an image in a mirror. Points where waves cancel are called nodal points. Two-dimensional waves, such as water waves, cancel in lines called nodal lines. Waves in three dimensions cancel in planes called nodal planes. Nodal lines are hyperbolas, and nodal planes are hyperboloids. Special conditions must exist for cancellation to take place.

For a simple example, imagine two waves moving toward one another on a string. If the two waves both have positive amplitudes when they meet, the amplitude will be the positive sum of the individual amplitudes. If they have equal amplitudes, for example, the interfering amplitude will be twice that of either of the individual waves. This situation is called constructive interference. However, if one wave has a positive amplitude and one has a negative amplitude when they meet, the sum of the waves will be smaller than either of the individual waves. This situation is called destructive interference. A combination of constructive and destructive interference is called an interference pattern. The interference pattern is made of areas of constructive interference where the sound is louder and destructive interference where the sound is softer.

Sound and Temperature

All of these things occur because sound is a wave and the speed of sound is constant for a given temperature. The speed of sound changes with temperature. The speed of sound in air is about 331.5 m/s at 0°C and increases about 0.60 m/s for every degree Celsius increase in temperature. In the FPS system, the speed is about 1,087 ft/s at 32°F and increases 1.1 ft/s for every degree Fahrenheit increase in temperature. Because sound travels faster in air at warmer temperatures, for a given frequency the wavelength is slightly longer than at freezing temperatures.

Suppose the temperature is 70°F; then the speed of sound is 1,087 ft/s plus (70–32)F° (1.1 ft/s), which turns out to be about 1,129 ft/s. For a frequency of 256 cycles/s, the wavelength is about 4.2 ft at 32°F and about 4.4 ft at 70°F. Remember that even though the speed of sound increases with temperature, all of the phenomena discussed are still observed at a given temperature. Notice that the frequency of the sound did not change due to the change in temperature. It is determined by the frequency of the source. Following are a few simple problems that incorporate some of the ideas we've discussed so far. Give them a shot and see how you do.

Newton's Figs

The speed of sound is less at temperatures below 0°C or 32°F and greater at temperatures above 0°C or 32°F.

Physics Phun

1. How much time is required for sound to travel 6.78 km when the temperature is 5°C?
2. How many feet long is the wavelength of a wave generated by a tuning fork with a frequency of 256 cycles/s if the temperature is 86°F?
3. A man drops a stone into a well 155 m deep. The temperature is 8°C. How long does it take for him to hear the stone splash in the water below?

Resonance

Resonance is an interesting phenomenon of sound energy. The easiest way for you to understand what it is, is to provide you with an analogy first. Imagine yourself pushing a child on a swing. The child on a swing represents a form of pendulum and has a natural period of vibration. If you apply successive pushes to the swing at random intervals, you will often push the swing as it is moving back toward you and will cancel what motion it possesses, slowing it. By persisting, you will keep the swing moving in accordance with your pushes, but you will expend a lot of energy doing so. If, however, you timed your pushes to match the natural vibration of the period of the swing, you would push each time as the swing begins to move away from you, thus adding to its velocity and increasing it further with each swing and rhythmic push. At the expense of far less energy, you would get a far more rapid and extended swing. Energy is absorbed by a system in harmonic motion if the energy is input at the same frequency.

The situation is analogous for sound waves. The sound wave of a particular note would push another object with each region of compression and pull it with each region of rarefaction. If the rhythmic push and pull did not match the natural period of the receiving object, the forced vibration could only be obtained at the expense of considerable energy being used to overcome that natural period. If the frequency of the note just matched the natural period of vibration of the receiving object, however, the latter would begin to vibrate more and more. This is called resonance.

Any given sound wave would produce far more vibration in a resonating object than in any other kind; in fact, only the resonating object might produce sound waves strong enough to be audible. Suppose, for instance, you raise the top of a piano to expose the wires and step on the "loud" pedal to allow all those wires to vibrate freely. Now sing a short, loud note. Only those wires that vibrate at the frequency of that note will resonate, and when you stop singing, you will hear the piano answer back softly in the same note.

Musical instruments depend on the resonance of the material making up their structure to strengthen and add richness to the notes produced. Pianos have a "sounding board" just under the wires, and this device can resonate with the various notes. Without that board, the notes sounded by the wires would be quite weak.

The sounds we ourselves make produce resonances in the air filling the hollows in throats, mouths, and nasal cavities. The natural vibrations of the air depend on the shape and size of the cavities, and because in no two individuals are these cavities of precisely the same shape and size, voices differ in quality; we usually have no trouble recognizing the voice of a friend from among a large number of others.

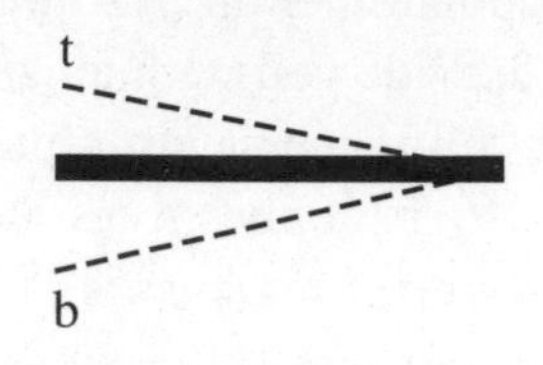

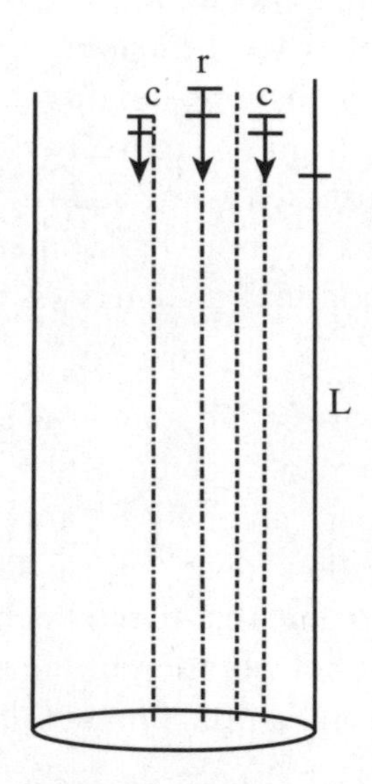

Figure 16.5

The conditions for resonance in a closed tube are shown by following the reflections of a compression.

Take a look at Figure 16.5. It shows that the reflections of a compression from each end of the tube are traced and the distance recorded. The vibrating source of sound is a tuning fork held above a tube. The correct length of an air column can be established experimentally by holding an open tube with one end in water. Hold the tuning fork just above the mouth of the open end, and then raise and lower the other end of the tube in the water until you hear an obvious increase in volume of sound. The increased volume of sound signals resonance; the column of air is vibrating at the fundamental frequency of the source or it is vibrating at some harmonic frequency of the source. Follow these steps to see how resonance in the closed tube is explained:

1. Assume that the vibrating source has completed a compression when the tip of the lower side of the source is at position *b* in Figure 16.5. A compression is sent down to the bottom at that time. In order to have resonance for the shortest length of the closed tube, the source must be creating a compression at the same time a reflection from the air at the top of the tube is a compression.

2. The compression travels to the bottom of the tube and is reflected as a compression.
3. The reflected compression travels to the top of the tube and is reflected as a rarefaction as shown. At this time, half a vibration of the source of vibration, the sound has traveled the length of the closed tube twice.
4. The rarefaction travels to the bottom of the closed tube, where it is reflected.
5. The reflected rarefaction travels back up the tube and is reflected at the top as a compression. At this time, second half of a vibration of the vibrating source, the sound has traveled the length of the closed tube twice.
6. At the instant the reflected compression is created and just starts its trip downward, the source must have created another compression that is traveling in the same direction at the same time to have resonance. That is, to receive just the right push at just the right time in just the right direction, the two compressions must satisfy all of these conditions.
7. The source will have completed one cycle to meet these conditions. That means the distance the sound has traveled in one cycle of the vibrating source is one wavelength, four times the length of the closed tube. The length of the closed tube is $\frac{\lambda}{4}$, which is one fourth of the wavelength of the vibrating source.

Notice that the conditions for resonance as outlined here are for the shortest length of the closed tube.

By varying the length of the closed tube, conditions for resonance can be established for the harmonics of the vibrating source. The lengths of a closed tube for the fundamental frequency and its harmonics are $\frac{\lambda}{4}, \frac{3\lambda}{4}, \frac{5\lambda}{4}....$

Newton's Figs

The shortest length of a closed tube that resonates with the source of sound is one quarter the wavelength associated with the fundamental frequency of the source.

The wavelength for the fundamental frequency is about four times the length of the tube or $\lambda = 4L$. The diameter of the tube does make a difference for the calculation of λ, so to make the calculation more accurate use $\lambda = 4(L + 0.4d)$, where d is the diameter of the tube. The tube can be glass, metal, or even a cardboard mailing tube. It would be fun to try this experiment and compare your calculation of the speed of sound with the value you calculate for the temperature that day. That is, because

$V = f\lambda$, and using the results of the resonance experiment $V = 4(L + 0.4d)$. Compare that value with $V = (331.5 + 0.6\Delta t)m/s$ or $V = (1087 + 1.1\Delta t)ft/s$.

And that brings our chapter on sound energy to a close. We hope that what you learned in this chapter resonates with everything else that you've learned so far.

Problems for the Budding Rocket Scientist

1. What is the speed of sound in super-heated 100°C air?
2. How fast does an airplane have to be moving to produce a sonic boom on a day when the temperature is 30°C?
3. A 1,000-Hz sound wave in air strikes the surface of a lake and penetrates into the water. What are the frequency and wavelength of the wave in the water? Assume that the speed of sound in water is 1,500m/s.
4. An underwater sonar source operating at a frequency of 60 kHz directs its beam toward the surface. What is the wavelength of the beam in the air above? What frequency sound due to the sonar source does a bird flying above the water hear? Assume that v in air = 330 m/s.

The Least You Need to Know

- The frequency of simple harmonic motion is the reciprocal of the period of motion.
- There are two basic types of physical waves: transverse and longitudinal.
- A longitudinal wave is a series of disturbances moving through a medium in which the particles of the medium vibrate in paths that are parallel to the direction of travel of the wave. Sound is a longitudinal wave.
- When two waves occupy the same physical space, the result is called interference.
- Resonance in sound is the increased amplitude of vibration of an object caused by a source of sound that has the same natural frequency as the object's frequency.

Part 5

The Anatomy of Atoms, Electricity, and Light

Have you ever thought about what an atom looks like? Or how electricity manages to do all that it does? Perhaps, like Albert Einstein, you wondered what it might be like to travel on a beam of light. Well, even if you've never asked these questions, this last part of our book provides you with the answers to some of them. By understanding the structure of atoms and molecules, you'll have the keys to knowing how electricity operates. And with this knowledge at your fingertips, you'll delve into the dual nature of light and get a glimpse at why it can be seen as composed of particles or why it behaves like a wave.

Chapter 17

Atoms, Ions, and Isotopes

In This Chapter

- Defining atoms, molecules, elements, and compounds
- Different models of atomic structure
- Isotopes and atomic number
- How molecules are formed

While we explore the atom and molecule as the building blocks of matter in this chapter, you'll find that there is much more to the story of what matter is made of if you continue your exploration and enter into the world of subatomic particles. There are several places you may want to visit on the Internet to find out more about the many particles with strange-sounding names. Here are some sites (there are many more): particleadventure_.org, whatisit.techtarget.com, and www.chemtutor.com.

A brief visit to any one of these sites and you'll see that our discussion of molecules, atoms, neutrons, protons, and electrons is about all that we can handle in this chapter. There are worlds within worlds at the atomic and subatomic levels of matter, and this chapter is meant to simply introduce you to the most fundamental aspects of the atomic level. But we urge you to explore the deeper levels of matter so you can see just how far down the rabbit hole in *Alice in Wonderland* really goes.

The Structure of the Atom

Imagine that you travel from your frame of reference at the normal physical level down into the insides of matter. The first particle that you'll see is fairly large as these tiny particles go. It is called a *molecule*, and molecules are made up of many different combinations of *atoms*. (We'll have more to say about molecules a little later in the chapter.) Within this context, there are also things called *elements*. An element, such as sodium or iron, is the simplest form that matter can exist in by itself. It was once thought that there were only four elements (the Greeks called them earth, air, fire, and water). Today there are 113 known elements. If you'd like to see what they are, check out the periodic table of elements at the following websites: www.webelements.com or www.chemicalelements.com. The second website is interactive.

def•i•ni•tion

A **molecule** is the smallest particle of a compound. An **atom** is the smallest particle of an element that exists alone or in combinations. An **element** is a substance such as copper, hydrogen, or neon that cannot, by chemical means, be broken down into other substances.

Atoms of these elements combine to form compounds, and a molecule of these compounds is the smallest particle of any particular compound. A compound is also the combination of two or more elements, in which the individual properties of the elements are lost but the combination has new properties. For example, common table salt is a compound of sodium and chlorine. Individually, sodium and chlorine are extremely poisonous, but together they form a compound we can't live without.

Thomson's Atomic Model

In 1897, a British physicist by the name of J. J. Thomson was working on identifying the particles that moved within a cathode ray tube (CRT) from the negatively charged terminal (cathode) to the positively charged terminal (anode) and lit up the tube where they hit. Thomson discovered that it was possible to bend these particles in their journey using a magnetic field, which meant that they had an electrical charge. He found that the charge was negative and these negatively charged particles came to be called *electrons*.

No one had any idea what an atom looked like. Some physicists thought that if there were negatively charged particles in an atom, there might also be positively charged ones, too. But how did it look? Some, like Thomson, felt it probably looked like raisin pudding, with the electrons stuck in the center. Others thought it was a spongy, gooey blob, or even something like a beehive with the electrons buzzing around it.

Thomson's model of the atom was really the first model that received any serious thought. He pictured it as something like a tiny bit of raisin pudding. Embedded within this pudding were tiny electron raisins. The number of electrons depended on what the substance was. For example, Thomson thought that a hydrogen atom, which was the simplest atom known, had one electron raisin, which had a negative electric charge. This charge was somehow balanced out by a positive electric charge.

def•i•ni•tion

The **electron** is a negatively charged particle that has a mass of 9.1×10^{-31} *kg*.

An atom in its normal state is electrically neutral. So in addition to the negative electric charge of the electrons, there must also be an equal amount of positive charge somewhere in the atom. Thomson said that the additional positive charge in the atom was distributed in the "pudding part" of the atom. No one had discovered any positively charged components within an atom, though. The mass of the electrons was found to be negligibly small in comparison to the total mass of the atom. So it followed that the positively charged part must represent almost all of the atom's mass, and thus should be most of its bulk as well.

Thomson's idea of raisin pudding as a model of the atom was difficult for physicists at that time to accept, because they didn't believe that there could be anything smaller than an atom. There was no experimental evidence to support Thomson's theory, at least not until Ernest Rutherford came along.

Alpha Particles and Protons

Ernest Rutherford (1871–1937) was a New Zealander who came to England in 1895 to work at the famous Cavendish laboratory. Three years later, he left for Montreal where he worked for the next nine years. In 1907, he returned to England, which is where he made most of his significant discoveries.

While in Montreal, Rutherford performed a number of experiments on radioactivity, that strange and unusual phenomenon discovered by Marie Curie and her husband Pierre. Rutherford named the two types of radiation given off by radium alpha and beta. He soon realized that the beta radiation consisted of the electrons that Thomson had recently proposed. But he had no idea what alpha radiation was. After a series of experiments, he found that alpha radiation was composed of positively charged particles. These particles could penetrate through thin layers of solid material, such as the glass walls of a tube. With this information, he decided to investigate the internal structure of atoms by shooting these tiny, positively charged particles through them.

In 1907, with the help of his assistant Hans Geiger (name sound familiar, as in _____ counter?), he developed an apparatus to study the effects of alpha particles on various substances. They bombarded a thin piece of gold foil with alpha particles emitted by a piece of radium. Behind the gold foil they placed a glass plate coated with a chemical that would emit a flash of light every time it was hit with a particle. Geiger later refined this apparatus into the first successful detector of alpha particles. You know this instrument as a Geiger counter. (More on Rutherford's experiment when we discuss radioactive particles later in this chapter.)

After accumulating an abundance of data, Rutherford realized that the entire atom wasn't positively charged, as in the pudding model. Instead, the positive charge appeared to be concentrated in a very tiny central part of the atom called the *nucleus*. This nucleus was much smaller than the entire atom, yet astonishingly, it accounted for almost all of the atom's mass. As a matter of fact, the nucleus is only one ten thousandth the size of the whole atom. Rutherford named the positively charged particles in the nucleus *protons*. Protons along with electrons were the second building block of the atom.

The third and last building block of the atom was predicted by Rutherford back in 1920. He thought that another particle would be found inside the nucleus. Sure enough, in 1932, the *neutron* was discovered. The neutron had no charge; it was electrically neutral. But it did help account for the mass of the nucleus.

The model put forth by Rutherford was known as the planetary model, because the structure of the atom was described as being similar to that of our solar system. The nucleus is like the sun, and the electrons orbit the nucleus like the planets. However, the atom is almost entirely empty space. If you were to take an atom and blow up its nucleus to the size of a golf ball, the atom would be the size of the earth, with an electron the size of a grain of rice in orbit. Matter is mostly just empty space. It appears solid only because the group of electrons orbiting each atom resists intrusion by the electrons of adjacent atoms. So each atom has its own space. The speed at which the electrons orbit the nucleus is also what gives matter its solidity. At least this is what was thought at the time.

The Atomic Model of Niels Bohr

Enter Niels Bohr (1885–1962). While working for Rutherford as his assistant, the two decided that they needed a better picture of the dynamics of atomic theory for

Rutherford's model to be fully accepted. (There were some problems with his theory, such as stability and the way atoms released energy.)

Bohr spent a lot of time studying the periodic table of elements, to see whether he could recognize a correlation between the structure of atoms and the periodicity of the properties of elements. Periodicity is another way of saying the position of an element in the periodic table. He found that electrons arrange themselves in orbits, or shells, one outside the other. As each shell becomes full, the next electron occupies the next-furthest shell. Bohr developed the following sequence based on the maximum number of electrons in each shell.

Bohr's Electron/Shell Configuration

Shell number	1	2	3	4	5
Number of electrons	2	8	18	32	50

So Bohr's Electron/Shell Configuration model used the nucleus of the Rutherford atom and proposed discrete orbits for the electrons. Without going into a deeper analysis of Bohr's model, which would involve knowledge of quantum mechanics, we can summarize by saying that the current atomic model is defined as having a nucleus, with the electrons orbiting the nucleus, not in discrete orbits, but rather in nebulous areas more like a cloud.

def•i•ni•tion

The **nucleus** of an atom is considered the core of the atom and is made up of protons and neutrons. The **proton** is a positively charged particle with a mass of about $1.7 \times 10^{-27} kg$. The **neutron** is a particle with no charge and about the same mass as the proton.

Isotopes and the Same Element

The model of the atom that we're using here has two important numbers associated with it: the *atomic mass number* designated by the symbol A, and the *atomic number* that identifies the number of protons in the nucleus, designated by the letter Z. These numbers are used to identify the element and to indicate the number of nucleons.

Let's suppose that you are given an atom with 4 protons and 12 nucleons. That means Z = 4 and A = 12, and because A is the number of nucleons, then A – Z = the number of neutrons in the atom. The atom has 12 nucleons made up of 4 protons and 8 neutrons. The symbol for an atom is written ${}^{12}_{4}Y$, where Y is the symbol of the element that can be found on a periodic table of elements. In general, the symbol for an atom is written ${}^{A}_{Z}Y$. As you can see, Z identifies the element even though the symbol for the element is included in the notation.

def•i•ni•tion

The **atomic mass number** of an atom is the number of nucleons (collectively the protons and neutrons) the atom contains. The **atomic number** is the number of protons in the nucleus of an atom. **Isotopes** are atoms of the same element that have different numbers of neutrons. The **unified atomic mass unit,** u, is the unit of mass used to stipulate nuclear masses.

There are atoms of the same element that have a different number of neutrons. For example, all oxygen atoms have eight protons, but they can have a different number of neutrons. ${}^{16}_{8}O$, ${}^{17}_{8}O$, and ${}^{18}_{8}O$ are all examples of atoms of oxygen. Nuclei having the same number of protons but a different number of neutrons are called *isotopes*. The oxygen isotopes listed are stable isotopes of oxygen.

All isotopes of an element are not equally abundant. Some isotopes can only be produced in the laboratory. Many isotopes occur naturally, such as the abundant isotope of carbon, carbon 12 or ${}^{12}_{6}C$. Did you know that the isotope ${}^{14}_{6}C$ is used for carbon dating? No, not that kind of dating; it is the process of checking the age of ancient objects that were alive at one time.

The unit of measurement for nuclear mass is the *unified atomic mass* unit or u. Carbon 12, ${}^{12}_{6}C$, was given the exact value of 12.000000 u on this scale. On this scale, a proton has an atomic mass of 1.007276 u, a neutron has a mass of 1.008665 u, and a hydrogen atom has an atomic mass of 1.007825 u. Let's use our knowledge of unit analysis to define the mass in grams of 1 u. Earlier we mentioned that a hydrogen atom has a mass of $1.673653 \times 10^{-27} kg$, which has a unified atomic mass of 1.00725 u. We can write that as a defining equation, then divide both terms by 1.00725 to find 1 u:

$$1.673653 \times 10^{-27} kg = 1.00725u$$

$$1.66054 \times 10^{-27} kg = 1u$$

You may use this last result to express nuclear mass in kilograms given the nuclear mass in u. The atomic mass number is the whole number closest to the atomic mass of an atom. That is, given that the atomic mass of oxygen is 15.9994 u, the atomic mass number is 16. That means that this isotope of oxygen has 16 nucleons, and because the atomic number of oxygen is 8, there are 8 protons and 8 neutrons in the nucleus. See if you can work the example yourself before looking at our solution.

Problem:

The periodic table gives the following information about an element: the symbol is Fe (iron), the atomic number is 26, and the atomic mass is 55.847. What is the mass number of Fe? Find the number of protons and neutrons in the nucleus.

Solution:

The atomic number, 26, is given, and that is the number of protons in the nucleus. The mass number is 56 because that is the number nearest the atomic mass of the element. That means there are 30 neutrons in the nucleus, because the sum of the neutrons and protons must be 56.

You can include more information about the iron atom because you know the mass of u in kilograms. The mass of an iron atom in kilograms is given by: $(55.847)(1.66054 \times 10^{-27} kg) = 9.2736 \times 10^{-26} kg$.

Instead of working with the mass of one atom, many times the scientist deals with the mass of a *mole* of atoms. A mole of anything is Avogadro's number (6.02×10^{23}) of the items. (You met that number way back in Chapter 1.)

The mass of a mole of a substance turns out to be numerically equivalent to the atomic mass of the smallest particle of the substance. The mass of a mole of iron in grams, assuming that the smallest particle of iron is one atom, is 56 g. You may check our previous answer by dividing the mass of a mole by Avogadro's number: $\frac{.056kg}{6.02 \times 10^{23}} = 9.3023 \times 10^{-26} kg$, which is essentially the same as before when we used the atomic mass of iron instead of the atomic mass number. So try some problems on your own and see how you do.

def•i•ni•tion

A **mole** is Avogadro's number of items (6.02×10^{23}). Another way to think of it is as a basic unit of quantity.

Note: tables usually express the atomic mass of an atom to six significant figures. Use a calculator and express your answers to reflect the same precision.

Physics Phun

1. The atomic number of an element is 47, and the symbol is Ag (silver). What is the mass number if the atomic mass is 107.868 u? How many neutrons and protons are in the nucleus? What is the mass in grams of a mole of silver given that the atom is the smallest particle of silver? What is the mass in grams of an atom of silver?
2. What is the atomic number of a lead (symbol Pb) atom if it is made up of 82 protons, 125 neutrons, and 82 electrons? What is the mass number of the atom? What is the mass in grams of a mole of these atoms? What is the mass in grams of one of these atoms?
3. What particles, and how many of each, make up an atom of gold (symbol Au), atomic number 79, mass number 197?

The Formation of Ions

The atomic model we've been discussing features a nucleus in the center where most of the mass is found surrounded by mostly empty space except for a cloud of electrons. In Bohr's electron configuration model, the electrons occupy regions of space called energy levels (we also called them shells earlier). Only a certain number of electrons can be in a particular energy level. For some atoms—such as argon, helium, krypton, and neon—all of the energy levels that have any electrons in them are filled. These elements are stable; that means they do not combine with other elements. They are found on the far right side of the periodic table.

Some atoms have outer energy levels that are incomplete. That may mean that the outer energy level of that type of atom may contain eight electrons, but that particular atom only has six in its outer energy level. It may mean that that type of atom has all its inner energy levels filled, but the outer energy level has only one electron in it even though it needs eight to complete its complement of electrons to become stable. It may help to refer back to the table in the section where we gave you Bohr's electron configuration model. If you look at the shells (energy levels) listed, you can see that only the elements that have electrons filling the exact amount needed in each shell including those before it are stable. That means that there are very many more unstable elements than there are stable. Electrons occupying an incomplete outer energy level are called *valence electrons*. The valence electrons are the outer electrons that largely determine the chemical nature of an element.

def•i•ni•tion

A **valence electron** is an electron in an incomplete energy level. Some atoms can lose their valence electrons and become stable. However, for different kinds of atoms stability is acquired by gaining electrons until the total number of outer electrons is eight. These loosely bound electrons are largely responsible for the chemical behavior of an element. An **ion** is a charged particle that is the result of an atom gaining or losing electrons to exhibit an excess or a deficiency of electrons.

If an atom gives up a valence electron for some reason, because it has complete energy levels in its core, it may become a stable *ion*. It is no longer a neutral atom, but has become a positively charged atom called an ion because it now has one more proton than it has electrons. If an atom gains an electron for some reason, it becomes a stable negatively charged atom because it has one more electron than protons. Neutral atoms become negative ions if they gain electrons.

Another Source of Ions

An ion that might be familiar to you is an alpha particle. (That was one of the two particles that Rutherford discovered during his experiments with radium.) And even though an alpha particle is a positive ion, it quickly acquires two electrons from surrounding matter and becomes a neutral helium atom. And in that sense the alpha particle is the nucleus of the helium atom. Helium has atomic number 2 and mass number 4, so you know quite a bit about this guy given that information. You know that the nucleus has 2 protons and 2 neutrons. Because an alpha particle $^{4}_{2}He$ is just the nucleus of the helium atom, it must have a positive charge of plus 2.

Some types of atoms are radioactive. That means that the nucleus spontaneously disintegrates and emits *alpha particles* $^{4}_{2}He$, *beta particles* $^{0}_{-1}e$, or *gamma rays*. Radium $^{226}_{88}Ra$ is a radioactive element that can cause cancer if the human body is exposed to it for long. (It was discovered by Madame Curie and her husband. She later died from cancer due to her prolonged exposure to radium. It was also the element used by Rutherford in his early experiments.)

def•i•ni•tion

Alpha particles are helium nuclei that have a positive charge and are emitted from the nucleus of some radioactive elements such as radium. **Beta particles** are high-speed, high-energy electrons emitted from the nucleus of some radioactive elements, and they have a negative charge just like the electrons found in the energy levels around the nucleus. **Gamma rays** are very penetrating, short-wavelength electromagnetic radiation emitted from the nucleus of some radioactive elements. They have no charge, but do carry away energy from the nucleus.

Rutherford's Radioactive Particles

Earlier we discussed a little of Rutherford's experiments with his assistant Hans Geiger. They had set up an apparatus to study the effects of alpha particles on various substances. Let's pick up now where we left off.

Rutherford and his assistants found that most of the alpha particles passed right through the gold foil, indicating that the foil is very porous to alpha particles. This meant that many of the atoms of gold must have been empty space. Otherwise, the alpha particles would have been pushed from their paths by the charged particles within the gold atoms.

> **Newton's Figs**
>
> Particles emitted from radioactive atoms travel at very high speeds but not for very far. A sheet or two of printer paper stops alpha particles. A beta particle can travel for several meters in the atmosphere before being stopped but is absorbed by about a centimeter of aluminum. Gamma rays are not charged and travel very fast. They are much more difficult to stop.

Much to his surprise, a significant number of the alpha particles were deflected at large angles to their paths. Because the alpha particles travel at about 20,000 mi/s, that result suggested that the alpha particles were interacting with something charged positively, the same as his alpha particles. It also suggested that the positive charge of the gold atoms had to be in a relatively small volume. This led to the idea of a positive nucleus for the atom.

Almost unbelievable to Rutherford was the result that some of the particles bounced straight back, opposite the direction they were traveling initially. Do you remember the interaction of the billiard balls earlier in this book? We would have been surprised to see a billiard ball bouncing straight back from a collision. But unlike the relatively equal mass of the billiard balls that touched when they collided, the alpha particles must have collided with something very massive and charged positively. This confirmed the ideas suggested about the nucleus.

As we already mentioned, Rutherford called the thing that the alpha particles interacted with the *atomic nucleus*. Because the interactions were infrequent, he concluded that the nucleus was very small. The diameter of an atomic nucleus is about 10m to 14m, which is the order of magnitude of the diameter of an electron.

Just think about it for a moment: an alpha particle ${}^{4}_{2}He$, traveling at 20,000 mi/s, colliding with the nucleus of a gold ${}^{197}_{79}Au$ atom! No wonder there was such a change in direction for the alpha particles. Of course, Rutherford did not know their relative sizes then, but his experiment laid the groundwork for our level of understanding of the nucleus of the atom.

The Formation of Molecules

Alpha particles are not the only ions that are familiar to you. There are some ions that you've run into every day, especially if you enjoy good food. Sodium has a single valence electron in its outer energy level that the sodium atom is just itching to give up. If the outer electron left, the sodium atom would then have eight electrons just like the noble gases found at the far right of the periodic table. (Noble gases are another name given to the inert gases such as argon and krypton. Inert gases are simply gases that are so stable they don't interact with anything around them.) Chlorine is lacking one electron in its outer energy level and would be delighted to fill that shell and become more stable also. When sodium and chlorine atoms are brought into the vicinity of each other, sodium quickly gives up an electron to chlorine and becomes a positive ion. Chlorine now has one more electron than it has protons so it becomes a negative chloride ion.

Electron transfer, electron sharing, and electron exchange are key ideas for remembering how molecules are formed. You find that, in nature, unlike charges attract each other, so sodium and chloride ions attract each other and cling together to become one happy unit of a crystal of table salt. Sodium is no longer a neutral atom, nor is chloride. They are both happily combined ions. Stability is happiness to an atom. This is one way that atoms combine to form stable units. It is called *ionic bonding*. In ionic bonding, one or more electrons are transferred from the outer energy level of one atom to the outer energy level of another atom. If the ions pull toward each other, the resulting ionic units are more stable than the original separated atoms. Since energy is given off during the electron transfer and ion attraction process, the resulting product is more stable. This is like a ball rolling downhill. Once it is down, it is likely to stay there. It is energetically stable.

def•i•ni•tion

Ionic bonding is the formation of a stable unit by the transfer of an electron from one atom to the other, creating two ions that attract each other. They are more stable than they were as atoms before the transfer. The two newly formed ions attract each other, due to opposite electrical charge, thus forming a stable unit. **Covalent bonding** is the combination of two atoms to form a molecule by sharing a pair of electrons.

Other Ways to Form Molecules

When two atoms share a pair of electrons to form a molecule, the combination is called *covalent bonding*. Atoms such as hydrogen, chlorine, and oxygen exist in nature in pairs of atoms. The atoms establish a more stable particle by sharing electrons to complete an energy level. Because the first energy level is full if it contains only

Newton's Figs

Hydrogen and chlorine are two elements that occur in nature as molecules. Each molecule of these elements is made up of two atoms that share two electrons in a covalent bond.

two electrons, two hydrogen atoms can share their electrons to complete the first energy level for each and become a more stable hydrogen molecule with symbol H_2.

Similarly, two chlorine atoms share two electrons, making both atoms appear to have complete outer energy levels in that configuration as a chlorine molecule with symbol Cl_2. The more an atom's ion's electron configuration resembles the electron configuration of an inert gas, the more stable the combination becomes. Two oxygen atoms combine by making the same covalent bonding by sharing two electrons to become a more stable molecule of two atoms. You can visualize the sharing of pairs of electrons by referring to Figure 17.1. It is not intended to suggest that the atoms look like the diagram.

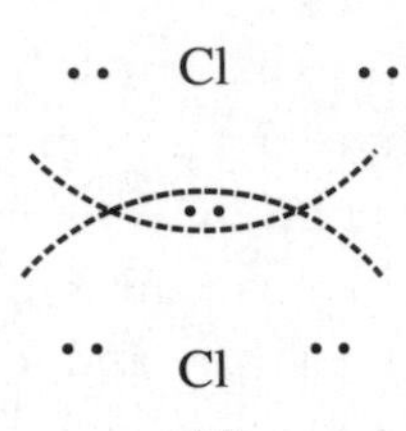

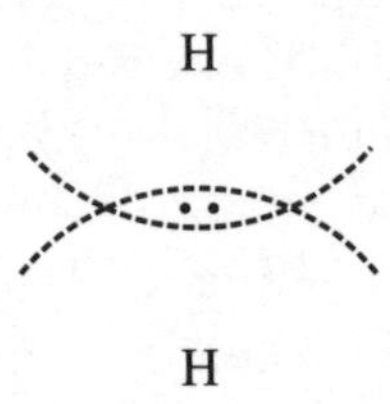

Figure 17.1

This diagram, of two hydrogen atoms and two chlorine atoms sharing a pair of electrons, helps to visualize the formation of certain molecules.

It may help you to see that two atoms can share a pair of electrons so that each atom in the combination may experience the stable configuration of an inert gas neighbor. The configuration of the electrons in atoms like these is the most stable configuration that the atoms may achieve. Some molecules are made of different atoms that share electrons, but they do not share evenly. One atom might control the electron more than the other. This uneven sharing is called polar covalent bonding. In either case, the sharing of electrons makes more stable electron configurations possible in the formation of a molecule by covalent bonding.

Metals Exchange Electrons

The atoms of solid metals are very close together. They are so closely packed (each atom of the element is surrounded by 12 other atoms) that electrons are shared freely throughout the atoms.

The reason is that the electrons in the outer energy levels are loosely held and easily moved from one atom to the next. The sharing of electrons in this setting enables the atoms to attract each other. This type of bonding is called *metallic bonding*. Later, when we look at how electric current moves through a metal wire, you'll see that the loosely held electrons that are continually exchanged are subject to being moved from one part of the solid to another by a special force. And although our discussion of how molecules are formed has only pertained to two atoms combining in some fashion to form a molecule, the same processes can involve several atoms in the formation of molecules.

def•i•ni•tion

Metallic bonding is the name of the attraction atoms of solid metals have for each other in a closely packed arrangement due to the continuous exchange of loosely held electrons in the outer energy levels of the atoms of the metal. Metals are found on the left and bottom side of the periodic table.

Movement of Atoms and Their Components

In the last few sections of this chapter, we've discussed the movement of some atoms and their components, such as the nucleus of the helium atom. Alpha particles emitted by radioactive atoms travel at about 20,000 mi/s. The beta particles are high-speed electrons, meaning they travel at speeds around 80,000 mi/s and up. The vapor trails of these particles can be observed easily in a *cloud chamber*. And it's always good to remember that the speed of atoms and molecules of matter depends upon the state and the temperature of the material.

At S.T.P., nitrogen molecules travel at about 500 m/s. Molecules of solids vibrate about an equilibrium position with amplitudes of about .01 nanometers with frequencies of 1,013 cycles/s. The atoms have diameters of about 0.2 nanometers, and molecules range in diameter from a few nanometers to hundreds of nanometers depending on the number of atoms in the molecule. The particles of matter race around, bounce off containers, and vibrate at incredible speeds. The type of motion depends on the state of matter, and the speeds are increased by increased temperature.

It is interesting to note that not only does radioactive material emit alpha particles but also beta particles! There are no electrons in the nucleus, and yet beta particles traveling at high speed and tremendous penetrating power have their origin in the nucleus.

The emission of a beta particle occurs when a neutron decays into two particles—a proton that stays in the nucleus and a beta particle that flies out of the nucleus. Let's consider the following reactions, one for alpha decay and one for beta decay:

$${}^{226}_{88}Ra \rightarrow {}^{222}_{86}Rn + {}^{4}_{2}He + energy$$

$${}^{60}_{27}Co \rightarrow {}^{60}_{28}Ni + \beta$$

The alpha decay yields a daughter nucleus with a mass number that is four less than the parent nucleus and an atomic number that is two less than the parent nucleus. The beta decay yields a daughter nucleus that has the same mass number as the parent, and the atomic number of the daughter nucleus is one higher than the parent nucleus. Radium has a *half-life* of about 1,620 years. That means that in 1,620 years a sample of 1.0 g of radium will become a mixture of 0.5 grams of unchanged radium and 0.5 grams of original radium sample that will have decayed to other atoms.

def•i•ni•tion

The **half-life** of carbon 14 is about 5,730 years, and the process of carbon dating is fairly accurate up to about 60,000 years.

And that's a wrap for your introduction to the atomic and subatomic world. We encourage you to explore these levels of the physical world in more detail, because so much of what goes on at these levels pertains not only to physics and chemistry, but also to some of the most profound questions about the origins of life, the state of our global environment, and the way in which energy will be used in the very near future.

Problems for the Budding Rocket Scientist

1. What is the number of neutrons in the nucleus of $^{22}_{11}Na$?
2. How many protons, neutrons, and electrons are there in (a) ^{3}He, (b) ^{12}C, (c) ^{206}Pb? Hint: you'll need to refer to a periodic table of elements to find out the atomic number of each element.
3. Carbon has two stable isotopes. Natural carbon is 98.89 percent carbon 12 and 1.11 percent carbon 13. Calculate the average atomic weight of carbon. (The atomic weight of ^{12}C is exactly 12.)

The Least You Need to Know

- An atom in its normal state is electrically neutral.
- The model put forth by Rutherford was known as the planetary model, because the structure of the atom was similar to that of our solar system.
- The atomic mass number and the atomic number are the two numbers used to identify an element and to indicate the number of nucleons it has.
- Electron transfer, electron sharing, and electron exchange are key ideas for remembering how molecules are formed.

Chapter 18

Electricity at Rest

In This Chapter

- Electrical charge at rest
- Substances that carry a charge
- Two ways to charge an electroscope
- Charging and discharging capacitors

As you walk down the sidewalk on your way to class, or perhaps on your way to go shopping, your cell phone vibrates. So you take the earpiece from the headphones to your MP3 player out of your ears and answer it. It's your friend calling about going to see a movie tonight. So you whip out your PDA and check to see if you're free tonight. All the while, you've been standing on the corner waiting for the light to change, and as you hang up the phone you notice the quiet hum of a car going by powered by an electric motor. The pedestrian light finally changes and as you look down the street you see lightning flash in the distance. Hmm, did you remember to bring your umbrella? You check in your backpack, but find only your digital camera, laptop computer, and some batteries. Oh well, it's pretty warm, what's a little rain? As a matter of fact, it's so warm you decide to take your sweater off. In doing so, you feel the crackle in your hair and wonder if you look like Einstein now.

Electric charge, electric currents, batteries, and electromagnetic signals are so much a part of our daily lives that we barely notice them anymore. Most of the time, we are surrounded by electrical phenomena in both the natural world and our constructed world. We've learned about the presence of electricity through experimentation and observation, and we've learned how to harness it as well to control and store electrical power for our use. We live in an electric world.

Early Ideas About Electricity

As with many fields of study in physics, we find that the ancient Greeks were the first to spend time considering and thinking about natural phenomena. Electricity is no exception. They discovered that when a material called *amber* was rubbed with a cloth, it would attract small pieces of straw. In later centuries, it was found that other materials possessed the same characteristic. There seemed to be some property of cloth that was apparently transferred to the amber when the two were rubbed together. That property enabled the amber (for a brief time) to attract small pieces of straw, or paper.

Since there seemed to be some sort of attractive force exhibited by the amber, early studies confused this characteristic with magnetism. William Gilbert, physician to Queen Elizabeth I, was the first person to conduct experiments to see if there was a connection between the two.

Gilbert wrote the earliest work on magnetism, *The Magnet.* (He probably spent all of five minutes coming up with that title.) In it, he discussed how the small spheres of magnetite (a form of iron ore that exhibits magnetic properties) that he experimented with were similar to the magnetic nature of Earth. He thought that the earth was just a huge magnet and was one of the greatest contributors to the understanding of the earth's magnetic field. It gave a rational explanation for the attraction of compass needles to the north and south poles. The publication of this work gave him such a great reputation that, in his honor, the unit of magnetic strength is called, you guessed it, the gilbert. However, this name didn't last very long as a standard unit of measure, and was replaced with the current standard for representing magnetic field strength, the tesla. (Named after Nikola Tesla, who invented the means to apply alternating current in practical, large-scale ways. It's what runs in our homes and businesses.)

But besides his theory of the magnetic nature of the earth, Gilbert also noted many other substances that, when rubbed, behaved just like amber. The word *electricity* comes from *elektron*, the Greek word for "amber." Gilbert used the word *electrum*, which is the Latin word for "amber," as the basis for a word that he invented, *electrica*.

He used this word for all substances that behaved like amber. His studies ultimately showed that magnetism and static electricity were different.

Static Electricity

All of us have experienced the phenomenon of static electricity. If you run a comb through your hair in dry weather (or remove a sweater on a dry day), you will sometimes hear a crackling noise. When you've been sitting in a car with cloth seats, you might get a slight electric shock when you touch the metal door as you slide across the seat to get out. Static electricity is simply the presence of electric charge (positive or negative) on the surface of the material. The branch of physics that studies the nature of charge that is not moving is called electrostatics.

Any two substances that are rubbed together can potentially become charged. If the substances are initially electrically neutral (as most substances are), a transfer of charge will give one substance a net positive charge and one a net negative charge. Electric current results from a movement of electric charges.

Experiments in the eighteenth century showed that a stick of hard rubber could be electrified by rubbing it with a piece of fur, and that likewise, a glass rod or tube could be electrified by rubbing it with a piece of silk. Like amber, both of these rods will attract small pieces of paper (just like the straw that the Greeks used); however, the hard rubber and the glass rod aren't identically electrified. Further experimentation showed that two electrified pieces of hard rubber will repel each other, but an electrified piece of hard rubber and an electrified piece of glass will attract each other.

What had been discovered by these experiments was a new force. Like gravity, this electrical force was exerted between two objects; unlike gravity (at least as far as we currently understand it) the electrical force can be either attractive or repulsive. Objects with the same charge (both positive or both negative) repel one another, and objects with opposite charge (one positive and one negative) attract one another. Benjamin Franklin was the first to name the two opposite charges positive and negative. He called the charge found on the glass rod positive (+) and the charge on the hard rubber negative (–).

Electric Charge and Atomic Structure

The nature of electrical force is fundamental to the understanding of chemistry, molecular structure, and the structure of the atom. Experiments to probe the strength of the electrical force have found that, over short distance, it is far stronger than the force of gravity and it is intimately related to the generation of electromagnetic waves.

As we learned in Chapter 17, all atoms consist of a nucleus that is made up of protons (+) and neutrons (no charge) (except for hydrogen, which has no neutron), as well as the negatively (–) charged electrons that surround the nucleus. Under normal conditions, atoms are neutral, with the positive charge of the nucleus (from the protons) being equal to the negative charge of the electrons. Such atoms are electrically neutral. And if you remember, an atom that has an excess or deficit of electrons is called an ion.

The electrons that surround the nucleus are indistinguishable and are held to the nucleus by an electrical force that decreases with the distance squared, much like the gravitational force.

By understanding the electrical structure of the atom, we'll be able to see how materials can end up with a net positive or negative charge. The outermost electrons in many materials are relatively loosely bound (because the electrostatic force drops off with the distance squared), and they can be easily knocked loose. When some materials are rubbed together, one material may be better able to hold on to extra electrons than the other. This material accepts extra electrons and the other material loses some electrons, giving a net negative and positive charge to the two substances. The nuclei of the atoms in the glass, rubber, silk, and fur are unaffected; only their outer layers (energy shells) where the electrons reside are in contact.

Conductors and Insulators

Materials that easily allow charged particles to pass through them are called *conductors*. Metals are good conductors of electricity and thermal energy. Recall the metallic bonding in which electrons are attracted to several different nuclei at the same time, thereby making them fairly mobile rather than localized as is the case in most bonding situations. Those electrons can be pushed around the metal easily.

Conversely, materials that do not easily allow charged particles to pass through them, such as the glass and hard rubber in our example of charge transfer, are called *insulators*. If a charged body is to retain its charge, it must be supported by a good insulator. If it is not, all the extra charge that builds up on it will simply flow out, like water out of a drain. Insulators act like drain plugs, holding in (or out) excess charge.

def•i•ni•tion

Conductors are materials that allow electrons to move freely throughout the material. An **insulator** is a material that does not allow free movement of charge.

Charging an Object

The first law of static electricity is: like charges repel each other, and unlike charges attract. That means that protons repel protons and attract electrons. In general, a body that has a deficiency of electrons is charged positively and an object with an excess of electrons is charged negatively. Two bodies attract each other if one is charged positively and the other has a negative charge.

> **Newton's Figs**
>
> The magnitude of the charge on the proton is the same as that on the electron. The charge on the proton is called positive and the charge on the electron is called negative. Electrons are free to move in a metal conductor while the protons remain in the relatively stationary nuclei of the atoms. Both positive and negative ions are free to move in a solution.

Suppose that you charge a hard rubber rod negatively by rubbing it with fur. Friction helps you to deposit electrons on the insulator by rubbing them off the fur. The insulator is now negatively charged because it has an excess of electrons after being rubbed by fur. Now you may take the negatively charged insulator to the location of another insulator, rub some of the electrons off the first insulator onto the second, and then they both will be negatively charged. The second insulator has been charged by contact.

By examining in detail the procedure of charging an object we can see that there are three characteristics of charging by contact. One is that the charged body must touch the charging body. The second is that the charged body ends up charged with the same kind of charge as the original charging body. Finally, the body doing the charging ends up with less magnitude of charge than it had before it shared its charge with the second body. The implication is that there must be at least one other way of charging a body. There is, and you'll see that method later in the chapter.

Let's suppose that you rub a hard rubber rod with fur. The rubber rod insulator would be negatively charged. If you take a piece of wire, a conductor, and hold it in your other hand, you could discharge the insulator by rubbing the wire across the insulator several times. The wire discharges the insulator faster if you actually drag every inch of the insulator across the wire. You essentially rake the electrons onto the conducting wire and into your body, which acts as a large reservoir, or ground.

> **Johnnie's Alert**
>
> Even if an isolated conductor has just one electron in excess, the conductor is negatively charged. It would be negatively charged regardless of how many electrons it had in excess. The difference is in the magnitude of the charge, not the type of the charge.

Johnnie's Alert

Ground or zero potential energy has the same meaning for electrons as it does for objects when considering their gravitational potential energy with respect to the surface of the earth. Electrons at ground have zero electrical potential energy just like an object lying on the surface of the earth (ground) has zero gravitational potential energy.

The wire is a good conductor, so the electrons move freely from the charged insulator through the wire to your hand. Your body is a fair conductor and provides a path for the electrons to move to the earth or ground zero unless you are standing in well-insulated shoes. The earth is a humongous reservoir of electrons as well as an endless source of electrons. A conductor that provides a path for excess electrons or a deficiency of electrons discharges the charged insulator.

The charges in the discussion so far are small; only a relatively few electrons are involved in charging and discharging the insulators.

Coulomb's Law

The SI unit of charge is the *coulomb* (c), and a charge of one coulomb is equal to the charge of 6.25×10^{18} electrons. Taking the inverse of this number reveals that the charge of a single electron is very small: $1.6 \times 10^{-19} C$.

Coulomb's law describes the electrical force between two charged particles separated by a distance r as varying inversely with the square of the distance, exactly like the gravitational force that we have discussed. Charles Coulomb (who else?) (1736–1806) discovered this relationship, which can be stated generally as

$$F_c \propto q_1 q_2 / r^2$$

This proportionality becomes exact if we write it with the constant

$$F_c = k \frac{q_1 q_2}{r^2}$$

where F_c is the electrical force (called the coulomb force), k is the coulomb constant, q_1 and q_2 are the charges of the two objects exerting a coulomb force on each other, and r is the separation between the two objects. In SI units,

$$k = 8.9875 \times 10^9 \frac{N \cdot m^2}{C^2}$$

You might wonder why the atom doesn't collapse because of the mutual attraction of the electron and the proton, and why the nuclei of atoms don't fly apart, filled as they are with positively charged nuclei. Don't they all repel one another?

It turns out that other forces are at work holding the nucleus of the atom together, and that electrons can be thought of as having angular momentum that keeps them in their orbits.

def•i•ni•tion

The **coulomb** is the fundamental unit of charge in the MKS system of measurement. The MKS system is the practical system for the study of electricity.

Detectors of Static Charge

The gold leaf electroscope or the aluminum foil electroscope is a very sensitive instrument for detecting electrical charge. It consists of a metal knob attached to a metal rod that has two leaves attached to the other end of the metal rod. The metal rod passes through an insulator that is embedded in a metal frame with transparent glass sides. The leaves are viewed easily through the glass and the insulator isolates the metal rod with the knob at one end and the leaves at the other.

The major components of the electroscope are shown in Figure 18.1. The neutral electroscope is shown in 18.1 (a) where the leaves hang limp with equal positive and negative charges. The electroscope indicates that a charge has been placed on the knob and conducted to the rest of the conducting parts, including the leaves, when the leaves spread apart.

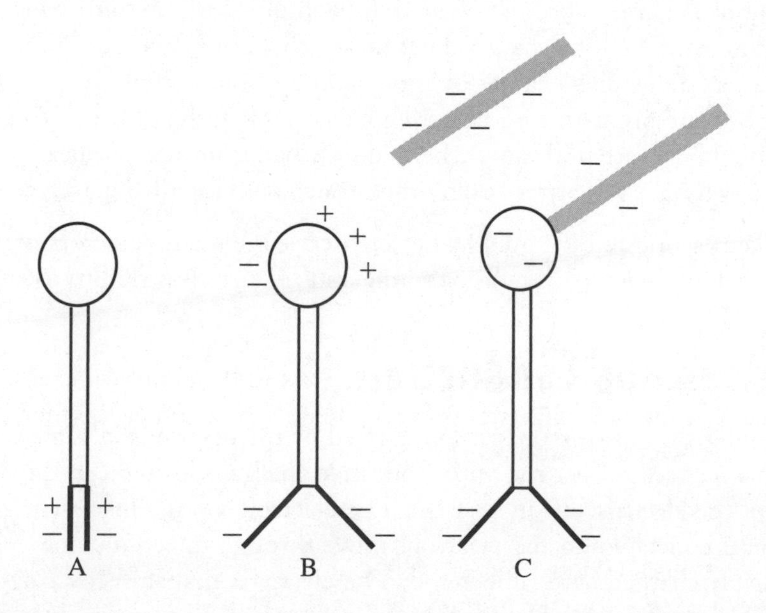

Figure 18.1

The gold leaf electroscope is a sensitive detector of electric charge.

Another good detector of static charge is the *pith-ball* electroscope. In such an electroscope, small balls of *pith* are suspended on insulating strings. Charge can be deposited on the pith-ball by touching it with a charged insulator. If an object of the same charge is brought near the charged pith-ball, it is repelled dramatically some distance from the charged object. Today pith-balls are not really made of plant stem tissue anymore. They're made of Styrofoam, and sometimes they are spray-painted with aluminum.

def•i•ni•tion

Pith is a very light, dry, fibrous material. Pith is the central column of spongy cellular tissue in the stems and branches of some large plants. The **pith-ball** electroscope gives you an idea about the force that has been alluded to pushing the pith-balls apart. The force is very much like the gravitational force except its source is charge. The force is called the coulomb force and is calculated as $F_c = k\frac{q_1q_2}{r^2}$, where q_1 and q_2 are the charges on the two bodies, *r* is the distance between the charges, and *k* is a constant that depends on the kind of insulator between the two charged bodies. Often it is a vacuum or air.

If an object with an unlike charge is brought near a charged pith-ball, the pith-ball swings dramatically toward the unlike charge. That means that once charged, the pith-ball electroscope can detect like charges by moving away from a similarly charged object. There is one problem with detecting a charge with the pith-ball electroscope that you should be aware of. If an uncharged pith-ball is brought near a charged pith-ball, the two balls will attract each other, touch, and then fly apart.

Both the pith-ball electroscope and the gold leaf electroscope are good detectors of charge, but the gold leaf electroscope is more versatile and requires less skill to use.

Charge by Induction and Conduction

A procedure for charging the electroscope is summarized by the diagrams in Figure 18.1 (b) and 18.1 (c). The hard rubber rod with a negative charge is brought in the vicinity of the knob of the electroscope in 18.1 (b). The electrons on the insulator repel the electrons on the metal knob and electrons move through the conductor away from the negatively charged hard rubber rod. The electrons that shift toward the leaves of the electroscope cause the leaves to repel each other. That means that the leaves have like charges, and because you know that the insulator has a negative charge, then the charge on the leaves is negative as long as an excess of electrons remains in the leaves. The knob on the top of the electroscope would be deficient in electrons, so it would be positive until the electrons move back to their original configuration. Keep in mind that the number of protons and electrons in the electroscope has not changed. The electroscope as a whole remains neutral.

We say as long as an excess of electrons remains in the leaves, because if the insulator is taken away in Figure 18.1 (b), the leaves will hang limp again because the electrons will shift back toward the knob rather than stay near each other in the leaves. Remember that the electrons are repulsive to each other. The negatively charged insulator is allowed to touch the knob in Figure 18.1 (c), enabling electrons to actually move onto the metal knob. The electrons move to the knob because as the charged insulator approaches the knob of the electroscope, the shift in electrons discussed above occurs. The apparent positive charge on the knob that occurs as the charged insulator comes in the vicinity of the knob attracts electrons from the insulator. The electroscope is no longer neutral as a whole. It now actually has an excess of electrons. Refer to Figure 18.2 to see what happens when the insulator is taken away. Can you explain why the leaves would behave as they do?

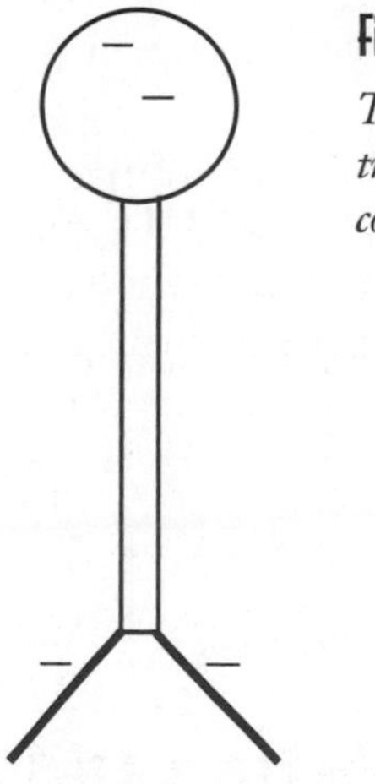

Figure 18.2

The negatively charged electroscope results from direct contact.

The knob collected some electrons when the insulator was in contact with it, leaving a net negative charge on the electroscope. The leaves of the electroscope spread apart, indicating that they have a net charge on them.

If you bring a positively charged rod, one that has had electrons removed by friction, near the knob of the charged electroscope in 18.2, as indicated in the diagram of Figure 18.3 (a), the electrons move toward one side of the knob.

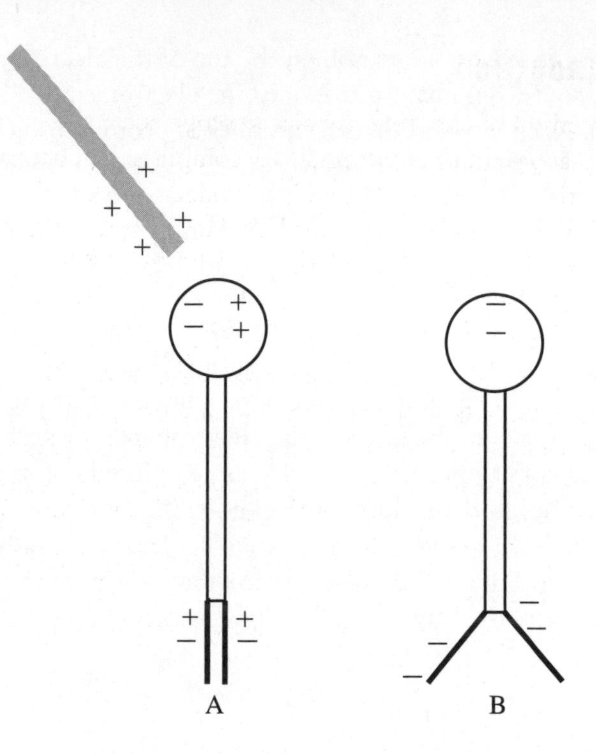

Figure 18.3

The electroscope detects charge if it is in the vicinity of the metal knob of the electroscope.

They are attracted to the positive charge on the insulator. The electrons throughout the electroscope are attracted to the positive charge that is in the vicinity of the knob. Because the leaves, connecting rod, and knob of the electroscope are all metal, the electrons move toward the area of attraction near the knob.

Johnnie's Alert

Keep the leaves of the electroscope hanging freely so that the tiniest charge will affect the movement of the leaves. Make sure not to touch the leaves with your hands.

The leaves collapse, indicating that the net charge on the electroscope is apparently neutral. If the positively charged insulator is removed, as shown in the diagram of Figure 18.3 (b), the electrons flow back. The electrons move throughout the conductor, repelling each other until they are equally distributed. Then the leaves open to their original position, indicating that there is no net loss or gain of electrons due to the presence of the positive charge as long as it did not actually touch the electroscope.

The procedure outlined here is called charging the electroscope by direct contact. The charging body actually touches the electroscope.

Charging by Induction

Another method of charging the electroscope is charging by induction. An important principle is applied in this process to accomplish the charging of the electroscope. At the proper time, a path to ground is provided for a supply of electrons. The path to ground is the body of the experimenter. Simply touch the knob of the electroscope and electrons can either flow off to ground or travel from ground onto the knob. Remember that by definition, a ground is a charge reservoir to which or from which electrons can move easily. Refer to Figure 18.4 for a step-by-step procedure of charging by induction. The electroscope is shown in Figure 18.4 (a) in its neutral state. The charging body is brought in the vicinity of the knob in Figure 18.4 (b), causing the leaves to spread apart because there is an excess of electrons on them. The excess electrons are the ones that were repelled down from the knob.

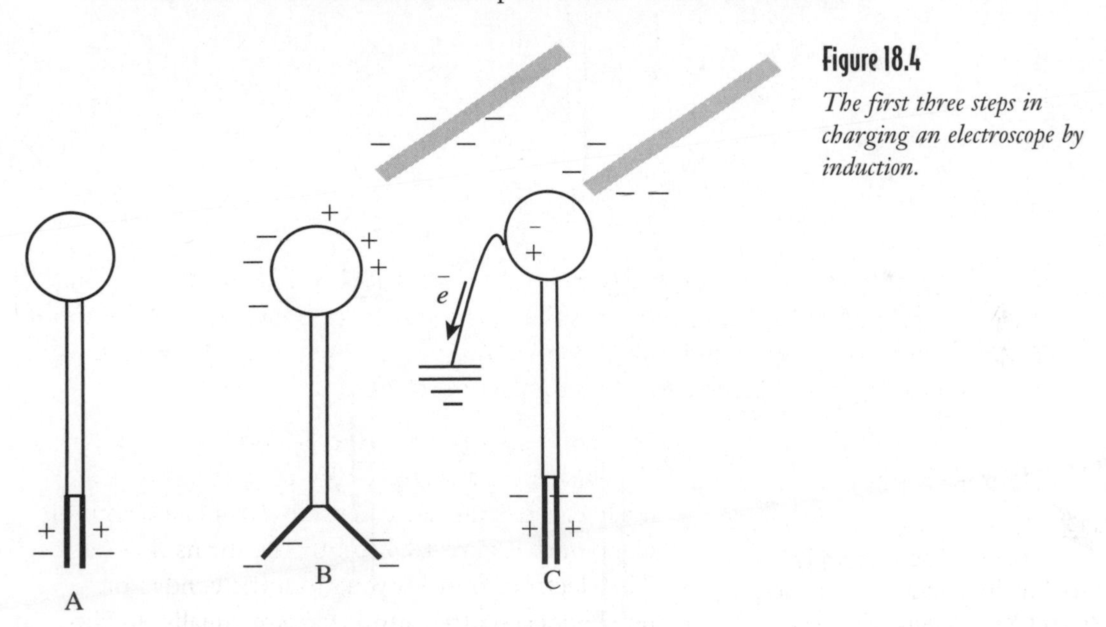

Figure 18.4

The first three steps in charging an electroscope by induction.

The ground connection is made in Figure 18.4 (c) while the charging body is still in the vicinity of the knob. This means that you touch the knob of the electroscope with your finger. Excess electrons are repelled off the electroscope to ground in the direction of the arrow. They are moving away from the negative insulator. Notice that the negative insulator can affect the electrons in the electroscope even though it is not touching the electroscope. The ability to affect a body at a distance is part of the idea of force fields. We have discussed gravitational and electric force fields so far without actually giving the idea a name. And we talk a bit more about the electric field in the next section.

> **Johnnie's Alert**
>
> It's a good idea to keep the source of charge far enough away from the electroscope so that it won't leave a residual charge by accident.

Electrons are represented symbolically as $\overline{e}$. Enough electrons are repelled to ground to make the leaves of the electroscope appear neutral because they drop down. Next, the ground connection is removed from the knob in Figure 18.5 (a), while the charging body is still in the vicinity of the knob and the leaves of the electroscope appear to be neutral. The leaves are spread apart in Figure 18.5 (b) when the charging body is removed, showing that overall the electroscope does have a charge on it. The electrons distribute themselves evenly throughout the electroscope, but there is now a deficiency of electrons. If you want to check the charge on the leaves of the electroscope, bring the negatively charged body near the knob and it will repel electrons to the leaves, causing them to start to drop.

Figure 18.5

The final steps in charging by induction show that the electroscope is charged opposite the charging body.

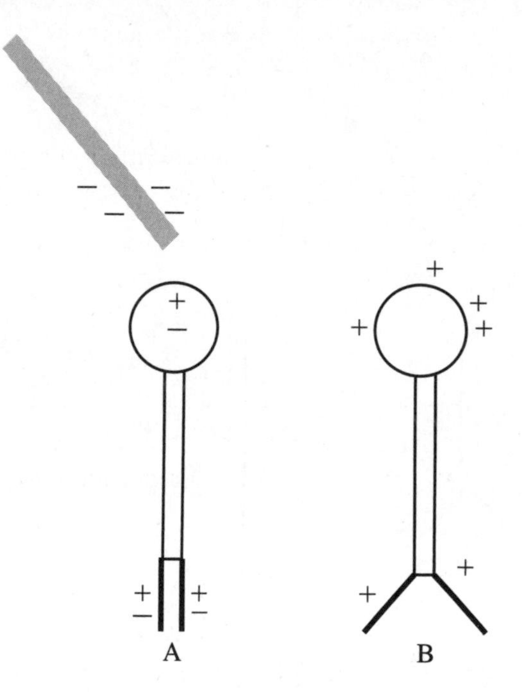

Because some electrons were repelled off to ground and the ground was removed so that they could not return, the electroscope has fewer electrons than it had initially. The electroscope is left charged positively, the opposite charge of the charging insulator. The charging body never lost any of its original charge and never touched the electroscope during this process. This process is involved in the theory of a *capacitor*, and the same process is used to explain the topic in the next section.

Electric Fields and Capacitors

We have seen that the electrical force, like gravity, can act between objects that are not physically in contact. In fact, any two charged particles will exert force on each other; but even a lone charged particle is surrounded by what we call an electric field that is a direct result of its net charge. The electric field (like the gravitational field of an object with mass) has both a magnitude and a direction; that is, it is a vector quantity.

def•i•ni•tion

A **capacitor** is an electrical device used to store an electrical charge. The capacitor is made up of two conductors separated by a good insulator. The conductors are often referred to as the plates of the capacitor.

The electric field is measured in units of force per unit charge, so that a charge q that experiences a coulomb force F is in an electric field of strength

$$E = F/q$$

where F is the coulomb force, and q is the charge of the particle. The SI unit for electric field strength is simply N/C (newtons divided by coulombs).

Because the electric field represents a vector quantity, we can draw field lines that represent the motion that positively charged particle would take if it were nearby. For this reason, positively charged particles have field lines emanating from them, and negatively charged particles have field lines that point toward them.

Electric field lines are always drawn as though they originate at positive charges and terminate at negative charges. Figure 18.6 shows an example of field lines around a positive charge, a positive and a negative charge, and two oppositely charged plates.

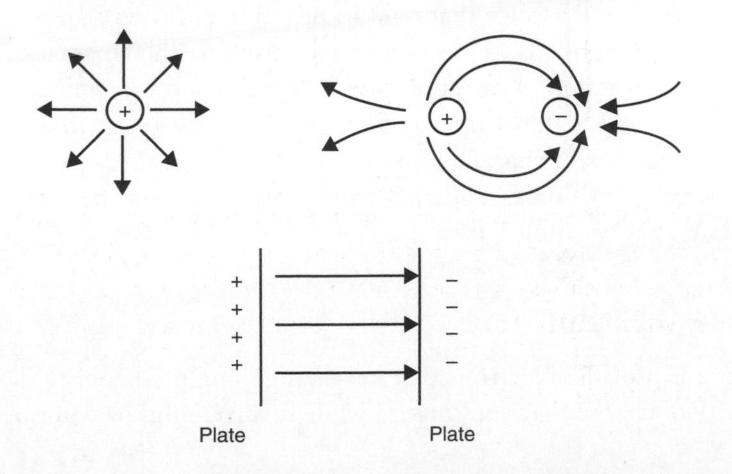

Figure 18.6

Examples of field lines around a positive charge, a positive and a negative charge, and two oppositely charged plates.

Generating Much Larger Amounts of Charge

The use of fur and an insulator to generate an electrostatic charge is fine when you are charging and discharging insulators used with the electroscope. There are times when you might want to generate a considerable amount of charge for some reason. The *Van de Graaff generator* is an instrument designed to generate high amounts of static charge. You may have seen demonstrations with that instrument.

> **def•i•ni•tion**
>
> A **Van de Graaff generator** is a motorized source of electrons of high potential energy.

> **Johnnie's Alert**
>
> The metal sphere of the generator must be clean and free of particles of any kind. Tiny particles can allow the charge to ionize the air and discharge the sphere so that it is unable to charge to capacity.

The generator is essentially a motorized version of the fur-and-insulator method of generating electrostatic charge. The major difference is that the charge is deposited on a spherical surface at the top of the generator. After the generator runs for a minute or so, there is a tremendous amount of charge on the spherical surface.

When a conductor (like the knob and leaves of the electroscope) is isolated by an insulator, the charge placed on the conductor distributes itself on the surface of the conductor. None of the electrostatic charge is on the inside of the conductor. It resides only on its surface. On a spherical surface, all of the charge is evenly distributed on the surface of the sphere, but appears to act as if all of the charge is concentrated at a point at the center of the sphere.

If you have a conductor with a surface that is not spherical, the electrostatic charges will repel each other in an attempt to become distributed evenly on the surface. They become very crowded on parts of the surface that tend to be pointed. If there are points on the surface, the crowding of charge becomes so intense that the air molecules become ionized. When ionization of the atmosphere near the point or points takes place, you may observe a bluish glow called a *corona* or *brush discharge* at those points. The charged object becomes discharged rapidly when ionization takes place. Sailors have observed a discharge like that emanating from sharp points on the masts of ships. They call the eerie glow St. Elmo's fire.

Charging and Discharging a Capacitor

The Van de Graaff generator provides electrons that have a high potential energy because of the high density of charge that you can use while it is running, or you can take electrons from the generator and store them on a capacitor. The *Leyden jar* is a good demonstration capacitor when it is used properly. As can be seen in Figure 18.7, the Leyden jar is a simple device, and yet it is a very dramatic piece of demonstration equipment. The Van de Graaff generator delivers electrons that have potential energy upward to 500,000 volts.

Newton's Figs

Capacitors do a lot of useful things for you including helping you to tune in your favorite radio station, AM and FM.

def•i•ni•tion

A **corona discharge** or **brush discharge** is a bluish glow of ionized gases formed at any sharp point of a conductor that is under the influence of a high concentration (or density) of electrons.

In the late 1700s, a group of scientists in Leyden, Netherlands, built an apparatus that could hold a huge charge. Their invention was known as a **Leyden jar.** This early capacitor would be later developed into a condenser, which is simply a very powerful capacitor that can also hold and store large amounts of electrical charge.

When the Leyden jar is charged, most of the charge can be delivered in a fraction of a second. The same is true of the charge on a television picture tube. A lot of charge delivered in a short time is the real danger even when the voltage is low. Now take a look at Figure 18.7. The symbol to the right of the Leyden jar represents a capacitor in the schematic diagram of some electrical circuits.

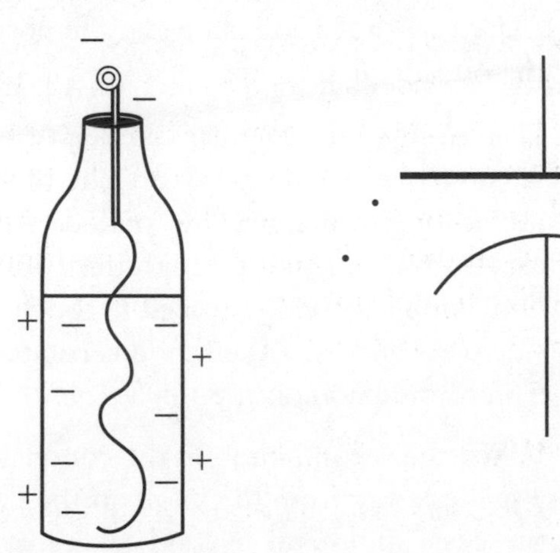

Figure 18.7

The Leyden jar and a symbol for a capacitor are diagrammed to emphasize a method for storing charge.

As you can see by looking at the figure, one side of the symbol for a capacitor is positive and the curved side of the symbol is negative. Both sides are conductors and you can see that they are separated by space. The space may be filled with any kind of good insulating material such as air, glass, or mica. Air is a good insulator for some applications of the capacitor, such as a tuning capacitor for a radio.

The Leyden jar is a thick glass jar with the lower portion covered with a thin layer of metal both inside and outside the jar. The glass, a good insulator, separates the metal conductors. The inside conductor may be connected to the metal knob outside the top of the jar by a chain that is long enough to reach the bottom of the jar. The chain is permanently attached to one end of a metal bar that is imbedded in a wooden top for the jar. Wood is also a good insulator. The other end of the metal bar contains the metal knob where connections to the inside conductor can be made. It has a round knob rather than a sharp point so that the high-potential electrons don't escape into the surrounding air. The metal sphere atop the Van de Graaff generator is first connected loosely to the knob or metal post of the Leyden jar. The generator is turned on and allowed to run for a short time. When the Leyden jar is fully charged, the generator is disconnected and the jar is placed in an isolated area. It now has many electrons packed on the inside surface of the jar, charging it negatively.

The negative charge on the inside surface of the jar induces a positive charge on the outside surface. The negative charge accomplishes that by repelling electrons from the outside surface to ground. The demonstration table supporting the jar is at electrical ground potential. The atoms of the insulator become aligned so that the electrons of those atoms are largely on the side away from the negatively charged plate. That leaves the positive side of the atoms of the insulator on the side near the negative plate. The atoms of the insulator in that state of stress help to hold the negative charge on the inside plate and the positive charge on the outside plate.

Johnnie's Alert

If you have the opportunity to work with a Van de Graaff generator, do not allow your feet to extend over the sides of the insulated stool, because dangling toes often discharge to the ground. You should not stand too close to the table that holds the generator because discharges from your body to the table can occur.

Discharging a Leyden jar can be pretty spectacular. It's done with a conductor attached to the end of the meter-long insulator. The conductor has two spherical knobs connected through the 10-inch conductor. One knob is carefully placed in touch with the outside plate and the other knob is rotated slowly toward the knob on the Leyden jar.

When the second knob of the conductor is from 6 to 4 inches away from the knob on the Leyden jar, there is a simultaneous loud crack like a large firecracker and a blue arc of discharging electrons! It's possible to obtain a second discharge. By repeating the same procedure another loud snap with a smaller bluish spark can be seen. This can be repeated until the jar is completely discharged.

Lightning and Charged Rain Clouds

Sharp objects like lightning rods can protect a house made of wood, bricks, or some other nonconducting material. The lightning rods can be placed at high strategic places on the building and, being well grounded, provide a path directly to ground for lightning. Any high object is a target for lightning. If you are caught out in a thunderstorm the best thing to do is get as low as possible, leaving higher targets for lightning. Any time a cloud in a rainstorm becomes charged, positively or negatively, it will induce the opposite charge on the earth.

As the intensity of the charge builds, the chance of a bolt of lightning grows. Lightning discharges can take place between clouds and from the earth to the cloud or from a cloud to the earth. It is said that one hundred bolts of lightning hit the earth each second. Any time charge is separated, for whatever reason, there is a good chance for a tremendous arc of energy that will occur to neutralize the charged bodies involved.

So now that you've had a good introduction to electricity at rest, our next chapter will cover electricity in motion. You'll learn all about current, voltage, resistance, and circuitry. Magnetism will be explored, and in the process you'll discover the close relationship between electricity and magnetism.

Problems for the Budding Rocket Scientist

1. What is the strength of the coulomb force between the two components—an electron and a proton—of a hydrogen atom? Is the force attractive or repulsive? The mean distance between the proton and electron is tiny, about $5.3 \times 10^{-11}m$. A proton has a charge of $1.6 \times 10^{-19}C$, and an electron has the exact opposite charge of $-1.6 \times 10^{-19}C$.
2. In the preceding problem, you calculated the coulomb force between an electron and a proton. Calculate the gravitational force between the same two particles using the universal law of gravitation.
3. Given the definition of the word *ion*, what would you say is required to "ionize" an atom?

The Least You Need to Know

- Objects with the same charge (both positive or both negative) repel one another, and objects with opposite charge (one positive and one negative) attract one another.
- The magnitude of the charge on the proton is the same as that on the electron.
- Electrons at ground have zero electrical potential energy just like an object lying on the surface of the earth has zero gravitational potential energy.
- The coulomb is the fundamental unit of charge in the MKS system of measurement. The MKS system is the practical system for the study of electricity.
- An electroscope can be charged either by conduction or induction.

Chapter 19

Electricity in Motion

In This Chapter

- Early discoveries in electricity
- Electromotive force and current
- Magnetic fields
- Resistance and Ohm's law
- Series and parallel circuits

The only kind of electricity that we've discussed so far is static, nonmoving electricity, whose charge can be expressed in coulombs. But the type of electricity that you're most familiar with is the kind you get by plugging an electrical device into a wall socket or using something that runs on batteries. That type of electricity will be the focus of this chapter. And to begin our discussion of moving electricity, let's go back and visit a few very electric Italians who laid the foundation for the age of electricity.

Volts, Amps, and Ohms

The story goes that one day while sitting in a restaurant, Luigi Galvani (1737–1798) noticed that the severed frog legs that were hanging from copper hooks jumped when they touched the iron balcony railing. He decided to conduct some experiments with frog legs. When he inserted a two-pronged fork made of one iron and one copper prong, he noticed that the frog leg moved or contacted. He called this animal electricity and thought that the nerve in the frog leg was responsible for producing this movement. He was wrong. In a moment you'll see why.

Newton's Figs

Luigi Galvani never had a unit of measure named after him, but the galvanometer, which is used to measure and detect small electric currents, and the word galvanize, which is a process used to plate metal with zinc, both use his name.

Another Italian by the name of Alessandro Volta (1745–1827) had been doing experiments using different types of metals. He thought that he could stimulate his sense of taste by placing two different metals on his tongue. When he placed a silver coin on the back of his tongue and touched the tip with a piece of tin, he perceived a strong acid taste. Touching other dissimilar metals to various parts of his body—don't ask where—caused a sensation of light in his eye. From these experiments, Volta concluded that the flow of electric charge was due to an electrical imbalance between two metals. The key to this flow was the interposition of a moist conductor between two dissimilar metals. The nerve in Galvani's frog legs simply acted as the conductor for the electrical charge.

Alessandro constructed what is known as a voltaic pile using a large number of alternating disks made of copper and iron or zinc separated by layers of cloth soaked in a salt solution. This pile was the prototype of today's modern electric batteries. In this process, he also discovered what is known as EMF, electromotive force, which is more commonly called volts. As you can see, the highest honor that a scientist can receive is to have a unit of measure named after him. And we'll have more to say about EMF in just a little while.

Electromagnetism and Amps

Early investigators of electric and magnetic phenomena knew that there was some sort of relationship between the two, but they couldn't quite get the connection. Magnets didn't influence electric charges, nor did charges affect magnets. The person who finally made the connection was a Danish physicist named Hans Christian Orsted (1777–1851).

Hans had built a voltaic pile to conduct experiments with. In 1820, while lecturing at the University of Copenhagen, he got an idea. Static electricity didn't seem to have any effect on magnets, but maybe the electricity moving through a wire connected to the two poles of his voltaic pile would. He connected a piece of wire across the two poles and placed a compass next to it. To his amazement, the compass needle pointed toward the wire, and no longer pointed to magnetic north. Orsted then reversed the wires and found that the compass needle reversed direction, too. He had found a relationship between electricity and magnetism and coined the term *electromagnetism.*

While in Paris, a French mathematician and physicist by the name of André-Marie Ampére (1775–1836) heard of Hans Orsted's famous discovery. Within weeks, he performed a number of his own experiments and found that not only does current (something we explain in more detail in the next section) traveling through a wire affect a compass needle, but currents traveling through two separate wires affect one another. If you place two wires next to each other and both are carrying electric currents, he found that one of two things happens:

- The wires are attracted to each other if their currents run in the same directions.
- The wires are repulsed from each other if their currents run in opposite directions.

In other words, the wires behave like magnets. Every time you replace a blown fuse, remember Ampére and the name given to the quantities of current, the amp. (Who would've guessed that's what they would be called?) If you blew a fuse, you tried to use too much current.

Being a mathematician, Ampére wanted to be able to quantify what he saw. He developed what's known as Ampére's law. This law states that the magnetic field at a given point in space is proportional to the current and inversely proportional to the distance. The important thing is that this law tells you exactly how much of a magnetic field is produced by a given current. So electric currents produce magnetic fields! More on this, too, later in the chapter.

Ohm's Law

The last physicist in this section contributed the third of the three most common units of electricity. You've met Volta and his volts and Ampére and his amps; now it's time to meet Ohm and his *ohms.*

The German physicist George Simon Ohm (1787–1854) was a schoolteacher in Cologne. He was interested in the effect different materials had on the flow of electricity. By using wire made of different substances and constructed of various thicknesses, he investigated whether the strength of the current depended upon these changes. He also wanted to know whether such changes influenced the amount of voltage needed.

He made two important discoveries:

- The strength of the current is directly proportional to the diameter of the wire and is inversely proportional to the wire's length. It is also dependent upon the material of the wire.
- The strength of the current is directly proportional to the voltage and inversely proportional to the resistance of the wire.

What all this means is that Ohm discovered and defined *resistance*, which is measured in ohms. He had also come up with Ohm's law, which establishes the relationship among voltage, current, and resistance. His law is at the heart of all electronics. Every technician knows Ohm's law by heart. Here's what the law looks like. (In the following equations, V equals volts, I equals amps or current, and R equals ohms or resistance.)

- Voltage is equal to the current times resistance. The formula is $V = I \times R$.
- Current is equal to voltage divided by resistance. The formula is $I = \frac{V}{R}$.
- Resistance is equal to voltage divided by current. The formula is $R = \frac{V}{I}$.

def•i•ni•tion

An **ohm** is a unit of electrical resistance. **Resistance** is the property that impedes the flow of current in a material. The opposite of resistance is conductance, which is the property that allows current to flow in a material. Because conductance is the exact opposite of resistance, the unit of measure of conductance is called the mho. Are you kidding? No, it's true!

There you have it. That's the relationship at the most basic level of everything that goes on in electric circuits. So let's spend some time going into these ideas in more detail.

Electric Potential Energy

Picture an old-fashioned shower: a bucket of water suspended over your head, perhaps from the branch of a tree, with a rope attached that you pull to release the water in a stream. The water flows downward because of the force of gravity. The water above your head has a type of potential energy that we've discussed in previous chapters called gravitational potential energy. The water got that potential energy through the work exerted in pulling the bucket up into the branch in the tree.

Newton's Figs

Remember that work is done when a force is applied and an object is displaced in the direction of the applied force or a component of the applied force. Work done on a charge requires a force to be applied in a direction opposite to the coulomb force. That is similar to work done by the force lifting a body that acts opposite to the direction of the force of gravity on the body.

We can think of electrical potential energy in the same way. In Chapter 18, we discussed static charge, charge that doesn't flow. However, if a charge is able to flow (because of the presence of a conductor), it will flow from a higher to a lower potential, in the same way that water flows downhill. In the production of lightning, for example, an *electric current*, or flow of electrons, results when the potential difference between the cloud and the ground gets sufficiently large.

When potential is equalized, because of the flow of electrons, the current will stop flowing. To keep a current flowing, there must be a way to artificially (or naturally) maintain a potential difference between two points. In the case of lightning, the potential difference is maintained by the internal dynamics of certain types of clouds. Rubbing a glass rod with silk transfers electrons from the rod to the silk, which creates a potential difference between these two materials. Touching the rod to another material causes current to flow from the object into the glass rod (because it has a deficit of electrons). When the rod has been discharged, though, current ceases to flow. We measure current as flow rate, in units of charge per unit time.

def•i•ni•tion

Electric current is the flow of charge past a point in an electric circuit for each unit of time. An electric circuit is the conducting path for the flow of charge. The **ampere** is the practical unit of electric current. The ampere is 1 coulomb of charge passing a point in an electric circuit in one second.

So as we mentioned, electric current is the rate of flow of charge or the flow of charge per unit of time. In symbols, $I = \frac{Q}{t}$, where I is current, Q is the charge in coulombs, and t is time. If the charge is 1 coulomb—remember, that is a lot of charge, 6.25 × 1,018 electrons—and the time is 1 second, then the current is defined to be 1 *ampere*. So $I = \frac{1C}{1s} = 1A$. And that means that if a current of 100 A is flowing (and this is a large current, by the way!), there are 100 C passing a given point in a conductor every second.

The ampere is a very large unit, so more often than not you use the milliampere, mA, or the microampere, μA, as a unit of current. You know that 1 mA = 10^{-3} ampere and 1 μA = 1.0×10^{-6} ampere. Some reference books use different symbols; one popular method is 1 mA = 10^{-3}A and 1 μA = 10^{-6}A.

def•i•ni•tion

A **schematic diagram** is a symbolic representation of electric circuits. That is, symbols representing components, conductors, and instruments of measurement are used instead of pictures of those items.

Now that you know the units of electric current, it's also nice to know about instruments used to measure current. The instrument used to measure current in basic circuits is the ammeter. Other instruments used for measuring current are the milliammeter, microammeter, and the multimeter. The multimeter is an instrument that provides you with the options of measuring any one of several quantities that are found in electric circuits. The use of any of these devices to measure current requires that you connect them in the circuit so that they sample all of the current in the circuit at the place where you want to measure it. That connection is illustrated in the *schematic diagrams* of electric circuits in Figure 19.2.

The electric currents that surround us in our daily lives are maintained by artificial means and are able to provide a more steady flow of electrons, a more steady current. We say that any device that maintains a potential difference provides an electromotive force. (And it's measured in what unit? Volts.) The batteries that power all the personal electronics that we use, and the electrical generators that provide the current available at the outlets in our homes, provide the potential difference that allows charge to flow.

There are two basic types of current, *direct current* (DC) and *alternating current* (AC). Direct current refers to electron flow in a single direction with time. Batteries of all varieties provide direct currents. Batteries have a positive and negative terminal, and are rated in the voltage (potential difference) that they can sustain. Car batteries, for example, are typically 12 V DC. Batteries that power your CD player may be 1.5 V DC, and you may need two of them to provide the power that your CD player requires.

Home appliances in the United States are powered by alternating current. Our most common house current is 110 to 120 volts AC. There are higher voltages, too, but they are all still in the form of alternating current (AC). The electrons that flow in alternating currents do not push electrons in a single direction but, rather, move back and forth, in a motion similar to the motion of a swing (simple harmonic motion). Because the electrons are constantly changing direction, this implies that the voltage of the EMF changes, too. The rate of change of the direction of the current and voltage in an AC circuit is measured in cycles per second (1/s) or hertz (Hz). In the United States, the current varies at a rate of 60 cycles per second, or 60 Hz, and maintains a voltage of 110 to 120 V.

def•i•ni•tion

Direct current, DC, is an electric current that travels in only one direction in the circuit. It travels from the negative side of the cell, through the circuit external to the cell, to the positive side of the cell. **Alternating current,** AC, is an electric current that travels in one direction and then the opposite direction with a fixed period. Your local electric company, through outlets in your home, provides AC. Although the current travels back and forth, seeming to get nowhere, the electrical energy still goes in a forward direction to your home.

Magnetic Fields and Induction

We mentioned earlier that there was a relationship between magnetism and electricity. It was discovered that when a current ran through a wire, a magnetic field surrounded the wire. But what, exactly, is a *magnetic field?*

A permanent *magnet* can exert force from a distance, the forces being exerted through what we call a magnetic field. You know this from experience if you played with magnets as a child. If you carefully bring the north pole of one magnet close to the north pole of another, the forces can be sufficient (depending on the mass of the magnets and the friction present) to push the other magnet away. And, of course, if you bring opposite poles within range of each other, the magnets "stick" together. Opposite poles attract, and like poles repel.

Although a magnetic field is strongest near the magnet, or magnets that cause it, it extends indefinitely into space. Within the magnetic field, force is exerted in curved patterns known as *magnetic lines of force.*

We can understand magnetic field lines in relation to the electric fields we discussed in Chapter 18. If you recall that in electric field diagrams, field lines are drawn from positive charge to negative charge, indicating the direction of motion of a positive test charge, then you can imagine how in magnetic field diagrams, field lines are drawn from the north pole (N) to the south pole (S). If you want to see what these lines of force look like, take a piece of paper, put a bar magnet under it, and sprinkle some iron filings on the paper. You will soon see a pattern take shape that shows the line of force created by the field of the magnet.

def•i•ni•tion

A **magnet** is an object made of iron or steel and is characterized by a north pole and a south pole each having the ability to strongly attract iron. The **magnetic lines of force** are invisible lines directed from the north pole through the space near the magnet into the south pole. The **magnetic field** is the region near the magnet containing all the magnetic lines of force. The magnetic field is to the magnet what the gravitational field is to the earth.

The lines of magnetic force that indicate the field due to a current in a straight piece of wire are found to be circles that go around the wire in one direction. The field is strongest near the wire and gets weaker with the distance from the wire in any direction. If the flow of current is reversed, the magnetic field lines again circulate around the wire but in the opposite direction. The strength of the magnetic field at some distance r from the wire is found to be directly proportional to the current and inversely proportional to the distance. You can determine the connection between current flow and the direction of magnetic field lines with the right-hand rule, which says that if the thumb of your right hand points in the direction of the current flow in the wire, the fingers will curl in the direction of the magnetic field lines that surround it. A great use for your thumb, other than for hitchhiking, although the way you hold it is basically the same.

Now there is something very cool that happens when you move a straight conductor through a magnetic field perpendicular to the magnetic lines of force. A current is induced in the wire. It is called induction, a concept that was explained by another famous physicist, Michael Faraday (1791–1867).

Faraday was the son of a poor blacksmith and had very little formal education. But his accomplishments reflected a brilliant, innovative mind and a strong belief in the fundamental unity of nature. He invented the first electric motor, showed that magnetism could be converted into electricity, and designed and constructed a dynamo. He also made the first electrical transformer and showed that magnetism could affect polarized light (or light that vibrates in only one direction, not all directions like ordinary light). Not bad for someone with an eighth-grade education.

Faraday knew that electricity could produce magnetism, so he wondered if he could use magnetism to produce electricity. During his construction of an electric motor, he had built wire coils of different sizes. Through experimentation, he found that if he passed a magnet through a coil of wire and hooked a meter up to the ends of the coil, the galvanometer (named after Luigi and his frog legs) showed that a current was created. Simply placing a magnet within the coil did nothing; the magnet had to be moving in and out of the coil to produce a current. The coils of wire broke the lines of force that were part of the magnetic field of the magnet. The more coils there were, the more lines of force were broken, and more current was produced. The changing magnetic field induced a voltage in the coil of wire.

Potential Difference

A few pages back, we introduced you to the voltaic pile and the concept of the EMF. Let's now look at this in more detail. In the same way that physical objects tend naturally to move from a region of high potential energy to low potential energy, charged particles move naturally from higher to lower electric potential energy states. To move a negatively charged particle along field lines requires work. And work, we're sure you recall, is defined as a force exerted over a distance. The force required to move a negatively charged particle along electric field lines is stored in the form of electric potential energy.

Doing this work results in a change in the potential energy of the particle. If a particle moves between two different states of electric potential energy, we refer to this as the *potential difference* between the two states. The potential difference, called voltage, is simply the change in potential energy between the two states, divided by the charge:

Potential difference = Change in electric potential energy/Charge

Potential difference and electric potential energy are both scalar quantities. And because the change in potential energy is measured in joules, and the charge is measured in coulombs, the SI unit of potential difference is J/C, where 1 J/C is also called 1 volt (V). And the force required to move a negatively charged particle along electric field lines is the electromotive force, which you already know as the EMF.

Let's calculate how much energy is required to move an electron (with a charge of $1.60 \times 10^{-19} C$) through a potential difference of 1 V.

PD = Change in electric potential energy/Charge

1 V = Change in EPE/$1.60 \times 10^{-19} C$, therefore

$\Delta EPE = 1.60 \times 10^{-19} C \cdot V$, *or* $1.60 \times 10^{-19}\ J$

This tiny amount of energy is sometimes called an electron-volt (eV). Other units of measurement you might see are these: 1 millivolt (mV) = 10^{-3} V, 1 microvolt (μV) = 10^{-6} V, and 1 kilovolt (kV) = 10^{3} V. The instrument that is used to measure potential difference in a circuit is called a *voltmeter*.

Resistance

Whenever an electric current is established in an electric circuit, it always experiences opposition. The *resistance* to the flow of current due to the components in an electric circuit can be thought to be like the resistance to the motion of a car due to friction.

def•i•ni•tion

The **potential difference** between two points in an electric field is the work done per unit charge as the charge is pushed by a force in a direction opposite to the coulomb force between those points. The **voltmeter** is the instrument used to measure potential difference in an electric circuit.

The **resistance** in an electric circuit is the opposition to the flow of charge or current in the circuit. A **resistor** is a component in an electrical circuit used to establish the amount of current and/or potential difference at different places in the circuit.

There is always some production of thermal energy associated with resistance in an electric circuit. This means that some of the electrical potential energy is converted to thermal energy. Sometimes sound is associated with resistance in a circuit. If the *resistor* is a filament in a lamp, there is some light in addition to much thermal energy associated with resistance.

Johnnie's Alert

A small amount of current can damage the human body. Around 50 mA causes considerable pain. Around 100 mA can cause problems with the heart and even stop an unhealthy heart. Anyone who tells you that DC will not hurt you needs to know about the Leyden jar, the wrong side of a television picture tube, and a lightning bolt.

Resistance and the Unit of Opposition

Resistance is measured in ohms, denoted by the symbol Ω. For a given potential difference, the greater the resistance in ohms the smaller the current in amperes. The resistance of a component is not measured, whereas the component is in a circuit. Instead, at least one end of the component to be measured must be disconnected.

The instrument used for measuring resistance is the ohmmeter. Usually the resistance of a resistor is marked by a number on the resistor or indicated with a color code. The color code is often explained in the package of resistors. Anyway, if there is ever any doubt, you can measure the resistance with an ohmmeter.

Components of an Electric Circuit

Given everything we've talked about up until now, you have at your fingertips all that you need to delve into the world of electric circuits. That is, you have a good basic understanding of resistance, potential difference, and current. Take a look at Figure 19.1 to see how these three components of every circuit are represented. The diagram in Figure 19.1 is a schematic diagram of a basic electric circuit. The break in the circuit is the symbol for a switch. No current is shown in the circuit because the switch is open or in the off position.

It's often a good idea to construct a circuit with a switch so that you can control the circuit, but it all depends on what you want the circuit to do. You know that current is in the circuit when the switch is closed, which makes the loop in Figure 19.1 complete. In that diagram, closing the switch causes a current in the circuit with a counterclockwise direction. Before reading any further, can you tell us why the current flows counterclockwise? The reason is that the electrons flow from the negative side of the potential difference around the circuit to the positive side. A single cell is represented by a short and long line, labeled V in the diagram.

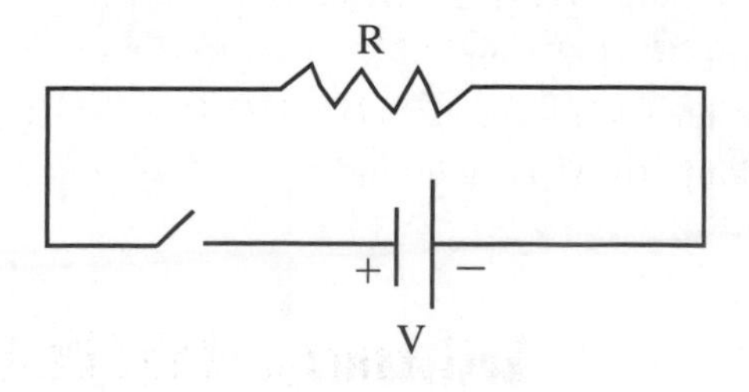

Figure 19.1

A simple electric circuit has a switch, a cell, and a load.

The current passes through a resistor, represented by a squiggly line and labeled R in the diagram, and of course through the switch when it is closed. The symbol for current is I. The current is not included in Figure 19.1 to emphasize that the switch is open. Future diagrams will indicate the direction of the current.

An electric circuit has at least one resistor of some sort, usually referred to as the load, symbol R_L, or total resistance R_T. And as previously mentioned, it's a good idea to have a switch if you want to control the circuit. It also has connecting conductors, symbolized by straight lines between components in Figure 19.1, and a source of continuous current providing a potential difference. The load resistor provides the output

Johnnie's Alert

Using a switch in an electric circuit is always a good idea. You can close it when you want the circuit to work and open it when you don't want current in the components. A switch can save components and measuring instruments when a connection is in error.

for the electrical energy; that is the purpose of the circuit. For example, the basic circuit can be a flashlight, so the load resistor is the light bulb that gives you the light for which you designed the circuit. The load may be made up of a string of colored lights where you can have many resistors. It can consist of several different devices, provided the source of continuous current and difference of potential (voltage) is sufficient. And it's pretty clear that a single cell cannot run the home refrigerator or a television set. These appliances do not use the same kind of current as is provided by a cell.

Series Circuits

The electric circuit in Figure 19.2 is called a series circuit because the resistors are connected end to end like links in a chain.

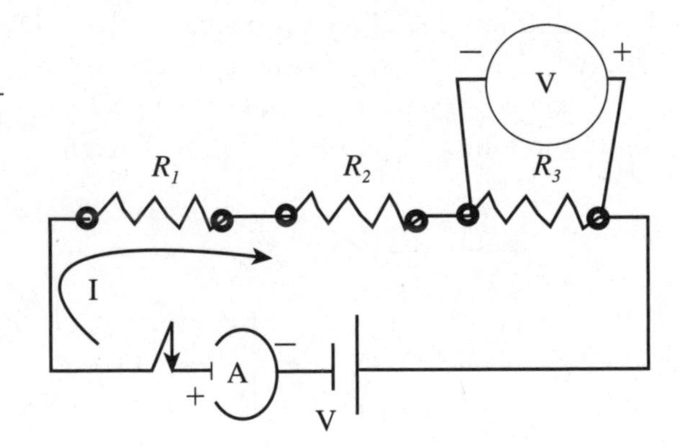

Figure 19.2

A series circuit contains resistors connected in a chain end to end so that the total current passes through each of them.

Resistors connected in a circuit in this way will all have the same amount of current pass through them. In Figure 19.2, the total current, I, in the circuit passes through each of the resistors, labeled R_1, R_2, and R_3. We connected the ammeter in series with the load, along with the voltmeter that is connected to measure the potential difference across R_3. Notice that the voltmeter is connected with its negative side closest to the negative side of the cell and its positive side closest to the positive side of the cell.

The ammeter is labeled the same way and is connected positive to positive and negative to negative. The meters are labeled in this way so that the DC current passes through the meter in the proper direction. Did you notice that the ammeter is not connected in the circuit in the same way as the voltmeter? We discuss that special connection of the voltmeter in just a little bit.

Johnnie's Alert

The positive and negative sides of the electrical meters must be connected properly to the circuit not only so that they will measure correctly but also to protect the meter. For example, the voltmeter is wired internally so that most of the current stays in the circuit and only a small amount passes through a properly connected voltmeter.

Earlier you were introduced to Ohm's law. Now we'll get a chance to put it into practice. But before doing so, let's cover just a few more key ideas. First, when the electrons of the current pass through a resistor they give up potential energy. It is much like the object falling from a point above the surface of the earth; as it falls, it gives up potential energy. The electrons give up energy as they travel around the circuit. For that reason, the decrease in potential energy or decrease of potential is called a potential drop.

def•i•ni•tion

A **battery** is a combination of cells. It is constructed to overcome the limitations of one cell. A battery provides a larger current, a larger potential difference, or both. A **branch** of a circuit is a division of a parallel part of the circuit.

The amount of the potential drop can be calculated using Ohm's law. One of the ways that it can be written is $V = IR$; the potential difference is equal to the product of the value of the resistance and the current through the resistor. The potential drop across a resistor is often called an *IR* drop. Another important idea is stated: the sum of the potential drops around a circuit is equivalent to the potential drop across the cell or *battery* in the circuit. Applied to the components in the circuit of Figure 19.2, this gives you $V = V_1 + V_2 + V_3 = IR_1 + IR_2 + IR_3$. This idea is known as Kirchhoff's second law. His first law is touched on shortly.

It's time for a short practice problem. Let's suppose V = 1.5V, $R_1 = 10.0\Omega$, $R_2 = 20.0\Omega$, $R_3 = 30.0\Omega$; calculate I, V_1, V_2, and V_3.

$$R_T = R_1 + R_2 + R_3$$

$$R_T = 10.0\Omega + 20.0\Omega + 30.0\Omega = 60.0\Omega$$

$$V = IR_T \; I = \frac{V}{R_L} = \frac{V}{R_1 + R_2 + R_3} = \frac{1.5V}{60.0\Omega} = 0.025A$$ By Ohm's law, $1V = (1A)(1\Omega)$.

That means the reading on the ammeter is 0.025 A. Because $I = 0.025\text{A}$ $V_1 = IR_1 = (0.025A)(10.0\Omega) = 0.25V$, $V_2 = IR_2 = (0.025A)(20.0\Omega) = 0.50V$, and $V_3 = IR_3 = (0.025A)(30.0\Omega) = 0.75V$.

We can check to see whether those three voltages added together gives us the original voltage given in the problem by using Kirchhoff's second law: $V_1 + V_2 + V_3 = 0.25V + 0.50V + 0.75V = 1.5V$, which gives us exactly what we started with.

Now let's take a look at Figure 19.3, where you find a diagram of three resistors connected in parallel. Notice that the current is not the same in each of the resistors but each *branch* has its own current.

Figure 19.3

The electric circuit with three resistors connected in parallel demonstrates branch currents.

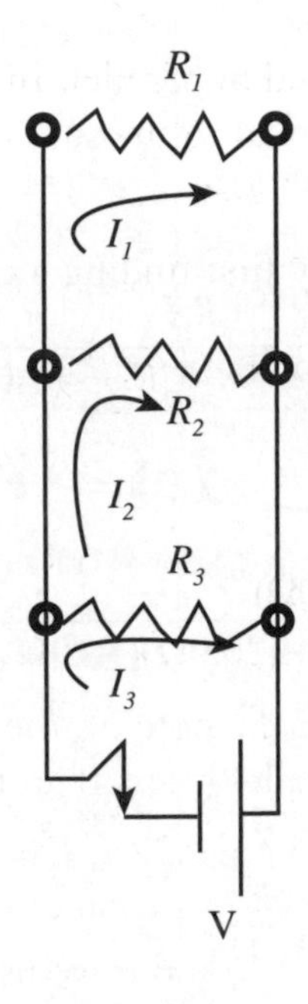

Parallel Circuits

Kirchhoff's first law states that the total current passing through an area of a circuit in which resistors are connected in parallel is equal to the sum of the currents passing through the branches of the parallel part of the circuit. That is essentially a statement of conservation of charge; no charge is gained or lost when circuits branch into different paths. Kirchhoff's second law is a statement of the conservation of energy; that is, no energy is gained or lost in electric circuits. The symbolic statement of that rule for Figure 19.3 is $I = I_1 + I_2 + I_3$.

Let's suppose the resistors and the potential difference all have the same values as in the last circuit. Calculate the total current and the branch currents in the diagram of Figure 19.3. We'll work this process together.

Notice that the potential drop across all resistors in this circuit is the same because each has one end connected to the negative side of the cell and each has the other end connected to the positive side of the cell. That means that the potential drop across each resistor is V and is stated symbolically as $V = V_1 + V_2 + V_3$, $I = I_1 + I_2 + I_3$, $\frac{V}{R_T} = \frac{V_1}{R_1} + \frac{V_2}{R_2} + \frac{V_3}{R_3}$, $\frac{1}{R_T} = \frac{1}{R_1} + \frac{1}{R_2} + \frac{1}{R_3}$ (dividing both members by V the potential of the cell).

This is a general result for resistors connected in parallel. In words, it states that the reciprocal of the load or total resistance is equal to the sum of the reciprocals of each resistor connected in parallel.

The terms in the right member are added by first finding a common denominator, which is $R_1R_2R_3$ for this circuit: $\frac{1}{R_T} = \frac{R_2R_3}{R_1R_2R_3} + \frac{R_1R_3}{R_1R_2R_3} + \frac{R_1R_2}{R_1R_2R_3} = \frac{R_1R_2 + R_1R_3 + R_2R_3}{R_1R_2R_3}$, $R_T = \frac{R_1R_2R_3}{R_1R_2 + R_1R_3 + R_2R_3}$.

The resistance of the load in Figure 19.3 is

$$R_T = \frac{(10.0\Omega)(20.0\Omega)(30.0\Omega)}{(10.0\Omega)(20.0\Omega) + (10.0\Omega)(30.0\Omega) + (20.0\Omega)(30.0\Omega)} = \frac{6000.0\Omega}{1100.0} = 5.45\Omega.$$

This is another general result that you can make note of: the total resistance or equivalent resistance of resistors connected in parallel is less than the smallest resistor of the parallel resistors: $I = \frac{V}{R_T} = \frac{1.5V}{5.45\Omega} = 0.275A = 0.28A$, $I_1 = \frac{1.5V}{10.0\Omega} = 0.15A$, $I_2 = \frac{1.5V}{20.0\Omega} = 0.075A$, and $I_3 = \frac{1.5V}{30.0\Omega} = 0.050A$. And if we use Kirchhoff's first law to check your result: $I = 0.15A + 0.075A + 0.050A = 0.275A = 0.28A$. Great—that's exactly what we expected!

Series-Parallel Circuits

We've progressed from a basic circuit to a series circuit to a parallel circuit. It is common to find a circuit that contains both series and parallel circuits, so let's take a look at one of those. We'll use the same components that we started out with, except that the potential difference of the cell has the same number of significant figures as the resistors—that is, V = 1.50V instead of simply 1.5V. So now we'll connect the resistors in a series-parallel circuit. The schematic diagram in Figure 19.4 is one possible arrangement.

To begin with, it's a good idea to tackle the most complicated part of a circuit.

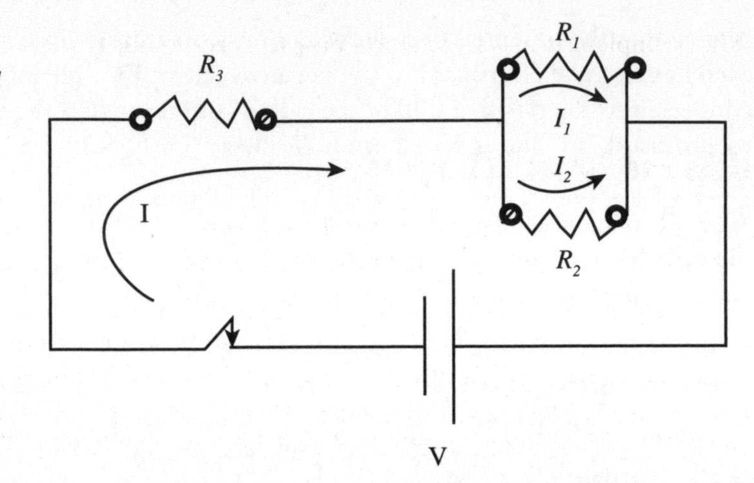

Figure 19.4

Three resistors are connected in a series-parallel circuit.

In this case, the parallel resistors are combined to find an equivalent resistor. Remember that the values of the resistors are the same as before. That is, $R_1 = 10.0\Omega$, $R_2 = 20.0\Omega$, and $R_3 = 30.0\Omega$. R_{eq} acts in the circuit like a resistor in series with R_3. We'll combine those resistors to find the load R_L. Using R_L and V and Ohm's law, we can find the current I. The IR drop across R_3 subtracted from V gives us the IR drop across the parallel resistors. Using the IR drop across the parallel resistors and Ohm's law, we can find the branch currents and then the problem is solved:

$$\frac{1}{R_{eq}} = \frac{1}{R_1 + R_2}$$

$$\frac{1}{R_{eq}} = \frac{R_2}{R_1R_2} + \frac{R_1}{R_1R_2}$$

$$R_{eq} = \frac{R_1R_2}{R_1 + R_2}$$

$$R_{eq} = \frac{(10.0\Omega)(20.0\Omega)}{30.0\Omega} = 6.67\Omega$$

$$I = \frac{V}{R_L} = \frac{1.50V}{36.7\Omega} = 0.0409A = 0.041A$$

$$V_3 = IR_3 = (0.0409A)(30.0\Omega) = 1.23V$$

$$V_1 = V_2 = V - V_3 = 1.50V - 1.23V = 0.27V$$

$$I_1 = \frac{V_1}{R_1} = \frac{0.27V}{10.0\Omega} = 0.027A$$

$$I_2 = \frac{V_2}{R_2} = \frac{0.27V}{20.0\Omega} = 0.014A$$

That completes the solution, and now to check using Kirchhoff's first law. $I = I_1 + I_2 = 0.027A + 0.014A = 0.041A$.

Resistors and Power Rating

There is one other thing that you should consider when you're experimenting with circuits. It's good to understand the heating effects of resistors. Resistors and other components have a power rating associated with them. Usually resistors have common ratings by physical size, such as ⅛ watt, ¼ watt, ½ watt, 1 watt, and 2 watt. The power rating insures that you use the correct component rated for the amount of current that it is supposed to handle. If the resistor has a power rating that is too small, the resistor will quickly burn out and act like a switch and open the part of the circuit in which it is connected.

Power can be calculated by finding the product of current and potential difference or by squaring the current and multiplying by the resistance: $P = \frac{W_k}{t}$ and $V = \frac{W_k}{Q}$ (so work equals voltage times charge), $P = \frac{VQ}{t} = \frac{Q}{t}V = IV$, and $P = I(IR) = I^2R$.

Checking the power rating needed by the resistors used in the circuits constructed is a great exercise. All of those are checked here for you.

The circuit that has three resistors, 10.0Ω, 20.0Ω, and 30.0Ω, connected in series with V = 1.50V; it has a current of 0.025A. The power for R_1 is calculated as $P_1 = (0.025A)^2(10.0\Omega) = .00625W$, R_2 is calculated as $P_2 = (0.025A)^2(20.0\Omega) = .0125W$, and R_3 is calculated as $P_3 = (0.025A)^2(30.0\Omega) = .0188W$. All of these resistors are safe if rated at ⅛ W or higher. Remember that ⅛ W equals 0.125 W, and each of these power values is well below that.

The circuit that has the same three resistors connected in parallel with the same potential difference has the following currents taken from the previous examples that we worked out, and the power is calculated here for you: $I_1 = 0.027A$, $R_1 = 10.0\Omega$, $P_1 = (0.150A)^2(10.0\Omega) = 0.225W$, $I_2 = 0.075A$, $R_2 = 20.0\Omega$, $P_2 = (0.075A)^2(20.0\Omega) = 0.112W$, $I_3 = 0.050A$, $R_3 = 30.0\Omega$,

Newton's Figs

The power rating of resistors used in a circuit, or any component for that matter, should be slightly larger than the calculated power to be generated in the component. The reason is that any change in current can burn out the component and open that part of the circuit.

$P_3 = (0.050A)^2(30.0\Omega) = 0.075W$. R_1 and R_2 should be at least ¼ W, and R_3 can be ⅛ W or higher.

The series parallel circuit with the same three resistors and the same potential difference has currents as listed and power as calculated in the same way as the example above: $I_1 = 0.027A$, $P_1 = (0.027A)^2(10.0\Omega) = 0.00729W$, $I_2 = 0.014A$, $P_2 = (0.014A)^2(20.0\Omega) = 0.00392W$, $I_3 = 0.041A$, $P_3 = (0.041A)^2(30.0\Omega) = 0.054W$. All of these resistors can be ⅛ W or higher.

Whew, that was a long chapter, but at least you now know the shocking truth about electric circuits and all that they entail. You're down to just two more chapters. And we promise that they will be most enlightening.

Problems for the Budding Rocket Scientist

1. How many electrons per second pass through a section of wire carrying a current of 0.7 A?
2. What is the current through an 8Ω toaster when it is operating on 120 V?
3. Determine the potential difference between the ends of a wire of resistance 5Ω if 720 C passes through per minute.
4. Three resistances of 12, 16, and 20Ω are connected in parallel. What resistance must be connected in series with this combination to give a total resistance of 25Ω?

The Least You Need to Know

- Electric currents produce magnetic fields.
- The three most common units of electricity are volts, amps, and ohms.
- Work done on a charge requires a force to be applied in a direction opposite to the coulomb force.
- Potential difference, called voltage, is the change in potential energy between the two states, divided by the charge.
- When analyzing a circuit, the best place to begin is in the most complicated part.

Chapter 20

Light as Particles in Motion

In This Chapter

- The dual nature of light
- Calculating the speed of light
- Reflection of light on flat and curved surfaces
- Refraction of light
- The nature of lenses

In this chapter and the next, we examine the properties of light. First, we study light as streams of particles that move in straight lines, reflecting off surfaces or bending through materials as they travel. In the last chapter, we look at the wave properties of light.

Is Light a Particle or a Wave?

The scientific study of light begins with you know who—yes, the ancient Greeks. Aristotle was concerned with vision and how our eyes perceived the world around us. He thought that light wasn't a substance and had no qualities that could define it. The eye was somehow thought to produce light that allowed perception to take place. Those ideas remained essentially the same for 1,500 years, with some insights coming through the study of lenses in the building of telescopes and microscopes in the fifteenth and sixteenth centuries.

Enter Sir Isaac Newton. We already know a lot about the contributions he made to physics, and with the publication of his book *Opticks*, Newton advanced the study of light to a whole new level. Here is a summary of his ideas:

- White light, or sunlight, is the mixture of all the colors of the rainbow. Prisms act to separate these colors.
- Light consists of particles (he called them corpuscles) that move in straight lines.
- When these corpuscles run into a surface, they set up vibrations in the particles of the surface, and we see the color produced by these vibrations. (The introduction of the concept of vibration was very important.)
- The particles that make up light move at a constant speed.
- Light was created at the beginning of the world. It is a unique substance, and it plays a special role in maintaining nature.
- Ether is the medium that carries light particles from one place to another. (This concept also played a big part in the study of light.)

Because these ideas were put forth by the great Isaac Newton, they went unchallenged for a little while, but experiments done by other physicists showed that light behaved more like a wave than a particle. And for a long time, Newton's ideas were thrown out the window; light was considered to be a wave and that was that.

But early in the twentieth century, the work of Albert Einstein again suggested that light behaved in some ways like a particle and carried energy in small packets, or quanta. Later, during the development of quantum mechanics, it was found that light and all forms of electromagnetic radiation had properties of both waves and particles. Light has since been defined as having a wave-particle duality, a very paradoxical property and a mystery that lies at the core of quantum mechanics. So let's go ahead and begin by examining the first half of the dual nature of light, how light behaves like a particle.

Johnnie's Alert

Newton was right, light does travel at a constant speed, but it does not travel at the same speed in all media. For example, light travels at a speed in water that is about three fourths the speed of light in a vacuum. The speed of light in air is slightly less than the speed in a vacuum.

The Inverse Square Law

If light is a particle traveling in straight lines radially outward at a constant speed, you might expect to detect a certain number of particles by surrounding the source with a spherical surface. That is the basis for experiments that have been done to see if light does travel in straight lines in all directions from the source. Let's take a look at some of the instruments that can detect light besides your eye.

A light meter is something that you may have used to see if there is enough light to take good photographs. (At least before the advent of digital cameras. Certain types of film could only be developed with good-quality pictures if the lens of the camera was adjusted to the brightness of the surrounding light, and that's where a light meter would come in.) The light meter can measure the amount of light falling on a unit area. Sources of light are rated in terms of *luminous intensity*. The *intensity of illumination* (*I*) is measured in lumens. If a light meter measures the intensity of illumination one meter from a source with a luminous intensity of 1 *candela*, the measurement is 1 *lumen*.

def•i•ni•tion

The **luminous intensity** is the strength of a source of light. The unit of luminous intensity is the standard candle or **candela,** abbreviated cd. The **intensity of illumination** (I) is the rate at which light energy falls on a unit area of surface. The unit of intensity of illumination is the **lumen.** One lumen is the rate at which light energy falls on 1 square meter of the surface of a hollow sphere having a light source at its center with luminous intensity of 1 candela.

Let's suppose you have a point source of light with particles traveling radially outward from that point as diagrammed in Figure 20.1.

Now imagine that we could have all the particles from the point pass through two concentric spheres. A cutaway diagram of those spheres, one with radius r_1 and the other one with radius r_2, has the point source at their center.

A cone of particles intersecting the surfaces of the spheres in areas that are directly proportional to the surface areas of the spheres A_1 and A_2 is shown with the dotted lines. Because we assume that the number of particles per unit of time remains constant at some number N as they travel outward, we can calculate the intensity of illumination (I) of each surface as follows:

$$\frac{I_1}{I_2} = \frac{\frac{N}{A_1}}{\frac{N}{A_2}} = \frac{A_2}{A_1} = \frac{4\pi r_2^2}{4\pi r_1^2}.$$

The surface area of a sphere is calculated using the formula $4\pi r^2$ and the ratio of the surface areas of the spheres is the same as the ratio of A_1 to A_2. So $I_1 = \frac{1.5V}{10.0\Omega} = 0.15A$. If $r_2 = 2r_1$—in other words, if the second surface is twice the distance from the source as the first—then $I_2 = \frac{r_1^2}{(2r_1)^2} I_1 = \frac{r_1^2}{4r_1^2} I_1 = \frac{I_1}{4}$. That means if you double the distance from the source, the intensity of illumination is ¼ the intensity of illumination on the first sphere. If the distance is tripled, then the intensity of illumination is 1/9 the intensity of illumination on the first sphere. In general, the intensity of illumination, *I*, is inversely proportional to the square of the distance from the source. That is, $I = \frac{k}{r^2}$, where *k* is the constant of proportionality that turns out to be the luminous intensity in candela of the source of light. This is known as the inverse square law.

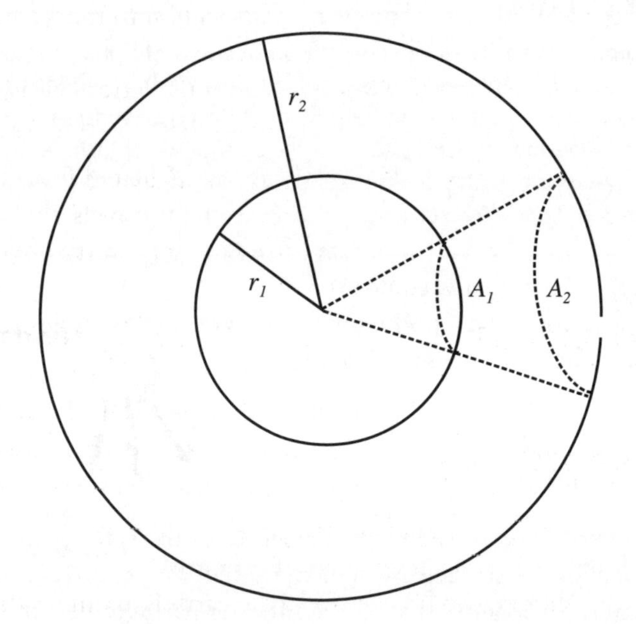

Figure 20.1

The illuminated surfaces are parts of two concentric spheres.

Direction and Speed of Light

The direction of the particles of light is radially outward from a point source, as you've learned. However, you may pick any one of those paths you want to help you to explain a particular behavior. The paths of particular particles are called rays, and a bundle of rays is called a pencil of light or a beam of light. The particles are so small that two beams or pencils of light can cross and the particles never deviate from their paths or bounce off each other. If there were any collisions, you would see bright spots or sparks where the beams cross, but from experience you know you see only two beams.

The first calculation of the speed of light was made by the Danish astronomer Olaus Roemer (1644–1710), quite by mistake. He was working on the orbits of the moons of Jupiter with the thought that precise periods of the moons' orbits might serve as a celestial timepiece. Roemer was surprised to notice that the moons appeared at different times than expected when Earth was in different points in its orbit (effectively at different distances from Jupiter). He calculated the speed of light by timing the appearance of one of the moons of Jupiter (as it emerged from behind the planet in its orbit) at different times in the earth's orbit. Using different points in Earth's orbit put the planets at different distances and allowed him to measure the different travel times. With a distance and a time, Roemer was able to calculate a velocity. His calculation of the speed of light (abbreviated with the letter *c*) was very close to what we know to be the correct speed of light (more than 186,000 miles per second).

In modern laboratories, the speed of light is calculated from the time is takes a beam of light to cover a known distance. Because light travels about 1,000 ft in one millionth of a second, the need for accurate timing is pretty obvious. The best current measurement of *c* (in a vacuum) is

$$c = 183{,}310 \text{ mi/s}$$

or

$$c = 299{,}776 \text{ km/s}$$

Newton's Figs

Light is different than sound in a lot of ways, but one huge difference is that light does not need a medium for transmission.

And just like Roemer, we can describe the motion of light with our knowledge of uniform motion. The distance to the sun from the earth is about 93,000,000 miles. How long does it take for light to travel from the sun to the earth? Using that same outline, we find:

$d = 93{,}000{,}000$ mi

$v = 186{,}000$ mi/s

$t = ?$

$$d = vt$$

$$t = \frac{d}{v}$$

$$t = \frac{93{,}000{,}000 mi}{186{,}000 mi / s}$$

$$t = 500s = 8.3 \text{ min}$$

Do you know what that means? The sun can drop from the sky or suddenly go out, and we will not know it for a little over eight minutes! (We use 186,000 mi/s for the speed of light by convention and also because it's close enough to the real speed and makes calculations easy.) Light is also significantly different from other types of waves we have studied so far. Unlike sound waves, or water waves, or human waves at a football game, light doesn't require a medium through which to travel. In the nineteenth century, physicists tried to identify such a medium, and even had a name for it: the ether. Measurements were attempted to determine how fast Earth was traveling through the hypothesized ether. In the middle of that century, James Clerk Maxwell (1831–1879) proposed instead that propagation, or movement of light waves, is the result of oscillations in electric and magnetic fields. Therefore, light can travel through a vacuum of space, because space is filled with electric and magnetic fields and not ether.

At this stage, the particle model is a good model to explain the direction and speed of light. It also explains that the intensity of illumination varies inversely with the square of the distance from the light source. If you think about it, those particles incident (the falling or striking of light on a surface) on the surfaces of the spheres suggest something else. If you have particles incident on an area, there should be pressure. The particle model predicts that there should be light pressure, and experiments have confirmed that light does exert pressure.

Reflection

When light strikes a barrier, or moves from one medium to another, part of the ray is reflected and part of it is transmitted. Depending on the properties of the material, more or less of the light may be reflected, or sent back in the direction from which it came. Some materials (mirrors) are more reflective than others (flat black paint) to visible light.

When light is reflected from a plane (flat) mirror, the incoming ray of light, called the incident ray, and the reflected ray of light, called the reflected ray, can be measured with respect to the *normal*, a line perpendicular to the plane surface. When any light ray strikes a plane, the angle of incidence, i measured from the normal to the incident ray, always equals the angle of *reflection*, r measured from the normal to the reflected ray, on the other side of the normal. This relationship is known as the law of reflection, and can be used to determine the image that will be formed by any reflective surface, whether it is flat or curved. Stated simply, the law of reflection says $i = r$.

The incident ray, reflected ray, and normal are all in the same plane; that is, a flat sheet could be made to pass through all three simultaneously with the flatness being distorted.

def•i•ni•tion

A **normal** is a line perpendicular to a line or to a surface. **Reflection** of light is the process of changing its direction when it strikes a smooth surface, causing the light to bounce off the surface.

Actually, it is rare to find a truly flat surface. Most surfaces have small unevenesses even when they appear flat. A beam of light, made up of parallel rays, would not display the same angle of incidence throughout. One ray might strike the surface at a spot where the angle of incidence is 0°, another might strike very close by, where the surface is at an angle of 10° to the light; elsewhere it is 10° in the other direction, or 20°, and so on. The result is that an incident beam of light with rays parallel will be broken up on reflection, with the reflected rays traveling in all directions over a wide arc. This is diffuse reflection.

Almost all reflection we come across is of this type. A surface that reflects light diffusely can be seen equally well from different angles, because at each of the various angles numerous rays of light are traveling from the object to the eye.

If a surface is extremely flat, virtually all the parallel rays of an incident beam of light will be reflected still parallel. As a result, your eyes will interpret the reflected beam as they would the original.

For instance, the rays of light reflected diffusely from a person's face make a pattern that the eyes transmit and the brain interprets as that person's face. If those rays strike an extremely flat surface, they're reflected without mutual distortion; when they strike your eye, you'll still interpret the light as representing that person's face.

Your eyes can't, however, tell the history of the light that reaches them. They can't, without independent information, tell whether the light has been reflected or not. Because we're used from earliest life to interpreting light as traveling in straight, uninterrupted lines, we still do so now. The person's face as seen by reflected light is seen as if it were behind the surface of reflection, where it would be if the light had come straight at you without interruption, instead of striking the mirror and being reflected to you. To get a better idea of what we're discussing now and in the upcoming paragraphs, take a look at Figure 20.2.

The face that you see in a mirror is an image. Because it doesn't really exist in the place you seem to see it (look behind the mirror and it's not there), it is a virtual image. (It possesses the "virtues," or properties, of an object without that object actually being there.) It is, however, at the same distance behind the mirror that the reflected object is before it, and therefore seems to be the same size as the reflected object.

Long ago, virtually the only surface flat enough to reflect an image was a sheet of water. Such images are imperfect because the water is rarely undisturbed, and even when it is, so much light is transmitted into the water and so little is reflected that the image is dim and obscure. Under such circumstances primitive peoples might not realize that it was their own face staring back at them.

A polished metal surface will reflect much more light, and metal surfaces were used throughout ancient and medieval times as mirrors. About the seventeenth century the glass-metal combination became common. Here a thin layer of metal (usually mercury) is spattered onto a sheet of flat glass. This, as you know, is a mirror or looking glass. A Latin word for "mirror" is *speculum*, and for this reason the phrase for the undisturbed reflection from an extremely flat surface is specular reflection.

Newton's Figs

A real image is formed when real reflected or refracted rays of light actually intersect. A virtual image is formed when virtual extensions of reflected or refracted rays of light only appear to intersect.

Johnnie's Alert

To understand how an image is formed, it's helpful to draw a ray diagram to locate the image formed by choosing rays of light from a point on the object that you understand. You know that a ray perpendicular to a plane mirror will reflect straight back the way it came. Draw any other ray that strikes the mirror, construct a normal at the point of incidence, and then construct the angle of reflection equal to the angle of incidence. Two rays are required to find the image of a point. Extend the reflected rays back until they intersect. That point of intersection is the image of the point source of light on the object. Take a look at Figure 20.2 if you need help carrying out this construction.

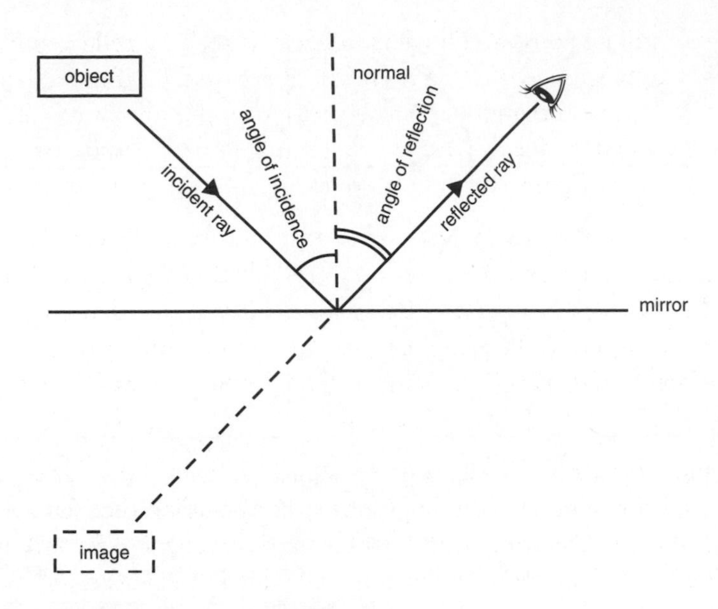

Figure 20.2

The image formed by a plane mirror is located by reflected rays.

Reflection and Spherical Mirrors

Light obeys the same laws of reflection any time it is incident on a smooth surface. Even if the surface is curved, the same laws will enable you to locate the image formed by a curved mirror. Let's suppose that a mirror is cut from a portion of a hollow metal sphere in which the inside surface is polished. That mirror is called a *concave spherical mirror.* If the outside surface is polished to reflect, it is called a *convex spherical mirror.* If you've ever looked at the outside of a shiny Christmas bulb or spherical lawn ornament (called a gazing ball), those would be examples of convex spherical mirror surfaces.

The side-view mirror on your car is a convex mirror, and the surveillance mirror in some stores is a convex mirror, too. The convex mirror forms only virtual images close to the surface of the mirror no matter how far away the object may be. The concave mirror also has many applications. One is that it can act as the collector for a solar oven in needy countries where there is plenty of sun. You may have seen a concave mirror on the flip side of makeup or shaving mirrors. It is the side that magnifies. We sketched one for you in Figure 20.3.

def•i•ni•tion

A **concave spherical mirror** is a segment of a sphere with the inside surface polished to reflect light. A **convex spherical mirror** is a segment of a sphere with the outside surface polished to reflect light.

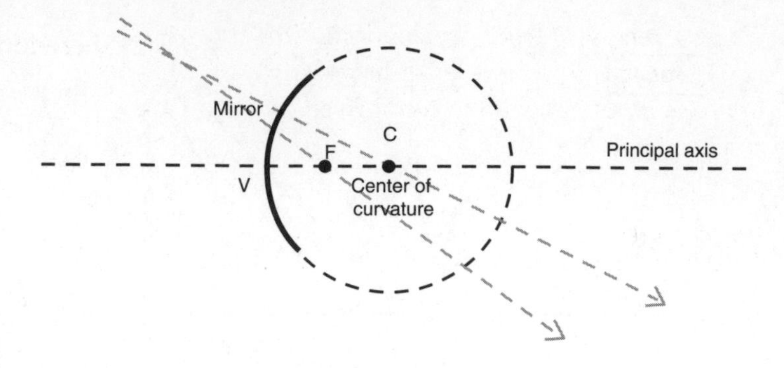

Figure 20.3

A concave spherical mirror.

def•i•ni•tion

The **center of curvature** of a mirror is the center of the sphere of which the mirror is a part. The **principal axis** of a concave spherical mirror is the line that passes through the center of curvature and intersects the mirror at a point called the vertex of the mirror. The **focal point** of a concave spherical mirror is a point on the principal axis where all rays intersect after they are reflected from the mirror, provided they are rays that are parallel to the principal axis and near the principal axis. The focal point is halfway between V and C. Therefore, the distance from V to C is 2f, where f is the focal length of the mirror, the distance from V to F.

The center of the sphere of which the curved mirror is part is called the *center of curvature*, C. A line connecting the center of curvature with the midpoint of the mirror is the *principal axis* of the mirror.

All of the rays of light that are close to and parallel to the principal axis are reflected back through a point called the *focal point*, F, of the mirror according to the laws of regular reflection. The object may be placed in front of the mirror in one of six regions that should be familiar to you. Those regions are as follows:

- Very far away
- A finite distance greater than C
- At C
- Between F and C
- At F
- Between F and V

That's a relatively basic explanation of the main aspects and principles of how light operates in relationship to a concave spherical mirror.

Newton's Figs

A light ray that leaves the object on a line that passes through the center of curvature travels along a radius of the sphere of which the mirror is a segment. The radius of a sphere is perpendicular to the surface of the sphere at the point of intersection of the radius and the surface. That means the angle of incidence is 0°, so the angle of reflection is 0°.

Refraction

In the previous section on reflection, light bounced back from a surface if it encountered it. There are two other possibilities: the light can be absorbed or transmitted. If light is incident on a transparent surface, some of the light is reflected according to the laws of regular reflection about which we have just talked, but part of it may be refracted. *Refraction* is our main focus at this stage of our discussion. We start with tracing the path of a ray of light as it travels from air into glass and is transmitted through a rectangular glass prism, as shown in Figure 20.4.

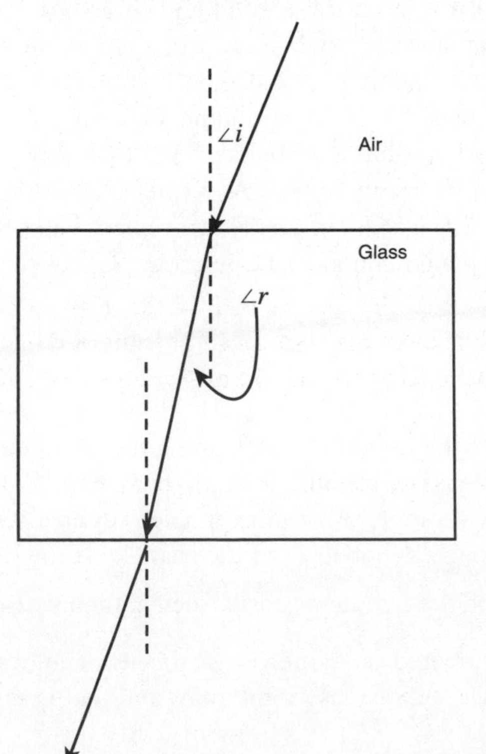

Figure 20.4

A light ray passing from air into glass bends toward the normal.

def•i•ni•tion

Refraction is the change in direction of light as it leaves one medium and enters a different medium. Any time light is incident on a transparent material, you'll observe both refraction and reflection. You're probably aware of this at night if you look outside through a window. You not only see the bushes outside but also a reflection of yourself and things in the room. The reflection of the room's contents and yourself form virtual images that appear to be on the other side of the window where the bushes are really located.

Again we have a normal (perpendicular) to the surface, but this time it is the surface of a glass prism. The angle between the normal and the incident ray in air is the angle of incidence, $\angle i$, and the angle between the normal and the refracted ray in glass is the angle of refraction, $\angle r$. The angle of incidence measures about 20° and the angle of refraction is about 13°.

You can see that the angle of refraction in glass is smaller than the angle of incidence in air without measuring the angles. One way to describe this fact is that the light bends toward the normal when it leaves air and enters glass. That means that if you were to imagine extending the original ray straight into the glass, the actual ray would be closer to the normal line than the imaginary extended ray. At one time, all of the known angles of incidence and corresponding angles of refraction for different substances were recorded in volumes of books. That is similar to Tycho Brahe's records of the locations of the heavenly bodies. As Kepler's laws replaced all of those volumes, the mathematical determination of Snell's law replaced all of the records of angles of incidence and corresponding angles of refraction. So let's see what Snell's law is all about.

Mathematics of Snell's Law

Snell's law requires that you know a small amount of mathematics beyond algebra and geometry. (Just a little bit of trigonometry, no big deal.) What you need to know is that the sine of an acute angle of a right triangle is defined as the ratio of the side opposite the angle to the hypotenuse of the triangle. If the side opposite the acute angle is labeled a, and the hypotenuse is labeled c, then $\sin\theta = \frac{a}{c} = \frac{\textit{side opposite the angle}}{\textit{hypotenuse}}$.

The 90° angle is not treated the same way as the acute angles in a right triangle, but there is one value that you may use right away, and that is $\sin 90° = 1$.

The simple relationship that replaced many volumes of books containing observations of angles of incidence and corresponding angles of refraction is called Snell's law and is written $\frac{\sin \angle i}{\sin \angle r} = n$, where n is a constant, called the index of refraction, and is associated with the type of matter (or medium) the light is entering, assuming that it is leaving a vacuum.

The index of refraction of a vacuum is defined as 1, and for air it is almost 1. The index of refraction of glass is $\frac{3}{2} = 1.50$ and the index of water is $\frac{4}{3} = 1.33$. (Any transparent matter has an index of refraction, but we use only these three in this book. You can look up others in reference books if you'd like.)

You can check the behavior of the ray of light entering glass in Figure 20.5 by measuring the angle of incidence and using Snell's law to calculate the corresponding angle of refraction. The angle of incidence is about 20°, so using Snell's law you find that the angle of refraction is calculated as follows:

$$\frac{\sin \angle i}{\sin \angle r} = n_g$$

$$\sin \angle r = \frac{\sin \angle i}{n_g}$$

$$\sin \angle r = \frac{\sin 20°}{1.5}$$

$$\sin \angle r = 0.2280$$

Most calculators have sine functions on them. Have the calculator determine the sine of 20°, and then divide that number by 1.5. The answer of 0.2280 tells you that there is some angle that has a sine value of 0.2280. Find the angle whose sine is 0.2280 by using the inverse key and then sin or by using the arcsin key or the sin–1 key. When we calculate the value, we get about 13°.

Did you notice in Figure 20.5 that when the light ray leaves the bottom of the glass prism, it bends away from the normal and enters air at an angle of 20°?

That makes the ray leaving the prism parallel to the ray that entered the prism at the top. Behavior like that happens when the sides of the prism are parallel. If you apply Snell's law to the ray inside the prism at the bottom, you can calculate the size of the angle of refraction into air. It's important to realize that the process of a light ray leaving the prism is just the inverse of the process of a light ray entering the prism, so that Snell's law states:

$$\frac{\sin \angle i_{glass}}{\sin \angle r_{air}} = \frac{1}{n_g}$$

$$\sin\angle r_{air} = n_g \sin\angle i_{glass}$$

$$\sin\angle r_{air} = (1.5)(\sin 13°)$$

$$\sin\angle r_{air} = 0.3374 \quad \sin\angle r_{air} = 20°$$

The Critical Angle of Incidence

You have just used the inverse of Snell's law. However, it works only for all angles of incidence in glass such that the ray leaves the glass and enters air or a vacuum. If you increase the angle of incidence in glass, the ray entering the air bends farther away from the normal. The last ray that will leave the glass and enter the air leaves at an angle of 90° in air. The angle in glass that would cause that angle of refraction is called the critical angle for glass. Any transparent substance has a critical angle relative to air, the last angle of incidence that has an angle of refraction in air equal to 90°.

You may use Snell's law to calculate the critical angle because it is calculated the same way as before. That is, Snell's law states $\frac{\sin \angle i_{glass\ crit}}{\sin 90°} = \frac{1}{n_g}$ or $\sin\angle i_{glass\ crit} = \frac{1}{n_g}$ because the value of sin 90° is 1. That means that $\angle i_{glass\ crit} = \sin^{-1}(\frac{1}{n_g})$ or $\angle i_{glass\ crit} = \arcsin(\frac{1}{n_g})$. Both types of notation mean the critical angle of glass is equal to the inverse sine of the reciprocal of the index of refraction of glass. You can get a numeric answer using that result: $\angle i_{glass\ crit} = \sin^{-1}(\frac{1}{1.5}) = 41.8°$. That means that any angle of incidence in glass less than or equal to 41.8° will leave the glass and enter the air. Any angle of incidence in glass greater than 41.8° will be totally reflected back into glass. That is called total internal reflection. The ray is not able to leave the glass if the angle of incidence is greater than the *critical angle*. Because it stays in the glass, the laws of reflection now take over as if it bounced off a mirrored surface. Actually, such reflections are even more efficient than mirrored reflections.

def•i•ni•tion

The **critical angle** of a substance is the angle of incidence in the substance that has an angle of refraction of 90° in a second substance.

The last result applies to any transparent substance. The critical angle may be calculated as

$\theta_{cs} = \sin^{-1}(\frac{1}{n_g})$, as long as the light is entering air or a vacuum, and is read: the critical angle of the substance is equal to the arcsin of the reciprocal of the index of refraction of the substance. We can calculate the critical angle for light as it tries to leave water and enter air: $\theta_{iw} = \sin^{-1}\left(\frac{1}{\frac{4}{3}}\right) = \sin^{-1}\left(\frac{3}{4}\right) = 48.6°$. The last ray that will leave the surface of water is incident at the angle of 48.6° and will enter air at an angle of refraction of 90°. Light enters water by following the inverse path. The last ray that enters water from air is at an angle of incidence of 90°, and it refracts at an angle of 48.6° in water. That is why fish see the world through a cone of light defined by the critical angle. To a fish, the world must appear as a circle above their eyes! If you want a fish-eye view of the world, the next time you go swimming stand on the bottom of the pool and look up at the surface. There is a circle through which you may see people standing at the side of the pool. Beyond that circle the surface of the water is a mirror and reflects all of the light back into the water.

Johnnie's Alert

If you trace a ray of light from a medium that has an index of refraction greater than the index of refraction of the medium the ray is entering, there will always be a critical angle. You will never have a critical angle going from a medium whose index of refraction is less than the index of refraction of the medium it is entering, such as going from air into glass.

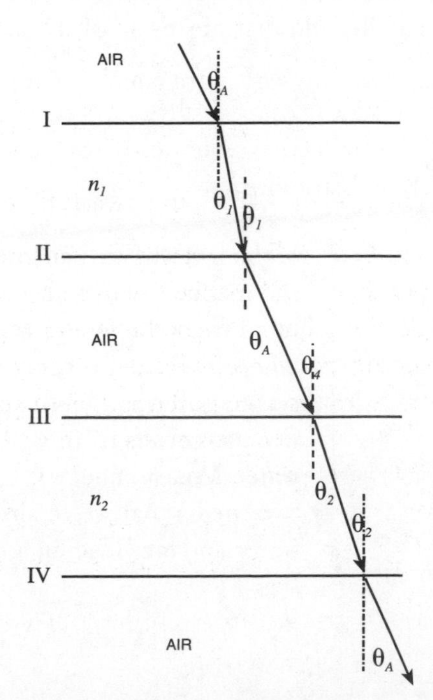

Figure 20.5

A ray of light travels from medium 1 into medium 2 according to the expression $n_1 \sin\theta_1 = n_2 \sin\theta_2$.

Snell's Law for Two Other Media

So far we've used only air as the medium to apply Snell's law. What happens if we trace a ray of light from water into glass? Even though there is no obvious answer, we can use information we have to develop an answer to the question. Suppose we have two prisms with parallel sides separated by a thin layer of air. Let's trace a ray of light from air through the prisms and the layer of air as shown in Figure 20.5.

Start with the surface at I and end at surface IV.

$$\frac{\sin\theta_A}{\sin\theta_1} = n_1, \text{ at I}$$

$$\frac{\sin\theta_1}{\sin\theta_A} = \frac{1}{n_1}, \text{ at II}$$

$$\frac{\sin\theta_A}{\sin\theta_2} = n_2, \text{ at III}$$

$$\frac{\sin\theta_2}{\sin\theta_A} = \frac{1}{n_2}, \text{ at IV}$$

Both the first and second equations show that $\sin\theta_A = n_1\sin\theta_1$ and both the third and fourth equations show that $\sin\theta_A = n_2\sin\theta_2$. If you look at Figure 20.6, you'll see that all of the normals are parallel and that means all of the angles in air, θ_A, are equal.

Therefore, all of the expressions, $\sin\theta_A$, are equal. That gives you:

$$\sin\theta_A = \sin\theta_A$$

$$n_1\sin\theta_1 = n_2\sin\theta_2 \text{ By substitution.}$$

Newton's Figs

Because the normals are all parallel because the faces of the prism are parallel, the angles in each medium are equal. The reason is that when parallel lines are cut by a transversal (the ray of light here), the alternate interior angles are equal.

If the layer of air becomes smaller and smaller, then the medium with index of refraction n_1 is not separated from the medium with index of refraction n_2. Therefore, the expression $n_1\sin\theta_1 = n_2\sin\theta_2$ enables you to describe quantitatively the behavior of light when it travels from water into glass or glass into water. We can check it to see whether it works for two media that we're already familiar with, such as water and air. Tracing a ray from air into water, the expression yields: $(1)\sin\theta_A = n_w \sin\theta_w$ because $n_A = 1$, $\frac{\sin\theta_A}{\sin\theta_w} = n_w$, and that is exactly the statement of Snell's law for that situation. That means that you may use the expression $n_1 \sin\theta_1 = n_2 \sin\theta_2$ as a generalization of Snell's law. It works for any transparent media.

We have concentrated on refraction in this section, but you know that light incident on the surface of transparent materials reflects at the surface as well as refracting into the material. You may be wondering at this time how some particles "know" to enter the water and some bounce off the surface. You know that the particles also change their average speed when they enter a second medium. Light travels at a constant speed, but the constant average speed is different for different media. Light does not require a medium for transmission; you know it travels through space, which is mostly empty and void of all atoms, to get to the earth from the sun.

Light travels slower in glass and water than it does in air. These are good ideas to consider as we continue to work with the particle model of light. It helps us to know how the particle model handles reflection and refraction at the same surface. Can it explain the change in speed when light travels from one medium to another? The ray diagram in Figure 20.6 shows how a single light ray *A* is partially reflected as *B* and partially refracted as *F*. The refracted portion is incident on the bottom of the glass prism and is partially reflected as *G* and partially refracted into air as *C*.

The ray *G* is incident on the upper surface of the glass prism where it is partially reflected back toward the bottom and partially refracted into air as *D*. The ray *D* emerging into air is parallel to the reflected ray *B*, and *E* emerges into air traveling parallel to *C*. Notice that part of the original ray emerges from the bottom of the glass prism traveling parallel to the original ray. It is offset a bit but leaves the bottom as if it traveled straight through the prism. If you observe this prism in the laboratory, you'll see that the rays emerging into air as *D*, *C*, and *E* are not nearly as bright as the original ray, *A*, incident at the top surface. It appears that no particles are lost and none are gained.

Let's trace a ray of light through at least one other shape of prism. Take a look at the glass prism in Figure 20.7. The ray *A* incident on the prism initially is parallel to the base of the prism. The ray *B* that emerges into air is obviously deviated from the original direction of *A* to a direction toward the base of the prism. This is an interesting outcome because the final direction of the emerging ray results from refraction at two surfaces.

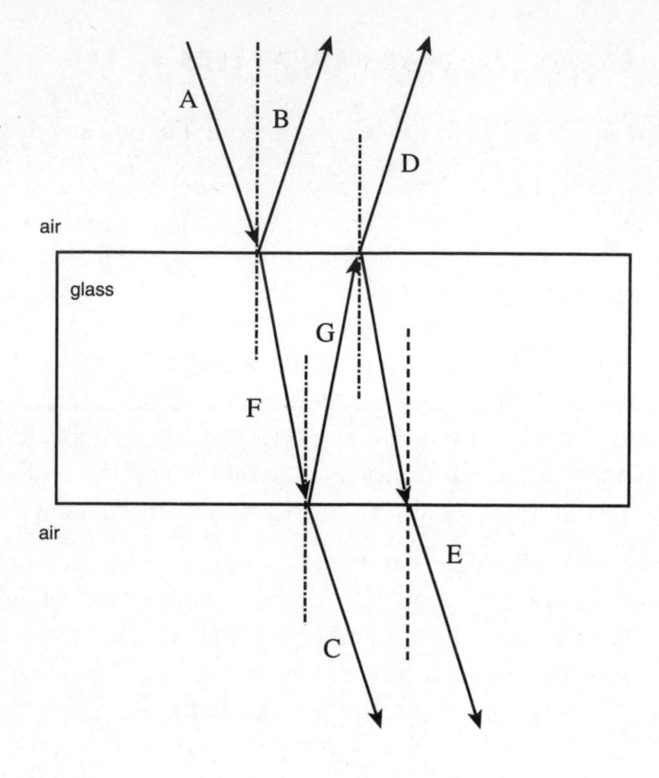

Figure 20.6

Reflection and refraction at the same surface at both parallel faces of a glass prism.

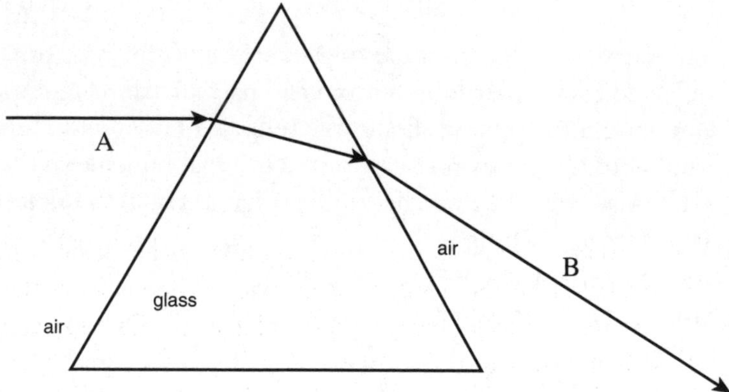

Figure 20.7

Tracing a light ray passing through a triangular prism of glass shows it emerging closer to the thicker portion of the prism.

Did you notice that the rays through prisms with parallel faces go practically straight through the prism? Also, incident rays of light parallel to a side of a triangular prism travel through triangular prisms made of the same medium and emerge traveling toward the thicker portion of the prism. That suggests an interesting combination of prisms.

The Double Convex Lens or Converging Lens

Let's suppose we make a new prism with nearly parallel sides in the middle and triangular-shaped edges. This type of prism is called a converging or double convex lens. That's because it converges rays of parallel light into a focus area. The lens in Figure 20.8 is such a prism and uses those features to form images.

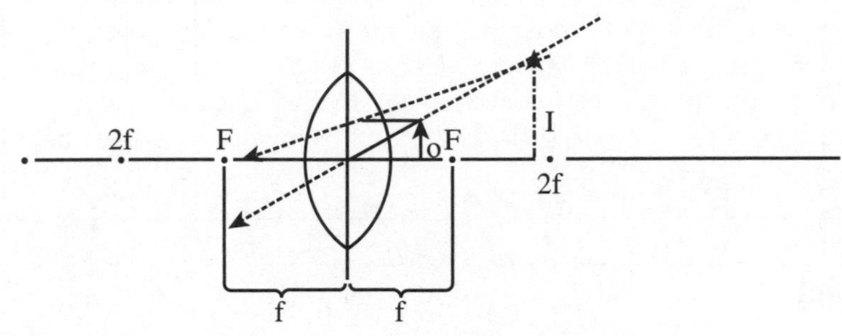

Figure 20.8

The image formed by a double convex lens can be a virtual image.

Do you recognize the image in the diagram? It's a magnifying glass. You'd be looking through the lens at the object from the left side of the lens. This is like holding the magnifying lens over words in a book and looking through the lens at the words. The ray diagram shows an image formed by light that is refracted through a prism (or lens) that has the features suggested earlier. Each side of the lens is convex so the lens does not have one center of curvature like the concave mirror. Distances from the middle of the lens, *F*, are measured in terms of the focal length, *f*, and twice the focal length, 2*f*, where *f* is the distance to the principal focus, *F*, on both sides of the lens.

Even though we know that refraction takes place at both surfaces, the ray diagram is drawn as if all of the bending takes place at one region, the middle, of the lens. Two rays are needed to locate the image of a point. One ray is parallel to the principal axis of the lens and near the principal axis. The other ray from the object is drawn through the region of the lens where the light travels practically straight through the lens without a net change in direction. It travels nearly straight because the edges of the lens are nearly parallel there. This is like the rectangular solid you saw earlier.

The images are formed where the refracted rays intersect or appear to intersect. The image of only one point is located because images of all other points are found the same way. The image of the base of the object is located on the principal axis if the object's base was on the principal axis. The object is drawn perpendicular to the principal axis so all other points of the image are drawn with the image being perpendicular to the principal axis, too.

Johnnie's Alert

The double convex lens or converging lens forms both real images and virtual images. Like the concave mirror, real images are formed where real light rays actually intersect and virtual images are formed where extensions of light rays seem to converge when your brain tries to find a place of convergence. Real images are always inverted and virtual images are never inverted. The light rays that form these images are rays refracted by the lens.

Testing the Particle Model

Why does it appear as though particles seem to "know" to change directions when traveling from one medium to another? Well, it has to do with refraction, and an experiment can give us a possible answer. The experiment involves treating light as a particle that is familiar and checking to see whether Snell's law is obeyed by the particle. The particle is a large steel ball.

The media are modeled by two horizontal surfaces separated by an inclined plane about a textbook thickness high. The particle is launched at the same speed on the upper surface and is allowed to roll down the inclined plane to the lower surface. The path of the ball on both surfaces is recorded by carbon paper tracks on plain white paper.

The ball is rolled over the top of the carbon paper several times at several different angles of incidence. The angles of incidence are matched up with the corresponding angles of refraction. Snell's law explains the change in direction of the ball on the lower surface compared to the upper surface. The angles on the lower surface were smaller than the angles on the upper surface, so the upper surface is a model of air and the lower surface water or glass.

Snell's law also explains the change in direction. However, you might recognize that there is a problem here even though the change in direction can be explained. You know that if the ball rolls down the inclined plane it travels faster on the lower surface than it does on the upper surface. That just doesn't agree with your experience because you know that light travels faster in air than it does in water.

The particle model for light doesn't do a complete job of explaining the behavior of light. As physicists we might have an inclination to toss it out because it can't explain refraction. But look at all of the things it does help to explain. The logical (and scientific) thing to do is continue using the particle model to explain things that it's good at explaining. Then look for a better model that can handle everything the particle model makes clear as well as things the particle model can't explain. We do exactly that in the next chapter as we continue to search for a more complete model for light.

Problems for the Budding Rocket Scientist

1. Assume that two mountains are separated by 50 mi. How long will it take for the light "released" when a lantern is uncovered on one mountain to reach the other one?
2. Light from a projection lantern provides an illumination of 12,000 lm/m^2 on a wall perpendicular to the beam and at a distance of 5 m from the source. What intensity must the source have to give this same illumination at a distance of 10 m?
3. You place an object 5 in. from a double convex lens with a focal length of 10 in. What type of image will it form? Will the image be larger or smaller than the object?

The Least You Need to Know

- The particle model is a good model to explain the direction and speed of light. It also explains that the intensity of illumination varies inversely with the square of the distance from the light source.
- The law of reflection states that the angle of incidence is always equal to the angle of reflection.
- Refraction is the change in direction of light as it leaves one medium and enters a different medium.
- The double convex lens or converging lens forms both real images and virtual images.

Chapter 21

Light as Waves

- The double-slit experiment
- What is color?
- The electromagnetic spectrum
- Diffraction and interference patterns

By the end of the eighteenth century, everyone had conceded that Newton's theory of light as a particle was wrong and that light was a wave, but what kind of wave? An ocean wave is not a thing, it's a property of water, something that water does. If there is no water, there is no wave. So if light was a wave, what was waving? That was the most urgent question physicists were asking.

To say that light consists of waves is not enough, for as you know, there are two important classes of waves, with significant differences in properties. Water waves are transverse waves, undulating up and down at right angles to the direction in which the wave as a whole is traveling. Sound waves are longitudinal, undulating back and forth in the same direction in which the wave as a whole is traveling. Which variety represents light waves? We reveal that enlightening answer shortly.

Thomas Young's Experiment

About 100 years after Newton's theory about the particle structure of light, a man by the name of Thomas Young performed a famous experiment in which he showed that light *propagated* as a wave.

def•i•ni•tion

Propagate is a term used both in biology and physics. In biology, it refers to the reproduction of a species, whereas in physics it refers to the transmission of sound waves or electromagnetic waves through air or water.

In Young's experiment, a light source was shown on a screen that had two holes a few millimeters apart. He put another screen behind the first, and the light coming through the two holes of the first screen illuminated this second target screen. As expected, two patches of light appeared. He then made the holes smaller and the correspon-ding patch of light became smaller, too.

But then something very unusual happened. When you made the holes very small, faint rings appeared around the patches on the target screen that actually made them bigger. Instead of the patches reducing in size to correspond with the smaller holes, they were larger. This couldn't happen if light were made of particles, because particles move in straight lines and wouldn't make these larger rings of light around the patches.

If he made the holes even smaller, the patches of light on the target screen began to overlap and became crossed with dark lines. These dark lines were caused by waves of light interfering with one another.

You can get a good idea of just how this works if you drop a couple of rocks in the water. The ripples sent out by each rock hitting the water interact with each other. Some cancel each other out and some amplify each other. The light and dark bands Young saw on the screen were the result of the light waves doing the same thing. The dark band was the absence of light, or when light waves canceled each other out, whereas the lighter bands were where light waves amplified each other. Such an interference pattern can only result if light is a wave, not a particle. And the type of interference pattern that was created also clarified for physicists that light was a transverse wave.

Based on the spacing of the interference bands of light and dark, Young was able to calculate the wavelength of light. He found that a single wavelength was fifty thousandth of an inch. It was also possible to show that the wavelengths of red light are about twice the length of violet light, which fit the requirements of wave theory.

Wavelengths and Frequency

In the metric system, it has proved convenient to measure the wavelengths of light in millimicrons ($m\mu$), where a millimicron is a billionth of a meter $F_c = k\frac{q_1 q_2}{r^2}$ or a ten millionth of a centimeter ($10^{-7}cm$). Using this unit the spectrum extends from $760m\mu$ for red light of the longest wavelength to $380m\mu$ for the violet light of shortest wavelength. The position of the wavelengths of any spectral line can be located in terms of its wavelength. Figure 21.1 shows how a prism breaks white light into the *spectrum* of colors.

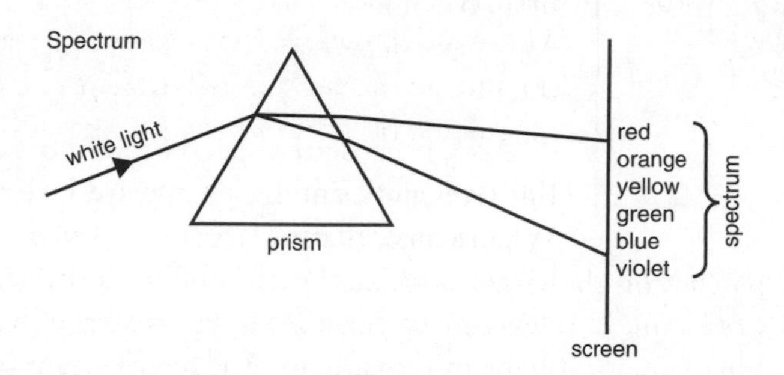

Figure 21.1

A prism breaks white light into a spectrum of colors.

One of those who made particularly good measurements of the wavelengths of spectral lines was the Swedish astronomer and physicist Anders Jonas Angstrom (1814–1874), who did his work in the mid-nineteenth century. He made use of a unit that was one tenth of a millimicron. This is called an angstrom unit (A) in his honor.

Before we go any further, a good question to ask at this time would be, what is color? Well, in terms of visible light, color is simply electromagnetic energy of a specific wavelength. Within the visible part of the spectrum (between 4000 A and 7000 A), the color perceived by the eye depends on the wavelength of the light. So objects that give off or reflect red light look red. Objects that give off or reflect blue light

def•i•ni•tion

When a beam of white light (sunlight) is refracted through a prism, and then allowed to strike a white surface, an extended band of colors displays. This rainbow of colors is called the **spectrum**.

Newton's Figs

An easy way to remember the order of the colors of the spectrum is simply to remember the name Roy G Biv. Each letter of the name is the first letter of a color. So, red, orange, yellow, green, blue, indigo, and violet.

look blue. The wavelength ranges for the different colors may be given roughly, for the colors blend into one another and there are no sharp divisions: red, 7600–6300 A; orange, 6300–5900 A; yellow, 5900–5600A; green, 5600–4900 A; blue, 4900–4500 A; violet, 4500–3800 A.

We can easily convert between wavelength and frequency of light by noting that the relationship among wavelength, frequency, and velocity for any wave is given by

$$v = \lambda f$$

where v is velocity, λ is wavelength, and f is frequency. Because the velocity of light, c, is constant in a vacuum, this relationship becomes

$$c = \lambda f$$

Also, because the speed of light c is a constant, wavelength and frequency are inversely proportional; that is, larger (longer) wavelengths are associated with smaller (lower) frequencies, and vice versa. And when we say that the spectrum is made up of Roy G Biv, these divisions are for convenience. The visible spectrum actually consists of an infinite gradation of color from red to violet.

Newton's Figs

In truth, nature is the ultimate crayon box, with colors varying infinitesimally from red through orange yellow to violet. Computers approximate this infinity of colors with varying degrees of accuracy. Older computer graphics cards produced 25 shades of color (called 8-bit color). Current graphics cards are 16 to 32 bit, offering 65,356 or more than 16 million colors, respectively.

Spectral Lines

Between 1814 and 1824, a German optician, Joseph von Fraunhofer (1787–1826), was working with particularly fine prisms and noticed hundreds of dark lines in the spectrum. He labeled the most prominent ones with letters from A to G and carefully mapped the relative position of all he could find. These spectral lines are, in his honor (it never stops), Fraunhofer lines.

Fraunhofer noticed that the pattern of lines in sunlight and the light of reflected sunlight (from the moon or from Venus, for instance) was always the same. The light of stars, however, would show a radically different pattern. He studied the dim light of heavenly objects other than the sun by placing a prism at the eyepiece of a telescope, and this was the first use of a spectroscope.

Different types of light sources produce different spectra, as can be seen with even a simple spectroscope. Objects such as a light bulb filament, the sun, or the heating element from a stove will produce a spectrum that appears to contain all of the colors of the rainbow, which is called a continuous spectrum. Many spectra that appear to be continuous actually contain gaps; that is, they are within the continuous spectrum wavelengths that are not represented, and appear as dark bands. The spectrum of the sun (examined in detail) contains many dark bands called absorption lines, which produces an *absorption line spectrum.*

Other spectra consist of only bright lines; this type of spectrum is called an *emission line spectrum.* Each of these types of spectra arises from different physical situations. For example, a continuous spectrum arises from any hot object (such as the sun or the heating element in a stove). The ideal version of such an object is sometimes called a blackbody. A blackbody is simply a perfect absorber and emitter of radiation. The wavelength at which the intensity of a blackbody peaks determines its temperature. Objects whose peak is closer to the red end of the spectrum are cooler, and those whose peak is closer to the blue end of the spectrum are warmer.

The dark lines in the solar spectrum seem to be an absorption spectrum. The blazing body of the sun is sufficiently complex in chemical nature to produce the continuous spectrum. As the light passes through the somewhat cooler atmosphere, it is partially absorbed. Those parts that would be most strongly absorbed, and would show up as dark lines in the spectrum, would correspond to the emission spectra of the elements most common in the solar atmosphere. For example, there are prominent sodium absorption lines in the solar spectrum (Fraunhofer labeled them the "D line"), and this is strong evidence that sodium exists in the solar atmosphere.

def•i•ni•tion

An **absorption line spectrum** arises when a hot object (such as the sun) is behind a cooler object (such as the outer layers of the sun's atmosphere). In this case, the intervening gas absorbs light at particular wavelengths, so these wavelengths get subtracted from the continuous spectrum. The exact colors (wavelengths) absorbed provide evidence about the atoms and molecules that are present in the outer layers of the sun. An **emission line spectrum** is produced when the gaseous form of an element is excited by some energy source. As the element returns to its nonexcited or ground state, it emits characteristic, signature wavelengths (and therefore colors) of light. The pattern of lines produced relates to the jumping of electrons from higher to lower energy levels within the atoms. Different jumps release different amounts of energy.

The pattern of emission lines produced is different for each element in the periodic table and each molecule that these elements can form. When we know the characteristic pattern of a given element, we can use that knowledge to identify elements present in any object from which we can gather light. This is what makes astronomy such a powerful investigative tool. We can determine the elements and molecules present in sources that are too distant to ever visit simply from the light that reaches us.

The composition of many distant sources can be determined using a spectroscope in conjunction with a radio telescope. This combination can detect the presence of complex molecules in distant clouds of gas from which stars are forming. Earthbound physicists and chemists know the lines that identify certain compounds, and detecting these same spectral lines from a source in space tells us that those elements are located in those regions, too.

Light and Paint

The results of mixing paints are altogether different from the results of mixing light. If you mix all the colors of light in the spectrum you produce white light. But you know that if you mix all the colors together in a paint box, you certainly don't get white paint, you produce a brownish-gray mess.

In terms of light or paint, virtually any color can be produced by mixing proper portions of the three primary colors: red, yellow, and blue. In fact, the colors on your TV screen are produced by clusters of these three colors covering the screen. If you project these three colors on a screen and let them overlap, you'll find that the region of overlap of all three colors is white.

Paints or pigments function by absorbing (removing) some wavelengths from white light striking the surface and reflecting the other wavelengths. Therefore, an object that looks red is one that absorbs all wavelengths except red, which it reflects. Its surface has a red color when illuminated by white light.

So when two colors of paint are mixed, the result is a color that is the combination of the two colors of that paint. Imagine that you mix together some yellow and some blue paint. We all know that the result is green paint. This is because the yellow paint absorbs blue wavelengths, and the blue paint absorbs red and yellow wavelengths, meaning that only green wavelengths aren't absorbed but reflected. Mixing together all colors in the paint set will mean that the resulting sludge will effectively absorb all colors and start to look gray or black (depending on how effectively the colors absorb across the spectrum).

Light and the Doppler Effect

With light viewed as a wave motion, it was reasonable to predict that it would exhibit properties analogous to those shown by other wave motion. The Austrian physicist Johann Christian Doppler (1803–1853) had pointed out that the pitch of sound waves varied with the motion of the source relative to the listener. If sound waves (and we covered this back in Chapter 16) were approaching the listener, they would be crowded together, and more waves would impinge upon the ear per second. This would be equivalent to a raised frequency, so the sound would be heard as being of a higher pitch than it would have been heard if the source were fixed relative to the listener. By the same reasoning, a receding sound source emits a sound of lower pitch, and the train whistle, as the train passes, suddenly shifts from treble to base.

In 1842, Doppler pointed out that this "Doppler effect" ought to apply to light waves too. In the case of an approaching light source, the wave ought to be crowded together and become of higher frequency, so the light would be come bluer. In the case of the receding source, light waves would be pulled apart and become lower in frequency, so the light would become redder.

When an object is in motion relative to Earth, the light that it emits is redshifted (to longer wavelengths) if it is moving away from us and blueshifted (to shorter wavelengths) if it is moving toward us. The shifts in wavelength are small but measurable. In fact, the change in wavelength is related directly to the speed of the source and the speed of light. The change in wavelength ($\Delta\lambda$) is related to the rest wavelength in the following way:

$$\frac{\Delta\lambda}{\lambda} = \frac{v}{c}$$

where v is the speed of the source relative to the observer, and c is the speed of light (3×10^8 *m/s*). Let's put this formula into practice by applying it to a problem.

The rest wavelength for a bright red line in the emission line spectrum of the hydrogen atom is 656.3 nm (nanometers). If an astronomer observes this line in a galaxy moving away from Earth at 100 km/s, what will be the change in wavelength, and will it appear as a shorter or longer wavelength?

First, if the source is moving away, the light will be redshifted, making the detected wavelength longer. So we use the formula

$$\frac{\Delta\lambda}{\lambda} = \frac{v}{c}$$

and multiply both sides by the wavelength to get

$$\Delta\lambda = \frac{\lambda v}{c}$$

Being careful to use the same units for the speed of light as our velocity (km/s), we get

$$\Delta\lambda = (656.3nm)(100km/s)(3 \times 105km/s) = 0.2nm$$

The Electromagnetic Spectrum

Our eyes are sensitive to a rather limited range of wavelengths or frequencies—namely 400 nm to 700 nm. They are insensitive outside of that range, and many physical processes in the world around us give rise to wavelengths that are outside that range as well. Wavelengths longer than 700 nm are referred to as infrared, and those that are shorter than 400 nm are referred to as ultraviolet.

X-rays have wavelengths still shorter than ultraviolet waves. As wavelengths get shorter and frequencies increase, the energy carried by light gets more and more intense. X-rays, with their high frequencies, are high-energy waves and are able to penetrate most materials, making them useful for imaging bones through skin, for example.

Gamma rays (such as those used by evil scientists in sci-fi movies) have even shorter wavelengths and are more energetic than x-rays. Gamma rays are given off in some of the most energetic events known, including the explosion of atomic weapons and the explosive collapse and death of a star, called a supernova.

Having wavelengths that are longer than optical or infrared radiation, radio waves are used to transmit information in television and radio broadcasts, as well as in cordless and cellular phones. Because radio waves are long wavelengths, they emit a low frequency, and typically are thought not to be harmful to humans.

All the types of waves described constitute the electromagnetic spectrum, and like all waves, they can be reflected, refracted, and diffracted. All these different types of waves also travel through a vacuum with the same speed (*c*). Look at Figure 21.2 to see just what the electromagnetic spectrum looks like and how small of a portion the range of visible light is.

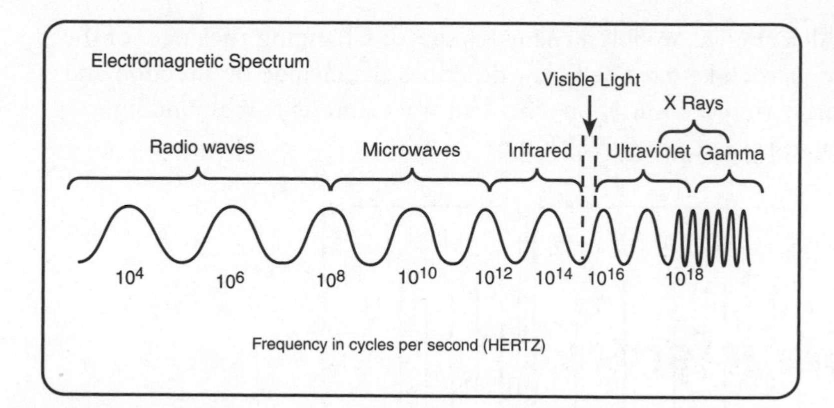

Figure 21.2

The electromagnetic spectrum.

Refraction

For the rest of this chapter, we use a general wave model to illustrate the characteristics of light waves. In the classroom, properties of light waves can be studied by understanding how other types of waves behave. One of the most common ways of doing that is to use what is called a *ripple tank.*

def•i•ni•tion

A **ripple tank** is a laboratory instrument used to study two-dimensional waves. It is a shallow tank that holds water about one to two centimeters deep. It has a glass or clear plastic bottom about 24 inches square.

The depth of water in the ripple tank can be changed so that there are extreme differences in depth by placing a glass or plastic plate just under the surface of the water. That was done for the diagram in Figure 21.3. The dark lines represent crests of straight waves moving from left to right in the diagram as indicated.

The line across the tank represents the change in depth of water from deep to shallow, reading left to right. The waves are periodic waves, so both the incident wave in the deep water and the waves in the shallow water have the same frequency. Different depths of water are different media. The wavelength in the deep water is greater than the wavelength in the shallow water, so $\lambda_d > \lambda_s$, $f\lambda_d > f\lambda_s$, and $v_d > v_s$.

Because the frequencies are the same and the wavelength in the deep water is greater than the wavelength in the shallow water, the speed is greater in the deep water. The speed is frequency times wavelength.

The waves in Figure 21.4 cross a barrier at an angle. The water is deep on one side of the barrier and shallow on the other. The diagram in Figure 21.4 shows clearly that there is a change in direction as well as a change is speed. Changing the angle of the barrier for several trials reveals that Snell's law describes the change in direction and more. Look at the blown-up version of one incident wave and its corresponding refracted wave in the lower left portion of Figure 21.4.

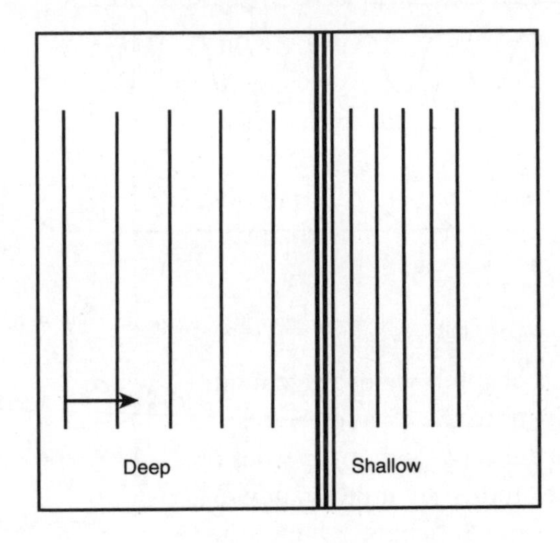

Figure 21.3

When straight waves change media, from deep to shallow water, the wavelength changes.

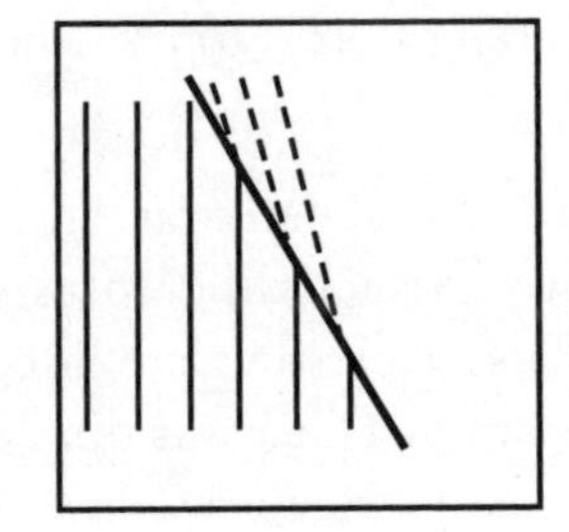

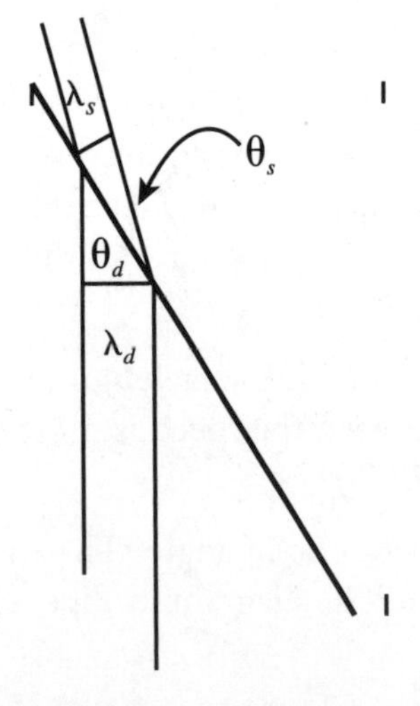

Figure 21.4

Straight waves change direction when they cross a barrier at an angle traveling from deep water to shallow.

The angles of refraction and incidence are measured the same way that the angles of reflection and incidence are measured. The angles to be measured are the smaller of the two angles between the wave and the barrier. We've drawn a line segment representing the wavelength of each wave along with a label. Two right triangles are formed; when that is done the triangles share one hypotenuse, which we'll call *H*. An expression for Snell's law may be written using the two right triangles:

Johnnie's Alert

The angle of incidence and the angle of refraction for waves is measured in a special way. The smaller of the two angles between the incident wave and the junction is the angle measured as the angle of incidence. The smaller of the two angles between the refracted wave and the junction is the angle measured as the angle of refraction.

$$\frac{\sin\theta_d}{\sin\theta_s} = \frac{\frac{\lambda_d}{H}}{\frac{\lambda_s}{H}}$$ By the definition of the sine of an angle.

$$\frac{\sin\theta_d}{\sin\theta_s} = \frac{\lambda_d}{\lambda_s}$$ Simplifying common factors.

$$\frac{\sin\theta_d}{\sin\theta_s} = \frac{\lambda_d f}{\lambda_s f}$$ Multiplying numerator and denominator by *f*.

$$\frac{\sin\theta_d}{\sin\theta_s} = \frac{v_d}{v_s}$$

The wave model predicts that since the ratio of the sine of the angle of incidence to the sine of the angle of refraction is greater than one, the ratio of speeds is greater than one.

That means that the speed in the second medium is less than the speed in the first medium when the angle in the second medium is less than the angle in the first medium. That is just the opposite of the prediction of the particle model. The wave model addresses this better than the particle model because waves crossing a junction between media change direction and have a correct change in speed.

Newton's Figs

The wave model correctly predicts the change in speed when the wave changes media. The change in direction is correctly explained with Snell's law.

Diffraction

Not only do waves change direction and speed when they change media but they also bend around corners. That is called *diffraction*. Straight waves are generated in the ripple tank as shown in Figure 21.5. A barrier with a slit smaller than the wavelength of the wave is placed in the path of the straight waves.

def•i•ni•tion

Diffraction is the curving of waves as they pass through an opening or pass by a sharp corner. The diffraction of light is an important property of light waves. Among other things, it provides the key to checking to see if you can measure the wavelength of light.

The waves pass through the opening and curve, bending around the edges of the slit. The slit acts very much like a point source of waves. It's like dipping your finger into the water and creating waves on the surface of the water.

The slit acts very much the same way if it is small compared to the length of the wave. If the slit is large compared to the length of the wave, the wave passes through the slit with slight curving at the edges. In fact, if the wavelength is very short compared to the width of the slit, the wave goes practically straight through the slit with very little curving.

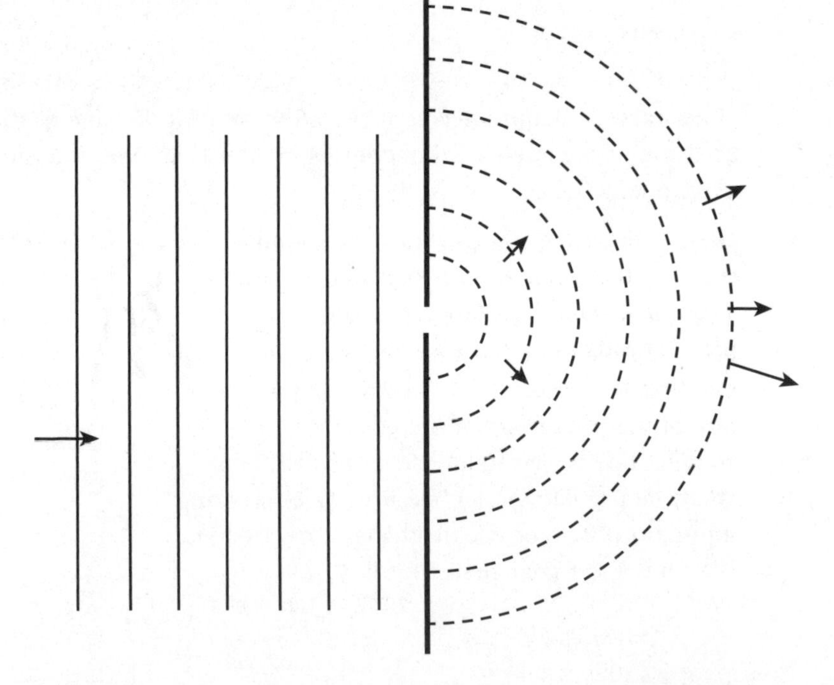

Figure 21.5

Straight waves passing through a slit bend sharply if the slit is small compared to the length of the wave.

The light passing through the slit with very little curving is similar to light passing by an object and casting a sharp shadow. That is a real possibility, but the behavior of very short wavelengths causing a sharp shadow suggests something else. If light is a wave, then the wavelength of light must be very short because of the sharp shadows cast by even small objects. If you think about that, light passes into our eyes through an opening of maybe half a centimeter. We see clear images of light bouncing off objects and entering our eyes. That means that if light is a wave, it must have wavelengths of much less than half a centimeter. Otherwise, we would have a large amount of curving and hence blurry images.

Newton's Figs

If you have a laser pointer, you can use it to observe the diffraction of light. In a darkened room shine the laser pointer at a white wall. You will see a spot of intense red light. Now, slowly raise the edge of a knife into the laser beam so that the edge of the beam strikes the knife. The pattern on the wall will broaden, indicating that the light is diffracting at this sharp edge.

Interference

Remember Thomas Young's double slit experiment? We briefly discussed what happened when light waves moved through a slit in a screen and observed the interference patterns indicating that light behaved like a wave. We can do a bit more of an in-depth analysis of the phenomenon of interference by examining two waves that interfere with one another in a ripple tank.

Dipping the finger into the water suggested that a point source of waves would dip into the water to cause circular waves. The ripple tank is equipped with two plastic beads mounted on a rocking arm that causes the beads to dip into the water in phase. The rocking arm is driven by a small electric motor that has a frequency that is the frequency of the dipping plastic beads. The frequency of the motor is measured with a calibrated strobe light. While the frequency is being measured, the wavelength of the water waves generated by the dipping beads is determined at the same time.

Newton's Figs

Viewing light reflected from the surface of a ripple tank when it reveals an interference pattern shows regions where the water is perfectly still. Regions of water that have maximum disturbance separate those regions. The regions of still water are the nodal lines, destructive interference, and the regions of maximum disturbance are regions of constructive interference.

The beads dip into the water, creating circular waves from two point sources. You know that the waves travel at the same constant speed in a single medium. The waves from the two point sources travel through each other and they create regions of higher waves and regions of perfectly still water. The total picture is called an interference pattern.

Lines of still water and lines of maximum disturbance, deeper troughs and higher crests, characterize the interference pattern. The lines of still water are called nodal lines. The nodal lines are the regions where a crest from one point source meets a trough from the other.

def•i•ni•tion

Waves **interfere** with each other when they move through each other and either reinforce each other or destroy each other. Waves **interfere constructively** when they interact to build each other up or reinforce each other. **Destructive interference** occurs when waves interact to cancel each other out or destroy each other.

Regions of maximum disturbance occur where a crest from one source meets a crest from the other source and a trough meets a trough. The drawing in Figure 21.6 helps you to visualize the interference pattern of two point sources, in phase and with the same frequency, in two dimensions. In Young's experiment, the dark and light lines represented the *interference* pattern of the light waves interacting with one another as they emerged from the two slits and hit the target screen. The light bands represented waves that were *interfering constructively* and the dark bands represented waves that were *interfering destructively*.

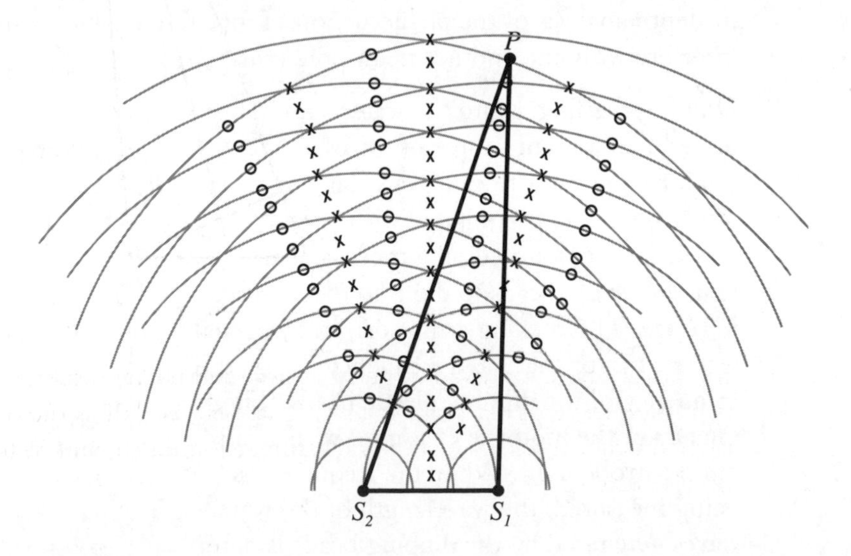

Figure 21.6

The nodal lines separated by regions of maximum disturbance of the medium characterize the two-dimensional interference pattern.

Calculating the Wavelength Using an Interference Pattern

> **Newton's Figs**
> The perpendicular bisector of the line segment joining the point sources is in the middle of a region of maximum disturbance.

Figure 21.7 is an enlargement of the dark triangle found in Figure 21.6. The perpendicular bisector of the line segment joining S_1 and S_2 is shown as a vertical dotted line. The distance between S_1 and S_2 is labeled *d* and the distance from the foot of the perpendicular bisector of *d* to *P* is labeled *L*. The distance from *P* to the perpendicular bisector of *d* is labeled *x*. Two acute angles of the two right triangles are labeled ϕ and ϕ'.

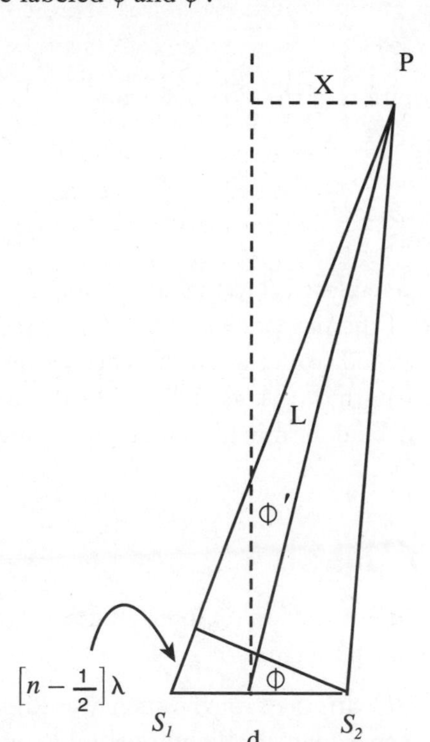

Figure 21.7
The geometry of the location of a nodal point enables you to calculate the wavelength of the waves.

If *P* is very far away from S_1 and S_2 compared to the distance between the sources, the two line segments drawn to *P* from S_1 and S_2, as well as the one from the foot of the perpendicular bisector of the line segment joining S_1 and S_2 to *P*, are parallel or nearly parallel.

Johnnie's Alert

The approximation that the three lines are parallel is good only if P is very far away from the point sources compared to the distance between them. Very far away would be at least 10 times the distance between the sources of waves.

That means that a perpendicular to PS_1 from S_2 cuts off a line segment that is the difference in path length, $PS_2 - PS_1$, and is labeled $(n - \frac{1}{2})\lambda$. That approximation is meaningless unless P is very far away from S_1 and S_2. The acute angles are equal because two angles are equal if they have their sides perpendicular right side to right side and left side to left side. Because the angles are equal, their sines are equal, so

$$\sin \phi = \sin \phi'$$

$$\frac{\left(n - \frac{1}{2}\right)\lambda}{d} = \frac{x}{L} \quad \text{Definition of the sine of an angle.}$$

$$\lambda = \frac{dx}{L\left(n - \frac{1}{2}\right)}$$

The interference pattern enables you to measure x and L, and you can determine n by counting on one side of the perpendicular bisector of d. The value for the wavelength of the wave determined from the interference pattern is the same as the value found by measuring the length of the wave directly with the strobe light. If light is a wave, the wavelength may be determined in much the same way using an interference pattern.

An Interference Pattern of Light

Young gave us the tool needed to view an interference pattern for light as well as the interference pattern in the ripple tank.

Two point sources in phase with the same frequency are needed to set up an interference pattern. In Young's experiment for light, one light source is used but it is viewed through two tiny slits very close together. The two slits act like two point sources just as you found in the diffraction discussion. The two point sources are in phase and have the same frequency because the same wave from a source passes through both slits at the same instant.

How about a little exercise of the imagination? Let's suppose you take the pattern in Figure 21.6 and rotate it about the line segment joining S_1 and S_2. Can you imagine nodal planes being generated by the nodal lines? If you view the nodal planes of an interference pattern of light, the nodal planes intersect the retina of your eye in lines and you see the lines of intersection. The lines are black because they are nodal planes that are regions of destructive interference. Bright regions, because of constructive interference of light, separate the black lines.

The diagram in Figure 21.8 helps to visualize the observed interference pattern and the experimental arrangement for viewing the nodal lines.

Johnnie's Alert

Young's experiment uses two tiny slits a small distance apart diffracting a sample of the same wave from the same source to create an interference pattern. The slits act like two point sources in phase with the same frequency.

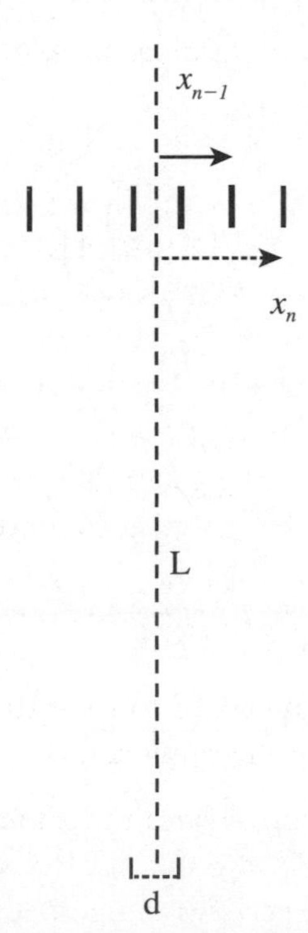

Figure 21.8

The nodal lines in the interference pattern for light are seen as black vertical lines separated by bright regions.

When two point sources generate an interference pattern in the ripple tank, it is found that the closer together the point sources are the fewer the number of nodal lines. That means that the point sources of light must be close together so that we can distinguish the nodal lines. It also means that the slits that serve as point sources are very narrow because you want as much diffraction by each slit as you can get.

Johnnie's Alert

The interference pattern of light that you see is actually formed on the retina of your eye. The measurements you make in the laboratory are in a triangle similar to the triangle in your eye.

The distance between slits is so small that the exact arrangement as used in Figure 21.7 is impossible to use. Even though you cannot measure x as you did in Figure 21.7, you can use that same information to calculate Δx, the distance between adjacent nodal lines. Let x_n represent the distance to the nth nodal line and x_{n-1} represent the distance to the nodal line next to it. Then

$$\Delta x = x_n - x_{n-1}$$

$$\Delta x = \frac{L\lambda\left(n-\frac{1}{2}\right)}{d} - \frac{L\lambda\left((n-1)-\frac{1}{2}\right)}{d}$$ By substitution, because the results of the analysis of Figure 21.7 shows that $x = \frac{L\lambda\left(n-\frac{1}{2}\right)}{d}$.

$$\Delta x = \frac{L\lambda}{d}\left(\left(n-\frac{1}{2}\right)-\left((n-1)-\frac{1}{2}\right)\right)$$ By factoring out the common factor of $\frac{L\lambda}{d}$.

$$\Delta x = \frac{L\lambda}{d}\left(n-\frac{1}{2}-n+1+\frac{1}{2}\right) = \frac{L\lambda}{d}$$ By simplification.

The expression from the analysis of Figure 21.7 simplifies to $\Delta x = \frac{L\lambda}{d}$ using the arrangement in Figure 21.8.

Johnnie's Alert

It's a good idea to go through each step of the algebra involved in the calculation of Δx until you can reproduce it on your own. As with anything, the more you practice the easier it gets.

A Final Word

The Copernican revolution was 2,000 years in the making. Classical mechanics was the result. Through the struggles and achievements of intellectual giants such as Galileo, Kepler, and Newton, who followed the right clues, a deeper understanding of the universe emerged. The nature of this new science was determined by the questions that it asked, and some of these questions were ones that we focused on in this book.

The great book of nature contains many more pages, full of new phenomena, new clues to the mysteries that remain unsolved; and new questions arise as each old question is answered, so the book of nature may not even have a final chapter. Throughout this book we've developed a conceptual as well as a mathematical background upon which we could study principles such as the conservation laws. This knowledge will serve as a trusty guide as ever-emerging questions together with new research take us beyond the confines of our present-day understanding of the universe.

Problems for the Budding Rocket Scientist

1. What is the frequency of radio waves, assuming that their wavelength is 21 cm?
2. If you are approaching a stoplight, how fast will you have to be traveling to make a red stoplight appear green? Hint: assume that the actual wavelength is 700 nm (red) and that you have to be going fast enough to shift it to 500 nm (green).
3. Find the angle of refraction of light incident on a water surface at an angle of 48° to the normal.
4. Light from a sodium vapor lamp (589 nm) forms an interference pattern on a screen 0.8 m from a pair of slits. The bright fringes in the pattern are 0.35 cm apart. What is the slit separation?

The Least You Need to Know

- Thomas Young was the first to show by experiment that light propagated as a wave.
- When an object is in motion relative to Earth, the light that it emits is redshifted (to longer wavelengths) if it is moving away from us and blueshifted (to shorter wavelengths) if it is moving toward us.
- The diffraction of light is an important property of light waves. It provides the key to measuring the wavelength of light.
- Waves interfere with each other when they move through each other and either reinforce each other or destroy each other.
- The wavelength at which the intensity of a blackbody peaks determines its temperature.

Glossary

acceleration The rate of change of velocity or the change in velocity divided by the time for that change to take place.

actual mechanical advantage The gain a user realizes from a machine because it includes the effects of friction. (A ratio between forces.) Ideal mechanical advantage is the theoretical gain a user expects from a machine because it does not include the effects of friction. (A ratio between distances.)

adhesive forces Forces of attraction between particles of different kinds of matter.

alpha particles ($^{4}_{2}He$) Fast-moving helium nuclei that have a positive charge and are emitted from the nucleus of radioactive elements such as radium.

alternating current (AC) An electric current that travels in one direction and then the opposite direction with a fixed period. Your local power company, through outlets in your home, provides AC.

ampere The practical unit of electric current. If 1 coulomb of charge passes through a cross section of a conductor at a point in an electric circuit in 1 second, then the current is rated at 1 ampere.

amplitude The maximum displacement (from its equilibrium position) of the object undergoing simple harmonic motion.

anode The positive electrode of a cell or battery.

atom The smallest particle of element that exists alone or in combinations.

atomic mass number The number of nucleons the atom contains.

atomic number The mass of an atom relative to that of the isotope carbon 12. The whole number nearest the atomic mass is the number of nucleons the atom contains.

average speed The magnitude of the change in position of an object that does not change direction, divided by the change in time required to make that change in position; or the distance an object travels divided by the time required to complete the trip.

battery A combination of cells. It is constructed to overcome the limitations of one cell. A battery provides a larger current, a larger potential difference, or both.

beta particles High-speed electrons emitted from the nucleus of radioactive elements. They have a negative charge.

branch (of a circuit) A division of a parallel part of the circuit.

British thermal unit (BTU) The quantity of thermal energy required to raise the temperature of 1 pound of water 1 Fahrenheit degree.

calorie The quantity of thermal energy needed to raise the temperature of 1 gram of water 1 Celsius degree.

candela The unit of luminous intensity.

capacitor An electrical device used to store an electrical charge.

cathode The negative electrode of a cell or battery.

center of curvature The center of the sphere of which the mirror is a part.

centripetal acceleration The center-seeking acceleration resulting from uniform circular motion. Furthermore, it is the only acceleration defined by uniform circular motion.

centripetal force The center-seeking force on an object at every instant that deflects the object into a circular path at a constant speed. At all times it is perpendicular to the direction of motion.

closed polygon method A way of adding vectors that allows any number of vectors to be added together by joining them together foot to head until the sum is complete. The resultant vector is found by closing the polygon with a vector drawn from the foot of the first vector in the sum to the head of the last, in that order.

cloud chamber A transparent container filled with alcohol vapor suspended between electrically charged plates. While a radioactive source emits alpha and beta particles, a source of light is directed at the alcohol-filled chamber. Trails much like vapor trails of jet planes are formed along the path where the radioactive particles have traveled. Their trails are easily viewed by the light reflected from the droplets of vapor in the cloud formed.

cohesive forces The attractive forces between particles of the same kind.

components (of a vector) Those parts whose sum is the given vector.

compression That part of a longitudinal wave where the particles of the medium are pushed closer together.

concave spherical mirror A segment of a sphere with the inside surface polished to reflect light.

condensation The process of a vapor or a gas changing to a liquid.

conduction The transfer of thermal energy within a substance from one particle to the next when the particles are not moving from one place to another. Thermal energy causes the particles to vibrate with greater energy, and the vibrating particles bump into neighboring particles, transferring the energy throughout the object.

conductors Materials that allow electrons to move freely throughout the material.

convection The transfer of heat by the movement of matter.

convex spherical mirror A segment of a sphere with the outside surface polished to reflect light.

corona discharge (brush discharge) A bluish glow of ionized gases formed at any sharp point of a conductor that is under the influence of high potential energy.

coulomb The unit of electrical charge. It is equal to the charge on 6.25×10^{18} electrons.

covalent bonding The combination of two atoms to form a molecule by sharing a pair of electrons.

crest That portion of the graph of a transverse wave that lies above the time axis.

critical angle The angle of incidence in the substance that has an angle of refraction of 90° in a second substance. (Only occurs with light as the light leaves a medium where it has a slower speed and enters a medium where it has a faster speed.)

cycle One complete trip for an object moving in a circular path at a constant speed as well as the corresponding trip of its projection on the diameter of the circular path.

defining equation A statement of a relationship between two units of measurement.

density The amount of matter in a unit volume. Because matter is measured in two different ways, there are two types of density. Mass density is the amount of mass in a unit volume of matter, and weight density is the amount of weight in a unit volume of matter.

derived quantities Physical quantities that are defined using two or more fundamental quantities or one fundamental quantity used more than once.

destructive interference Waves interacting to cancel each other or destroy each other.

deuteron The nucleus of an isotope of hydrogen made up of one neutron and one proton.

diffraction The curving of straight waves as they pass through an opening or pass by a sharp corner.

diffusion The movement of particles of one kind of matter into the empty space of a different kind of matter because of the random motion of the particles.

direct current (DC) An electric current that travels in only one direction in the circuit. It travels from the negative side of the cell, through the circuit external to the cell, to the positive side of the cell. Conventional current is in the opposite direction.

displacement A vector defined as a change in position.

Doppler effect The apparent shift in the pitch of a source of sound because of the relative motion between the source and the observer.

dyne The force required to accelerate a mass of 1 gram at a rate of 1 centimeter per second squared.

efficiency A fraction greater than zero and less than one that expresses what part of input work the output work amounts to. The efficiency is usually stated as a percent obtained by multiplying the fraction by 100 percent.

electric circuit The conducting path for the flow of charge.

electric current The flow of charge through a cross section of a conductor past a point in an electrical circuit for each unit of time. (The rate of flow of charge.)

electron A particle that has a charge of 1.6 × 10 – 19 coulomb and a mass of 9.1 × 10 – 31 kg.

element A type of matter like copper, hydrogen, or neon that cannot, by chemical means, be broken down into simpler forms of matter.

equal vectors Vectors that have equal magnitudes and the same direction.

equilibrium When the sum of the forces is zero. If one force is used to balance the effects of two or more other forces, then that force is called the equilibrant and is equal in magnitude and opposite in direction to the resultant of the other forces.

evaporation The process of a liquid changing to a gas or a vapor. It can happen at many different temperatures.

focal point (of a concave spherical mirror) A point on the principal axis where all rays intersect after they are reflected from the mirror. That is true if they are rays that are parallel to the principal axis and near the principal axis. The focal point is halfway between V and C. That means the distance from V to C is $2f$, where f is the focal length of the mirror, the distance from V to F.

focal point (of a double convex lens) A point on the principal axis of the lens where all rays intersect after being refracted by the lens. That is true if they are rays that are parallel to the principal axis and near the principal axis. Because refraction is reversible, and works in either direction through the lens, the lens has a focal point on each side that is the same distance from the center of the lens.

force of gravity The downward pull on any mass placed above or on the earth's surface. This is a pull at a distance because nothing is attached to the mass pulling it toward the earth.

freely falling body An object that is influenced by the force of gravity alone.

freezing point The temperature at which a liquid changes to a solid at standard pressure.

frequency The reciprocal of the period; given by $f = \frac{1}{T}$ and has units of $\frac{cycles}{s}$ or just $\frac{1}{s} = s^{-1}$ because a cycle is not a unit of measurement. Another common unit of measurement of frequency is the hertz, where $1Hz = 1s^{-1}$.

fundamental quantities The building blocks for the foundation of physics. Time, space, and matter are the quantities required to study that area of physics called mechanics.

fusion A name given to the process of changing from a solid to a liquid. It is more commonly called melting. Fusing means to bring together; when metals were melted, they were brought together to make alloys.

galvanometer A meter connected in series in an electrical circuit to indicate the presence of a current. It also indicates the direction of the current in the circuit.

gamma rays Very penetrating short-wavelength electromagnetic radiation emitted from the nucleus of radioactive elements.

graphic solution Achieved by using drawing instruments to construct a scale drawing. A ruler provides the measurement of magnitude to scale and the protractor enables you to measure angles.

gravitational field That region of space where the force of gravity acts on a unit mass at all locations on or above the surface of the earth.

half-life The time required for half of the atoms in a given sample of a radioactive element to decay.

heat The internal kinetic energy and potential energies of the particles of matter.

heat capacity The quantity of thermal energy required to raise the temperature of a body 1 degree.

heat of fusion The quantity of thermal energy required to change a unit mass or weight of a solid to a liquid at the normal freezing point without changing the temperature.

heat of vaporization The quantity of thermal energy required to change a unit mass or weight of a liquid to a gas or vapor at the normal boiling point without changing the temperature.

hydrometer An instrument used to measure the specific gravity of a liquid.

impulse A physical quantity that results from a force being applied to a body for a certain amount of time that is equal in magnitude to the product of their values.

inertial frame of reference A frame of reference that is at rest or moving at a constant speed. For our purposes, the earth is an inertial frame of reference.

instantaneous velocity The velocity of an object at any instant of its motion. The instantaneous velocity vector is tangent to the path of motion of the object at every point along the path.

insulator A material that does not allow freedom of movement of electrical charge.

intensity of illumination The rate at which light energy falls on a unit area of surface.

interfere Waves that move through each other and either reinforce each other or destroy each other. Waves interfere constructively when they interact to build each other up or reinforce each other.

ion A charged particle caused by an atom gaining or losing electrons to exhibit an excess or a deficiency of electrons.

ionic bonding The formation of a unit by the transfer of an electron from one atom to the other creating two ions more stable than the atoms were before the transfer. The two newly formed ions attract each other, due to opposite electrical charge, thus forming a stable unit.

junction Where two media are joined together. That is, a spring can be anchored at a junction, two springs may be attached at a junction, or a spring may be attached to a string at a junction, or oil may float on water forming a junction.

kilocalorie The quantity of thermal energy required to raise the temperature of 1 kilogram of water 1 Celsius degree.

kinetic energy The energy an object has due to the fact that it is in motion.

law of heat exchange The thermal energy gained by cold substances is equal to the heat lost by hot substances. (The first and second laws of thermodynamics.)

longitudinal wave A disturbance traveling through a medium in which the particles vibrate in paths parallel to the direction the wave is traveling.

loop (or anti-node) The region of reinforced amplitude in a standing wave pattern.

luminous intensity The strength of a source of light measured in a unit called the candela.

lumen The unit of intensity of illumination.

magnet An object made of iron or steel that is characterized by a north pole and a south pole each having the ability to strongly attract iron.

magnetic compass A device used to find direction on the surface of the earth. The tip of the needle points to geographic north and the other end points to geographic south.

magnetic field The region near a magnet containing all of the magnetic lines of force. The magnetic field is to a magnet what the gravitational field is to a planet.

magnetic force A force resulting from the attraction or repulsion of a magnetic pole.

magnetic lines of force Imaginary lines directed from the north pole through the space near a magnet into the south pole.

matter Anything that occupies space and has mass.

maximum height (of a projectile) The greatest distance the projectile reaches above the earth or the level of launch.

melting point The temperature at which a solid changes to a liquid at standard pressure.

metallic bonding The name of the attraction atoms of solid metals have for each other in a closely packed arrangement due to the continuous exchange of loosely held electrons in the outer energy levels of the atoms of the metal.

mole Avogadro's number of items (6.02×10^{23}) or a basic unit of quantity.

molecule The smallest particle of a compound.

momentum A physical vector quantity that has its magnitude determined by the product of its mass and velocity.

move An object experiences a change in position.

neutron A particle with no charge that has about the same mass as the proton.

newton The force required to accelerate a mass of 1 kilogram at a rate of one meter per second squared.

nodal lines The regions in an interference pattern where the waves from two point sources that are in phase and with the same frequency cancel each other out. The destructive interference takes place where a trough from one source meets a crest from the other source.

node (or nodal point) The points in a standing wave pattern where the superposition of the waves produces zero amplitude. Nodes are half a wavelength apart, as are the loops of a standing wave.

normal A line perpendicular to a line or to a surface.

nucleus The core of the atom that is made up of protons and neutrons.

parallelogram method A way of adding vectors that requires the feet of two vectors to be located at the same point. A parallelogram is then constructed by drawing a line parallel to one of the vectors through the head of the other. A second line is constructed parallel to the second vector passing through the head of the first. The resultant vector is found by joining the feet with the opposite vertex of the parallelogram, in that order.

period The time for the object in simple harmonic motion to complete one cycle.

phase The position and direction of movement of the object undergoing periodic motion.

pith A very light, dry, fibrous material. Pith is the central column of spongy cellular tissue in the stems and branches of some large plants. It is used with electroscopes to detect electric charge and is replaced by Styrofoam balls in some schools.

position A vector used to locate a point from a reference point or frame of reference.

potential difference (between two points in an electric field) The work done per unit charge as the charge moves between those points.

potential energy The energy an object has because of its placement in a force field.

pound The force required to accelerate a mass of 1 slug 1 foot per second squared.

pressure A quantity determined by the force on a unit of area.

principal axis (of a concave spherical mirror) The line that passes through the center of curvature and intersects the mirror at a point called the vertex of the mirror.

principal axis (of a lens) The line passing through the principal focus and the center of the lens.

propagation The sending out or spreading out of a wave from a source.

proportion An equation each of whose members is a ratio.

proton A positively charged particle having a charge of the same magnitude as that of an electron with a mass of about 1.7×10^{-27}kg.

pulse A wave or a disturbance of short duration.

quantitative description Involves actual measurements of such quantities as temperature, time, and current.

radian The central angle of a circle subtended by an arc that is equal in length to the radius of the circle.

radiation The transfer of thermal energy by having only the energy transferred. No substance or convection currents are needed.

range (of a projectile) The maximum distance traveled horizontally by the projectile. It is measured from the point of launch to the point of return to the same level.

rarefaction That part of a longitudinal wave where the particles of the medium are being spread apart.

reflection The change in direction of light when it strikes a boundary causing the light to bounce off the surface of the boundary so that the angle of reflection is equal to the angle of incidence.

refraction The change in direction of light when it leaves one medium and enters a different medium.

regelation The melting of ice under pressure and then freezing again after the pressure is released.

resistance (in an electric circuit) The opposition to the flow of charge or current in the circuit.

resistor A component in an electrical circuit used to establish the amount of current and/or potential difference at different places in the circuit.

resolution The name of the process of identifying the parts of a vector.

resonance (in sound) The increased amplitude of vibration of an object caused by a source of sound that has the same natural frequency.

resultant vector The name assigned to the vector representing the sum of two or more vectors.

ripple tank A laboratory instrument used to study waves in two dimensions. It is a shallow tank that holds water about 1 to 2 centimeters deep. It has a glass or clear plastic bottom about 24 inches square.

scalar A quantity that has magnitude only. Any units of measurement are included when we refer to magnitude.

scalar multiplication The product of a scalar and a vector that results in a vector with a magnitude determined by the scalar. The direction of the product is the same as the original vector if the scalar is positive and opposite the direction of the original vector if the scalar is negative.

schematic diagram A symbolic representation of electric circuits. That is, symbols representing components, conductors, and instruments of measurement are used instead of pictures of those items.

significant figures Those digits an experimenter records that he or she is sure of plus one very last digit that is doubtful.

simple harmonic motion The to-and-fro motion caused by a restoring force that is directly proportional to the magnitude of the displacement and has a direction opposite the displacement. Examples include the motion exhibited by a vibrating string or a simple pendulum.

simple pendulum A physical object made up of a mass suspended by a string, rope, or cable from a fixed support. It is called a simple pendulum because the string, rope, or cable has a negligible amount of mass compared to the mass of the object being supported. A physical pendulum is a physical object having the mass distributed along the full length of the pendulum and not having most of the mass concentrated in one place as in the simple pendulum.

solidification The process of changing from a liquid to a solid. Commonly known as freezing.

specific gravity The ratio of the weight density of the liquid to the weight density of water. It is also the ratio of the mass density of the liquid to the mass density of water.

specific heat The ratio of its heat capacity to the mass or weight of a substance.

spring balance A device containing a spring that stretches when pulled by a force. Amount of stretch is calibrated to correspond to force units, newtons, dynes, or lb.

standing wave The superposition of two waves of the same frequency moving in different directions in the same medium. A standing wave can be established in a medium of proper length. Resonance of sound in a closed pipe occurs for lengths of a quarter wavelength or odd integral multiples of quarter wavelengths.

S.T.P. Properties of gases are given in terms of S.T.P, or Standard Temperature and Pressure, defined as 760mm of mercury at 0°C.

sublimation The direct change of a solid to a vapor without going through the liquid state.

superpose When one wave adds onto the other. During the instant of superposition, neither pulse is recognizable individually but as one combination of pulses.

surface tension A condition of the surface of a liquid resulting from the attractive forces of the molecules of the liquid that causes the surface to tend to contract.

temperature The condition of a body to take on thermal energy or to give up thermal energy. Bodies with high temperature are in a condition to give up thermal energy to cooler bodies. Bodies with low temperature are in a condition to take in thermal energy from warmer bodies.

terminal velocity The maximum velocity of a falling object in air. The air friction on a falling body will increase with velocity until the drag of air friction is equal in magnitude to the force of gravity on the body.

thermodynamics The study of quantitative relationships between other forms of energy and thermal energy.

time of flight (of a projectile) The total time it is in the air.

trajectory The arced or curved path of a projectile.

transverse wave A disturbance traveling through a medium in which particles of the medium vibrate in paths perpendicular to the direction of motion of the wave.

trigonometry A branch of mathematics that deals with relationships of angles and corresponding sides of triangles.

triple beam balance A device for measuring mass by comparing an unknown mass with a known mass by balancing two pans. The two pans are attached to a beam that is supported by a fulcrum in much the same way as a seesaw or teeter-totter.

trough The portion of the graph of a transverse wave that lies below the time axis.

unified atomic mass unit The unit of mass used to stipulate nuclear masses. The atomic mass unit is equal to $\frac{1}{12}$ of the mass of an atom of carbon 12.

uniform motion Motion characterized by a constant speed. Whenever a situation in physics states or implies uniform motion, then you will know that the speed is constant. If the speed is constant, then you know that the object has uniform motion.

unit analysis The process of defining a disguise of unity (one) in terms of units of measurement that will enable you to change from one unit of measure to a larger or smaller unit of measure without changing the value of the measured quantity.

valence electron An electron in an incomplete energy level that an atom can lose and become more stable or for a different atom to gain to become more stable resulting in a more stable bond. These loosely bound electrons are largely responsible for the chemical behavior of an element.

Van de Graaff generator A motorized source of a high concentration of electrons with high potential energy.

vector A quantity that has magnitude, direction, and obeys a law of combination.

velocity The rate of change of displacement. Velocity is a vector quantity that has speed as its magnitude.

voltmeter The instrument used to measure potential difference in a circuit.

wavelength (of a transverse wave) The distance from the beginning of a crest to the end of an adjacent trough. Is made up of a crest and a trough. The distance from the beginning of a crest to the end of the same crest is one half of a wavelength.

weight (with respect to the earth) The force of gravity on any object. It is a vector quantity with direction radially inward toward the center of the earth. The direction as described locally is down.

Appendix B

Physics Phun Answers

Chapter 1

Problem Set I

1. 2
2. 3
3. 1
4. 4
5. 1

Problem Set II

1. 2.5×10^5mi
2. 4.83×10^6m
3. 2.3×10^7s
4. 8.76×10^{-4}kg
5. 1.0003×10^0cm

Problem Set III

1. 951.2 cm
2. 324.0 ft
3. 35879 m
4. 76.7 cm

Chapter 2

Problem Set I

1. 1.5×10^{11}m
2. 8.64×10^4s or 8.6400×10^4s because the digits in defining equations are all significant.

Problem Set II

1. $x = v_0 t + \frac{1}{2} at^2$

Chapter 3

1. 2.6 cm E30°S or S60°E
2. The difference vector $\vec{B} - \vec{A}$ is the opposite of $\vec{A} - \vec{B}$.

   $\vec{B} + (-\vec{A}) = \vec{B} - \vec{A} = 3.5$S10°W or 3.5W80°S

Chapter 4

1. 2.3 m/s
2. 5.5 s
3. 4.5 m

Chapter 5

1. $3.5 \times 10^1 N$
2. 68 ft/s^2
3. 16.4 N S36°W or W54°S

Chapter 6

1. 160 lb
2. 5.0×10^1kg

Chapter 7

Problem Set I

1. $y_1 = 57ft, y_2 = 82ft, y_3 = 75ft, Y = 83\text{ft}$
2. $Y = 39.4m, y_3 = 39.3m$

Problem Set II

1. $\text{T} = 5.10\text{s}, Y = 31.9\text{m}, X = 221\text{m}$
2. $t_{max\ ht} = 44.4s, a = 9.80\text{m/s}^2$

   $v = v_{0H} = 435\text{m/s}, F_{net} = 1.12N$

   $t_{max\ ht} = 44.4\text{s}, \text{T} = 88.8\text{s}, Y = 9650\text{m}$

   $\text{X} = 38{,}600\text{m}$

Chapter 8

Problem Set I

1. 1.0 s
2. 979cm/s^2 or $9.79\ \text{m/s}^2$
3. 24,000 dynes or 0.24 N
4. $\frac{2}{3}\frac{1}{s}$
5. The acceleration is maximum when the displacement is maximum.
   The acceleration is zero when the velocity is maximum.

Problem Set II

1. $2.56 \times 10^{-7}N$
2. 3.4 N, the scale would have a higher reading at the north pole because there is no centripetal force.

Problem Set III

1. 8.43 m/s^2
2. 443 N
3. 5.96×10^{24}kg

Chapter 9

Problem Set I

1. 334 ft-lb
2. 1,870 ft-lb
3. 2,780 joules

Problem Set II

1. (a) $W_{k\ out}$ = 1300 joules; (b) $W_{k\ in}$ = 1700 joules; (c) efficiency = 76%
2. A little less than 310 N

Chapter 10

Problem Set I

1. (a) 93 ft-lb
2. (a) 1.48×10^{5}kg ft-lb; (b) 176 lb; (c) 23.4 hp assuming one horsepower is 550 ft-lb/s
3. 4.56×10^{-10}kg ergs

Problem Set II

1. (a) 44.9 m/s; (b) 6880 joules; (c) You might expect the ball to make a dent in the earth.
2. (a) 242,000 ft-lb; (b) 242,000 ft-lb; (c) The brake did the work along with the tires sliding on the road.

Chapter 11

1. 52,100 ft-lb, 70.2 mi/hr or 103 ft/s
2. 7.67 m/s

Chapter 12

Problem Set I

1. 4360 lb
2. 15.0 N-s, 3.00 m/s
3. 8.90 N

Problem Set II

1. 0.247 m/s in the opposite direction of the object
2. 17,100 kg
3. 0.71 m/s in the direction opposite the bullet

Chapter 13

Problem Set I

1. 23.5 N

Problem Set II

1. 4.54×10^4 N/m^2
2. 9.8×10^4lb
3. 2.56 lb/in^2

Chapter 14

1. 2.78×10^4lb/ft^2 ; No; 193 lb
2. 7.5
3. 3.11 ft

Chapter 15

Problem Set I

1. 70°C
2. 2400 Btu

Problem Set II

1. 2.1×10^4 Btu
2. 34,600 cal
3. 104°F

Chapter 16

Problem Set I

1. 0.333 m, 0.0222 s
2. 60 Hz, 0.0167 s
3. 16.5 m, 0.0165 m

Problem Set II

1. 20.4 s
2. 4.48 ft
3. 0.9 s

Chapter 17

1. 108, 47 protons, 61 neutrons, 108 g
2. 82, 207, 207 g
3. 79 protons, 79 electrons, 118 neutrons

Appendix C

Solutions to Budding Rocket Scientist Problems

Chapter 1

1. (a) 6.274×10^2, (b) 3.65×10^{-4}, (c) 2.0001×10^4, (d) 1.0067×10^0, (e) 6.7×10^{-3}
2. (a) Convert kilometer to meters, then meters to millimeters, = 1×10^6, (b) Divide a centimeter (100) by a millimeter (1,000), = 0.1
3. (a) $(1.27 \times 10^7$ m$) \times (1{,}000$ mm/1m$) = 1.27 \times 10^{10}$ mm, (b) Divide 1.27×10^7 m by 1 km or 10^3 m = 1.27×10^4 km, (c) Then use (1 km/1000 m) $\times$ (1 mi./1.61 km) = 7.89×10^3 miles.

Chapter 2

1. Speed = distance divided by time = 250 mi/5 hr = 50 mi/hr.
2. Time = distance divided by velocity = (1 mi)/(10 mi/h) = 0.1 hr, or 6 min.
3. (a) 0 = 0 m/s + (9.8 m/s^2)(10 m)(60 s/min) = 5.9×10^3 m/s, or 5.9 km/s; (b) (5.9 km/s)(3600 s/hr) = 2.1×10^4 km/hr.

Chapter 3

1. A car speedometer only indicates speed. To specify velocity, we need an additional instrument that would indicate direction, like a compass for horizontal direction. If you said velocity instead of speed, you were either wrong or you have a unique instrument that can tell you speed and direction at the same time.
2. Let's label our three vectors A (60 mi at 30° N of E), B (30 mi due E), and C (40 mi at 30° W of N). Using a scale, I chose 1 cm = 22.5 mi; we can construct the vector sum R = A + B + C, by adding the vectors head to tail. By connecting the starting point to the final point of our construction, we obtain the resultant vector R. Measuring the length of R, we find it to be approximately 4 cm, which corresponds to 90 mi. The direction from the starting point, as measured with a protractor, is about 46° N of E.
3. Let's call the swimmer's velocity in still water v_{sw}, the velocity of the river water v_{rw}, and the swimmer's velocity across the river v_{sr}. These velocity vectors are connected by the vector equation $v_{sr} = v_{sw} + v_{rw}$.

   From this relation, we can solve for the velocity of the river current, $v_{rw} = v_{sr} - v_{sw}$.

   Using a scale (arbitrary) of 1 cm = 0.85 m/s. we construct the vector subtraction by the tail to head method. Measuring the length of the vector v_{rw}, we find it to be 1.4 cm, which corresponds to a speed of 1.4 × 0.85 m/s or 1.2 m/s. Therefore, the river current has a velocity of 1.2 m/s downriver.

Chapter 4

1. $a = \frac{v^2}{R} = \frac{(6m/s)^2}{1.5m} = 24m/s^2$
2. (a) 180°; (b) π rad; (c) 20 × 360° = 7,200°; (d) $20 \times 2\pi = 40\pi$ rad.
3. 2π rad/5 s = 0 + angular acceleration (5 s); angular acceleration = 0.25 rad/s^2

Chapter 5

1. We first find the force F using $F = ma$. F = (2 kg)(3 m/s^2) = 6N. Then solving for $a = F/m$, we have for the different masses m = 1 kg, a = (6N)/(1 kg) = 6 m/s^2; m = 4 kg, a = 1.5 m/s^2. So our answers are (a) 6 m/s^2, (b) 1.5 m/s^2, (c) 6N.

2. $F = ma$ and $\frac{F}{a} = m$, so $m = \frac{70N}{(20 ft/s^2)(0.305m/ft)} = \frac{70kg\ m/s^2}{6.10m/s^2} = 11.5kg$.
3. For the boy to be in equilibrium, the floor must push up on the boy's feet with a force F equal and opposite to the compound weight of the flour and the boy. Let m equal the mass of the boy and w the weight of the flour:
   $F = mg + w = 75(9.8) + 40 = 735 + 40 = 775N$

Chapter 6

1. Using the acceleration formula, $x = x_0 + v_0t + \frac{1}{2}at^2$, and input values $x_0 = 0m$, $v_0 = 10m/s$, $a = -2.5m/s^2$, and $t = 4s$ we find that the car will stop in a distance of $x = 0m + 10m/s(4s) + \frac{1}{2}(-2.5m/s^2)(4s)^2$, or $x = 40m - 20m = 20m$.
2. The resultant force on the man is the vector sum of the two forces; the weight $w = mg = (91.8\ kg)(9.8\ m/s^2) = 900\ N$ downward, and the 225 force upward. Then the resultant force is $900 - 225 = 675\ N$ downward.
3. Using $F = ma$, for the horizontal direction we get $150lb \times 4.45N/lb = m(3.0m/s^2)$, and $m = 223$ kg.

Chapter 7

1. Choose downward as positive with origin at the edge of the tabletop.

   $v_{0X} = v_0 = 20cm/s \quad v_{0Y} = 0 \quad a_Y = +g = +980cm/s^2 \quad a_X = 0$

   To find the time of fall,

   $y = v_{0y}t + \frac{1}{2}gt^2$, or $80cm = 0 + (490cm/s^2)t^2$; $t = 0.40s$

   The horizontal distance is gotten from

   $x = v_{0X}t = (20cm/s)(0.40s) = 8.0cm$

2. This is a good projectile problem with $v_0x = 100\ cm/s$. For the horizontal motion:

   $x = v_{0x}t \quad 30 = 100t \quad t = 30$ s

   For the vertical motion:

   $Y = v_{0y}t + \frac{1}{2}at^2 = 0 + \frac{1}{2}(980)(0.30)^2 = 44cm$ The height of the table.

3. In the horizontal problem, $x = v_x t$ gives $v_x = \frac{0.70}{t}$. In the vertical problem, choosing down as positive, $v_0 = 0$, $y = 0.7m$, and $a = 9.8m/s^2$. Plugging these values into $y = v_0 t + \frac{1}{2}at^2$, we get $t = 0.378$ s. Then $v_x = 1.85m/s$.

Chapter 8

1. $$F = \frac{(6.67\times10^{-11}\frac{N\cdot m^2}{kg^2})(80kg)(5.98\times10^{24}kg)}{(6.38\times10^6)} = 784N$$

2. We can set up the ratio of g_m to g_e:

$$\frac{g_m}{g_e} = \frac{GM_m / R_m^2}{GM_e / R_e^2} = \frac{M_m}{M_e}\left(\frac{R_e}{R_m}\right)^2$$

and inserting values of the ratios, we have

$g_m = g(0.1)(1/.05)^2 = 0.4g = 3.9m/s^2$.

3. In SI units the gravitational constant has a value $G = 6.67 \times 10^{-11}$ $N\cdot m^2/kg$

$$F = G\frac{m_1 m_2}{r^2} = (6.67\times10^{-11})\frac{90(90)}{(0.40)^2} = 3.38\times10^{-6}N$$

4. First find the spring constant k:

$$k = \frac{(50g)(980cm/s^2)}{4cm} = 12250dyn/cm$$

To find the period, use this:

$$T = 2\pi\sqrt{\frac{m}{k}} = 2\pi\sqrt{\frac{150}{12250}} = 2\pi(0.1107) = 0.695s$$

5. $T = 2\pi\frac{l}{g}$, so $(2.4)^2 = (4\pi^2)(l/9.8)$, and $l = 1.43s$

Chapter 9

1. $Work = (200N)(10M) = 2000N\cdot m = 2000\, joules$

2. (a) The lifting force is in the direction of displacement and just balances the weight. $F = mg = 39.2N$

$W = Fh = (39.2N)(1.5m) = 58.8\, joules$ (b) If the object is lowered, F is opposite to the displacement. $W = -Fh = (39.2N)(1.5m) = -58.8$

3. Work = force × distance = $(10{,}000)(9.8m/s^2)(0.5m) = 4.9 \times 10^4$ *joules;* Power = work/time = $(4.9 \times 10^4 J)/(30\text{min})(60s/\text{min}) = 27.2J/s = 27.2$ *watts*

Chapter 10

1. $KE = \frac{1}{2} mv^2 = \frac{1}{2}(950kg)\left(\frac{10^5 m}{3600s}\right)^2 = 3.67 \times 10^5$ *joules*

2. Truck: $KE = \frac{1}{2} mv^2 = \frac{1}{2}(5000kg)[(50km/h)(1h/3600s)(1000m/1km)]^2 =$ $48{,}000kg\ m^2/s^2$ *or* 4.8×10^5 *joules*

   Bullet:

   $KE = \frac{1}{2} mv^2 = \frac{1}{2}(0.1kg)(500m/s)^2 = 13{,}000kg\ m^2/s^2$ *or* 1.3×10^4 *joules*

3. $W = \Delta K = \frac{1}{2} mv^2 - 0 = \frac{1}{2}(1300kg)(20m/s)^2 = 260kJ$, and

   $W = F \times D,\ F = \frac{W}{D} = \frac{260kJ}{80m} = 3.25kN$

Chapter 11

1. $PE = (500kg)(9.8m/s^2)(10m) = 4.9 \times 10^4$ *joules* Anvil

   $PE = (3kg)(9.8m/s^2)(200m) = 5.9 \times 10^4$ *joules* Bowling ball has more. And the work required is equal to the PE of each object.

2. (a) Because the spring is linear,

   $k = \frac{\Delta F}{\Delta x} = \frac{(0.020kg)(9.8m/s^2)}{0.07m} = 2.8N/m.$

   (b) $T = 2\pi\sqrt{\frac{m}{k}} = 2\pi\sqrt{\frac{0.050kg}{2.8N/m}} = 0.84s$

3. Assuming all her potential energy is converted to kinetic energy, PE + KE(top) = PE + KE(bottom);

   $(2m)(9.8m/s^2)(25kg) + 0 = 0 + \frac{1}{2}(25kg)v^2; v = \sqrt{39.2m^2s^2} = 6.3m/s$

Chapter 12

1. (a) Impulse = $\Delta(mv) = 80(5) = 400\ N \cdot s$

   (b) $\overline{F} = \Delta(mv)/\Delta t = 400/0.3 = 1330N$

2. Conservation of momentum:

   $(0.4 \times 3) + 0 = 0.4v + 0.6V$ or $v + 1.5V = 3$

   Elastic: velocity of separation = –velocity of approach, or $-v + V = 3$.

   We solve by adding the two equations to yield

   $2.5V = 6 \quad V = 2.4m/s \quad v = -0.6m/s$.

3. Consider the system (block + bullet). The velocity, and hence the momentum, of the block before impact is zero. The momentum conservation law tells us that

   momentum of system before impact = momentum of system after impact

   (mass) × (velocity of bullet) + 0 = (mass) × (velocity of block + bullet)

   $(.008kg)v + 0 = (9.008kg)(0.40m/s)$

   where v is the velocity of the bullet. Solving gives $v = 450m/s$.

Chapter 13

1. Density = $(75kg)/(0.064m^3) = 1.2 \times 103kg/m^3$; she will sink.
2. $\rho = density = \frac{mass}{volume} = \frac{63.3g}{80.0mL} = 0.791g / mL = 791kg / m^3$

   $specific\ gravity = \frac{density\ of\ alcohol}{density\ of\ water} = \frac{791kg / m^3}{1000kg / m^3} = 0.791$

3. $P_i = 125kPa \qquad V_i = 4.5L$

   $P_f = 75kPa \qquad V_f = ?$

   $P_iV_i = P_fV_f$

   $125 \times 4.5 = 75 \times V_f \qquad V_f = \frac{562.5}{75} = 7.5L$

Chapter 14

1. $specific\ gravity = \frac{density\ of\ cork}{density\ of\ water}$

   $0.25 = \frac{density\ of\ cork}{62.5}$ and $15.6lb/ft^3 = density\ of\ cork$

   Next, $W = d_wV$, $4 = 15.6V$, and $V = 0.26ft^3$

2. The atmosphere exerts a force normal to any surface placed in it. Consequently, the force on the windowpane is perpendicular to the pane and is given by $F = pA = (100kN/m^2)(0.40 \times .80m^2) = 32kN$.

3. In terms of the weight density, ρg, of water,

$$p_{bottom} = \rho gy + p_{atm} = \rho gy + p\left(\frac{\rho_{Hg}}{\rho}\right)gh_{Hg} = \rho g[45.7 + (13.6)(0.762)] = 45.7\rho g + 10.4\rho = 56$$

For the bubble, Boyle's law states that pV = constant, assuming that the temperature remains fixed. Then, $V_{surface} = \frac{p_{bottom}}{p_{surface}} V_{bottom} = \frac{56.1\rho g}{10.4\rho g} \times 33 = 178cm^3$.

4. By Pascal's principle,

pressure under large piston = pressure under small piston $\frac{F_1}{A_1} = \frac{F_2}{A_2}$

$$F_1 = \frac{A_1}{A_2} F_2 = \frac{200}{5}(250N) = 10kN$$

Chapter 15

1. $100°F = \frac{9}{5}T + 32$; $T = 38°C$, $T(K) = 311K$

2. $Q = mc\Delta T$,

   $Q(AL) = (1kg)(920J/kg°C)(20°C) = 1.8 \times 10^4$ *joules*

   $Q(H_2O) = (.5kg)(4186J/kg°C)(5°C) = 1.1 \times 10^4$ *joules*, the aluminum requires more energy.

3. $Q = mc\Delta T$

   (a) $Q = (1)(1.0)(212°F - 68°F) = 144Btu$

   (b) $Q = (2)(0.36)(212°F - 68°F) = 104Btu$

   (c) $Q = (0.20)(1.0)(212 - 68) = 86Btu$

Chapter 16

1. Speed of sound $v = (331 + 0.6T)m/s = 331m/s + 0.6(100) = 391m/s$

2. $v = 33 + 0.6(30) = (349m/s)(1km)/(1000m)(3600s/h) = 1300km/h$

3. The number of complete waves passing any point in air and in water in unit time is the same, so f = 1000Hz for both media. Therefore,

$$\lambda_w = \frac{v_w}{f} = \frac{1500m/s}{1000s^{-1}} = 1.5m.$$

4. As indicated in the last problem, f remains constant—in this case at 60 kHz. The wavelength $= \frac{v}{f} = \frac{331}{60.\times 10^4} = 5.5m$.

Chapter 17

1. $A - Z = 22 - 11 = 11$

2. (a) The atomic number of He is 2; therefore, the nucleus must contain 2 protons. Because the mass number of this isotope is 3, the sum of the protons and neutrons in the nucleus must equal 3; therefore, there is 1 neutron. The number of electrons in the atom is the same as the atomic number, 2.

   (b) The atomic number of carbon is 6; hence the nucleus must contain 6 protons. The number of neutrons in the nucleus is equal to 12 – 6 = 6. The number of electrons is the same as the atomic number, 6.

   (c) The atomic number of lead is 82; hence, there are 82 protons in the nucleus and 82 electrons in the atom. The number of neutrons is 206 – 82 = 124.

3. The average atomic weight $= \frac{0.9889(12) + 0.0111(13)}{1.000} = 12.011$.

Chapter 18

1. Because the charges are opposite, the force is attractive. To calculate the force, we simply use the following:

   $F_c = k\frac{q_1 q_2}{r^2}$, so

$$F_c = 8.9875\times 10^9 N\cdot m^2/C^2 \frac{(1.6\times 10^{-19}C)(-1.6\times 10^{-19}C)}{(5.3\times 10^{-11}m)^2} = 8.2\times 10^{-8}N$$

2. $F_g = (6.67\times 10^{11} N\cdot m^2/kg^2)\frac{(9.11\times 10^{-31}kg)(-1.67\times 10^{-27}kg)}{(5.3\times 10^{-11}m)^2} = 3.6\times 10^{-47}N$,

   far smaller than the coulomb force.

3. Enough energy to remove one or more electrons.

Chapter 19

1. I = 0.7 A means 0.7 C/s. Dividing by $e = 1.6 \times 10^{-19}C$, the magnitude of charge on a single electron, we get a number of electrons per second = $\frac{0.7}{(1.6 \times 10^{-19})} = 4.4 \times 10^{18}$.
2. This is an application of Ohm's law: $V = IR$, $120V = I(8\Omega)$, and $I = 15$ A.
3. First we determine the current, $I = Q/t$, or $I = 720$ C/60s = 12 A. Then use Ohm's law, $V = IR$, or $V = (12\text{ A})(5\Omega) = 60V$.
4. The resistance R of the parallel combination is given by $\frac{1}{R_T} = \frac{1}{R_1} + \frac{1}{R_2} + \frac{1}{R_3} = \frac{1}{12} + \frac{1}{16} + \frac{1}{20} = \frac{20}{240} + \frac{15}{240} + \frac{12}{240} = \frac{47}{240}$ or $R_T = 5.11\Omega$. Then $R_X + R_T = 25$ or $R_X = 25 - 5.11 = 19.89\Omega$.

Chapter 20

1. Time = $(50mi)(268 \times 10^{-4})(3{,}600s/h) = .98s$
2. $I = \frac{k}{r^2}$, $k = Ir^2$, then $(12{,}000lm/m^2)(10m)^2 = 1.2 \times 10^5 cd$
3. You will form a virtual, upright image. The image will be larger than the object. This is a magnifying glass.

Chapter 21

1. Frequency = c/wavelength = $(3 \times 10^8 m/s)/(0.21m) = 1.4 \times 10^9$, or 1.4 GHz
2. Change in wavelength/wavelength = v/c; $v = 3 \times 10^8 m/s(200nm/700nm) = 8.6 \times 10^7 m/s$, or about 30 percent of the speed of light.
3. From Snell's law, $1\sin 48° = 1.333\sin\theta_w$ and $\sin\theta_w = 0.5575$; $\theta_w = 33.9°$.
4. The distance between adjacent fringes is as follows:

$$\Delta x = x_{n+1} - x_n = \frac{L\lambda}{d}(n+1) - \frac{L\lambda}{d}n = \frac{L\lambda}{d} \text{ so}$$

$$d = \frac{L\lambda}{\Delta x} = \frac{(0.8m)(589 \times 10^{-9} m)}{0.35 \times 10^{-2} m} = 1.35 \times 10^{-4} m = 0.135mm$$

Index

A

B

C

D

E

F

G

H

I-J-K

L

M

N

O

P–Q

R

S

T

U

V

W-X-Y-Z